Antigen Processing and Presentation

Methods and Protocols

Edited by

Efstratios Stratikos

Department of Chemistry, National and Kapodistrian University of Athens, Zografou, Greece

Humana Press

Editor
Efstratios Stratikos
Department of Chemistry
National and Kapodistrian University of Athens
Zografou, Greece

ISSN 1064-3745 ISSN 1940-6029 (electronic)
Methods in Molecular Biology
ISBN 978-1-0716-5091-2 ISBN 978-1-0716-5092-9 (eBook)
https://doi.org/10.1007/978-1-0716-5092-9

This Humana imprint is published by the registered company Springer Science+Business Media, LLC, part of Springer Nature.
The registered company address is: 1 New York Plaza, New York, NY 10004, U.S.A.

Preface

The human adaptive immune system relies on the biochemical pathway of antigen processing and presentation to be able to tell which cells are healthy and which are infected, diseased, or cancerous. This pathway starts from the proteolytic processing of protein antigens and generation of small peptide fragments which are loaded onto nascent Major Histocompatibility Complex (MHC) molecules and ends with the recognition of the peptide-MHC complexes by specialized immune receptors on T cells and Natural Killer cells. Studying this pathway has tremendous reprecursions on our understanding of human immunology and the basic mechanisms that underlie normal immune responses to infections and cancer as well as aberrant responses towards self that lead to autoimmunity. Many highly complex assays and techniques have been devised by research groups around the world to help pick apart molecular events along this pathway. Sharing detailed experimental protocols is critical to best coordinate experimental efforts that will carry the field forward, validate key findings, and help newcomers make meaningful contributions. This book focuses on exactly that. A series of cutting-edge protocols and methods are presented in detail to allow reproduction and empower future research. The book focuses on the analysis of key molecules and processes along the Antigen Processing and Presentation pathway, starting from how to make and analyze MHC molecules and their cargo (Chapters 1, 2, 3, 4, 5, and 6), analyzing the function of chaperones that regulate antigen presentation (Chapters 7 and 8) and of enzymes that generate antigenic peptides (Chapters 9, 10, and 11), how to make and analyze T cell receptors (Chapters 12, 13, and 14), and finally presenting a series of functional assays to follow adaptive immune responses to antigen presentation (Chapters 15, 16, 17, 18, and 19).

Zografou, Greece *Efstratios Stratikos*

Contents

Contributors

YOU MIN AHN • *Infection and Immunity Program, La Trobe Institute for Molecular Science (LIMS), La Trobe University, Bundoora, VIC, Australia; Department of Biochemistry and Chemistry, School of Agriculture, Biomedicine and Environment (SABE), La Trobe University, Bundoora, VIC, Australia*

MIGUEL ÁLVARO-BENITO • *Department of Immunology, Ophthalmology, ENT, School of Medicine, Universidad Complutense de Madrid, Madrid, Spain; Immunobiology, Health Research Institute Hospital 12 de Octubre (imas12), Madrid, Spain; Laboratory of Protein Biochemistry, Institute of Chemistry and Biochemistry, Department of Biology, Chemistry and Pharmacy, Freie Universität Berlin, Thielallee, Berlin, Germany*

A. W. PESHALA AMARAJEEWA • *Department of Endocrinology, The Second Affiliated Hospital of Xi'an Jiaotong University, Xi'an, Shaanxi, China; Core Research Laboratory, The Second Affiliated Hospital of Xi'an Jiaotong University, Xi'an, Shaanxi, China*

MYRTO ATHINA BALACHOUTI • *Department of Chemistry, Laboratory of Organic Chemistry, National and Kapodistrian University of Athens, Panepistimiopolis, Zografou, Athens, Greece*

LOMANE BERTHY • *CEA, INRAE, Médicaments et Technologies pour la Santé (MTS), SIMoS, Université Paris-Saclay, Gif-sur-Yvette, France*

LOUISE H. BOYLE • *Department of Pathology, University of Cambridge, Cambridge, UK*

DEMETRA S. M. CHATZILEONTIADOU • *Infection and Immunity Program, La Trobe Institute for Molecular Science (LIMS), La Trobe University, Bundoora, VIC, Australia; Department of Biochemistry and Chemistry, School of Agriculture, Biomedicine and Environment (SABE), La Trobe University, Bundoora, VIC, Australia; Department of Biochemistry and Molecular Biology, Monash University, Clayton, VIC, Australia*

LAURENT DEVEL • *CEA, INRAE, Médicaments et Technologies pour la Santé (MTS), SIMoS, Université Paris-Saclay, Gif-sur-Yvette, France*

IRINI EVNOUCHIDOU • *Centre de Recherche sur l'Inflammation, INSERM UMR1149, CNRS EMR8252, Faculté de Médecine site Bichat, Université Paris Cité, Paris, France*

ALEXANDER FISH • *Division of Biochemistry and Oncode Institute, Netherlands Cancer Institute, Amsterdam, Netherlands*

CHRISTIAN FREUND • *Laboratory of Protein Biochemistry, Institute of Chemistry and Biochemistry, Department of Biology, Chemistry and Pharmacy, Freie Universität Berlin, Thielallee, Berlin, Germany*

DORIANA FRUCI • *Bambino Gesù Children's Hospital, IRCCS, Rome, Italy*

MALGORZATA A. GARSTKA • *Department of Endocrinology, The Second Affiliated Hospital of Xi'an Jiaotong University, Xi'an, Shaanxi, China; Core Research Laboratory, The Second Affiliated Hospital of Xi'an Jiaotong University, Xi'an, Shaanxi, China*

DIMITRIS GEORGIADIS • *Department of Chemistry, Laboratory of Organic Chemistry, National and Kapodistrian University of Athens, Panepistimiopolis, Zografou, Athens, Greece*

STEPHANIE GRAS • *Infection and Immunity Program, La Trobe Institute for Molecular Science (LIMS), La Trobe University, Bundoora, VIC, Australia; Department of Biochemistry and Chemistry, School of Agriculture, Biomedicine and Environment*

(SABE), La Trobe University, Bundoora, VIC, Australia; Department of Biochemistry and Molecular Biology, Monash University, Clayton, VIC, Australia

LANCE M. HELLMAN • *Nevada State University, Physical and Life Sciences, Henderson, NV, USA*

JAKOB HOLZAPFEL • *Laboratory of Protein Biochemistry, Institute of Chemistry and Biochemistry, Department of Biology, Chemistry and Pharmacy, Freie Universität Berlin, Thielallee, Berlin, Germany*

EDWARD JAMES • *Centre for Cancer Immunology, Faculty of Medicine, University of Southampton, Southampton, UK*

KAMILA KRÓL • *Bambino Gesù Children's Hospital, IRCCS, Rome, Italy*

MARIE LAUNAY • *CEA, INRAE, Médicaments et Technologies pour la Santé (MTS), SIMoS, Université Paris-Saclay, Gif-sur-Yvette, France*

LIYING LU • *Department of Pathology, University of Massachusetts Chan Medical School, Worcester, MA, USA; Department of Biochemistry and Molecular Biotechnology, University of Massachusetts Chan Medical School, Worcester, MA, USA*

YELYZAVETA MAKEDON • *Constructor University, Bremen, Germany*

MANOUSOS MAKRIDAKIS • *Center of Systems Biology, Biomedical Research Foundation of the Academy of Athens, Athens, Greece*

HOLLY ANNE MARTIN • *Nevada State University, Physical and Life Sciences, Henderson, NV, USA*

FRANÇOIS-XAVIER MAUVAIS • *Université Paris Cité, Paris, France; Service de Physiologie-Explorations Fonctionnelles, AP-HP, Hôpital Universitaire Robert-Debré, Paris, France*

GEORGE MAVRIDIS • *Department of Chemistry, National and Kapodistrian University of Athens, Athens, Greece*

JACK L. MORLEY • *Department of Pathology, University of Cambridge, Cambridge, UK*

ANASTASIA MPAKALI • *Laboratory of Biochemistry, Department of Chemistry, National and Kapodistrian University of Athens, Athens, Greece*

LAWTON D. MURDOLO • *Infection and Immunity Program, La Trobe Institute for Molecular Science (LIMS), La Trobe University, Bundoora, VIC, Australia; Department of Biochemistry and Chemistry, School of Agriculture, Biomedicine and Environment (SABE), La Trobe University, Bundoora, VIC, Australia*

BIANKA NAGY • *Constructor University, Bremen, Germany*

MARTHA NIKOPASCHOU • *Department of Chemistry, National and Kapodistrian University of Athens, Athens, Greece*

ATHANASIOS PAPAKYRIAKOU • *Institute of Biosciences and Applications, National Centre for Scientific Research Demokritos, Agia Paraskevi, Greece*

HOANG ANH T. PHAN • *Center for Computational and Genomic Medicine, Children's Hospital of Philadelphia, Philadelphia, PA, USA; Department of Pathology and Laboratory Medicine, Children's Hospital of Philadelphia, Philadelphia, PA, USA; Department of Biochemistry and Biophysics, Perelman School of Medicine, University of Pennsylvania, Philadelphia, PA, USA*

KARTHIK M. RAMESH • *Department of Pathology, University of Massachusetts Chan Medical School, Worcester, MA, USA; Department of Biochemistry and Molecular Biotechnology, University of Massachusetts Chan Medical School, Worcester, MA, USA*

EMMA REEVES • *Centre for Cancer Immunology, Faculty of Medicine, University of Southampton, Southampton, UK*

SANDRA LLAMAS RIZO • *Department of Chemistry, Laboratory of Organic Chemistry, National and Kapodistrian University of Athens, Panepistimiopolis, Zografou, Athens, Greece*

GEORGIA ROUNI • *Biomedical Sciences Research Center "Alexander Fleming,", Institute for Bioinnovation, Vari, Greece*

ANKUR SAIKIA • *Constructor University, Bremen, Germany*

MARTINA SAMIOTAKI • *Biomedical Sciences Research Center "Alexander Fleming,", Institute for Bioinnovation, Vari, Greece*

LOREDANA SAVEANU • *Centre de Recherche sur l'Inflammation, INSERM UMR1149, CNRS EMR8252, Faculté de Médecine site Bichat, Université Paris Cité, Paris, France*

NIKOLAOS G. SGOURAKIS • *Center for Computational and Genomic Medicine, Children's Hospital of Philadelphia, Philadelphia, PA, USA; Department of Pathology and Laboratory Medicine, Children's Hospital of Philadelphia, Philadelphia, PA, USA; Department of Biochemistry and Biophysics, Perelman School of Medicine, University of Pennsylvania, Philadelphia, PA, USA*

SEBASTIAN SPRINGER • *Constructor University, Bremen, Germany*

LAWRENCE J. STERN • *Department of Pathology, University of Massachusetts Chan Medical School, Worcester, MA, USA; Department of Biochemistry and Molecular Biotechnology, University of Massachusetts Chan Medical School, Worcester, MA, USA*

EFSTRATIOS STRATIKOS • *Department of Chemistry, National and Kapodistrian University of Athens, Athens, Greece*

PETER VAN ENDERT • *Université Paris Cité, CNRS, Inserm, Institut Necker-Enfants Malades, Paris, France; Service d'Immunologie Biologique, AP-HP, Hôpital Universitaire Necker-Enfants Malades, Paris, France*

JEROME ZOIDAKIS • *Department of Biology, National and Kapodistrian University of Athens, Athens, Greece*

Chapter 1

High-Throughput Production of Class I MHC Tetramers for Immunological Studies Using Chaperone-Mediated Peptide Exchange

Hoang Anh T. Phan and Nikolaos G. Sgourakis

Abstract

The production of Major Histocompatibility Complex (MHC) Class I proteins bound to different peptides of interest is essential for many immunological studies. However, the preparation of MHC-I molecules, whether wild-type or engineered, requires in vitro refolding with a high-affinity peptide, which stabilizes the complex, leading to a sufficient protein yield. This presents a challenge for subsequent peptide exchange reactions, commonly used to generate pMHC-I tetramer libraries. Here, we describe an optimized protocol for chaperone-mediated exchange of a high-affinity placeholder peptide for an array of target peptides. Using this protocol, a pMHC-I tetramer library encompassing 135 different antigen specificities can be produced within 3 days.

Key words MHC class I, TAP-binding protein related (TAPBPR), Peptide exchange, Chaperone-mediated, Temperature-assisted, Tetramer library

1 Introduction

Multimerized MHC class-I reagents, especially tetramers, have been used to investigate T cell responses, to identify antigen-specific cells, and to characterize T cell receptor (TCR) repertoires [1, 2]. Given the large diversity and antigen cross-reactivity of polyclonal TCR repertoires, pMHC-tetramer libraries encompassing diverse antigen specificities have open new possibilities for cross-reactivity screening studies, for monitoring the dynamics of T cell responses to viral pathogens, and for the validation of new technologies and therapeutics [3–6]. Rapid and reliable production of pMHC-tetramer libraries is thus of high interest. However, given the inherent instability of peptide-deficient, or empty, MHC-I molecules, it is challenging to produce properly folded MHC-I molecules presenting different peptides. To address this problem, previous studies have designed conditional MHC-I ligands that are

Efstratios Stratikos (ed.), *Antigen Processing and Presentation: Methods and Protocols*, Methods in Molecular Biology, vol. 3007,
https://doi.org/10.1007/978-1-0716-5092-9_1,

UV-light labile, or sensitive to high temperature [7–9]. However, these approaches have limitations; they often require a UV source, additional design or optimization of the conditional ligands, a more elaborate protein purification protocol, and can result in low protein yields. To address these barriers, our group previously developed destabilizing placeholder (i.e., goldilocks) peptides and used them to isolate stable, peptide-deficient MHC-I molecules in complex with the molecular chaperone TAP-binding protein related (TABPR). This process allows chaperone-mediated peptide exchange of a placeholder for a desired peptide of interest to proceed at scale, and can ultimately be used for high-throughput pMHC-I tetramer library production [5]. More recently, we developed the open MHC-I system [10], which enables peptide exchange in the absence of chaperones, and adapted it to generate ready-to-load nanoparticles for in vitro T cell stimulation studies [11].

Herein, we elaborate an optimized protocol for rapid peptide exchange, using a chaperone-mediated approach (summarized in Fig. 1). The MHC I protein is first refolded and purified with a placeholder ligand, which can be an intermediate-affinity peptide (e.g., leading to a pMHC-I with a melting temperature value, T_m, of approx. 50 °C), designed as described previously [14] Tetramerization is then done in a large batch prior to the peptide exchange step. We also detail a convenient protocol for efficient preparation and quantification of a synthetic peptide library in a 96-well plate format, prior to exchange. Using this workflow, a peptide library of 135 peptides can be produced within 3 days (about 45 peptides per day).

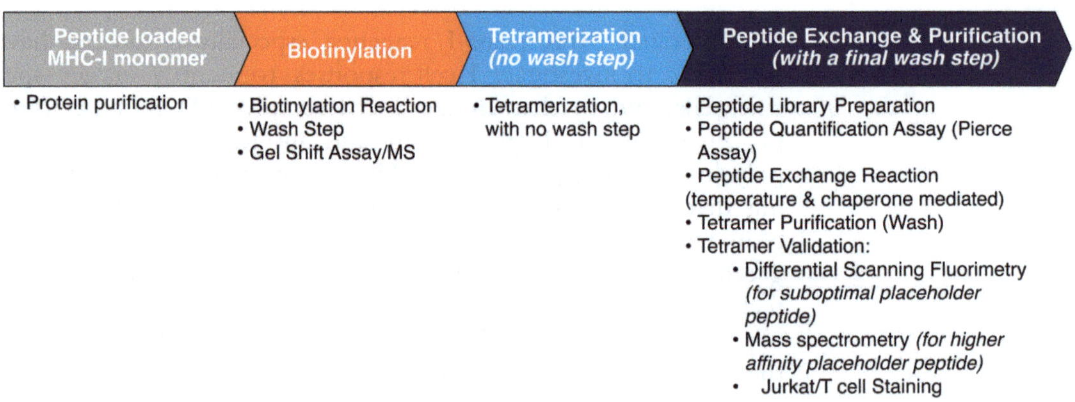

Fig. 1 General Workflow for Tetramer Library Production. Generally, the MHC-I monomer with a biotinylating tag (BirA) is refolded and purified with a placeholder peptide prior to biotinylation, and the efficiency of the biotinylation reaction is confirmed via a gel shift assay. The biotinylated pMHC-I is then tetramerized in a large batch (without a wash step). Afterward, the tetramerized pMHC-I will undergo peptide exchange, followed by a wash step and an optional validation step. This protocol focuses on the peptide exchange method

2 Materials

2.1 MHC Class I Tetramer—Batch Production

1. Biotinylated pMHC-I protein was prepared according to previously established protocols, and protocols outlined in the biotin-ligase kit provided by the manufacturer, Avidity, LLC (*see* **Note 1**) [10, 12].
2. Streptavidin-R-Phycoerythrin (Streptavidin-PE) (Agilent) (*see* **Note 2**).
3. Eppendorf Tubes.
4. Vortex Mixer.
5. Ice.

2.2 Preparation and Quantification of the Peptide Library

1. Peptides of interest in powder or lyophilized form.
2. Dimethylsulfoxide (DMSO), sterile.
3. Eppendorf tubes.
4. RNAse-/DNAse-/Protease-free water (*see* **Note 3**).
5. 96-well Clear Flat Bottom Not Treated Cell Culture Microplate, with Lid (*see* **Note 4**).
6. Pierce Quantitative Fluorometric Peptide Assay kit (*see* **Note 5**).
7. 96-Well Plates for Pierce Quantitative Fluorometric Peptide Assay (*see* **Note 6**).
8. Plate Reader (with fluorescence reading capability at 390 nm excitation and 475 nm emission).

2.3 Peptide Exchange

1. Tetramerized pMHC-I molecules from 2.1.
2. Chaperone-TAP-binding protein related (TAPBPR) (*see* **Note 7**). Either human hTAPBPR, chicken chTAPBPR, or engineered TAPBPR^{HiFi} can be used; protocols to express and purify these proteins can be found in previous publications [5, 13–15].
3. Stocks of peptides of interest for exchange in water from 2.2.
4. RNAse-/DNAse-/Protease-free water.
5. 1X PBS, pH 7.2, sterile.
6. Ice.
7. Heat block.

2.4 Tetramer Purification

1. Centrifugal filters: Amicon Ultra Centrifugal Filter, 100 kDa MWCO (*see* **Note 8**).
2. Eppendorf tubes.
3. 1X PBS, pH 7.2, sterile.

3 Methods

3.1 MHC Class I Tetramer Production

The following tetramer protocol is adapted from the NIH Tetramer Core Facility's Protocol for Class I MHC Tetramer Preparation [16]. This takes approx. 2 h.

1. Assuming 100% biotinylation efficiency, calculate the amount of biotinylated pMHC needed for the whole library: 490 pmol/reaction * *number of total reactions*. We will need a minimum of 490 pmol (about 24 μg) of pMHC per reaction, so make sure that the calculation accounts for extra reactions for potential pipetting errors (*see* **Note 9**).
2. Calculate the total amount of Streptavidin-PE you should add to the reaction. Please note that the NIH Tetramer Core Facility's standard protocol for tetramerization uses 159 μL Streptavidin-PE at 2 mg/mL (about 318 μg) per 200 μg pMHC protein. Below is an example calculation to figure out how much Streptavidin-PE to add to 25 μL of 2 mg/mL pMHC (or 2 μg/μL pMHC):

   Total amount of pMHC:

   25 μL*(2 μg/μL pMHC) = 50 μg pMHC

   Total amount of Streptavidin-PE to add:

   (159 μL Streptavidin-PE)/(200 μg pMHC) = (X μL Streptavidin-PE)/(50 μg pMHC)

   → X = 40 μL Streptavidin-PE.
3. Add Streptavidin-PE to the pMHC in 10 intervals, every 10 min (*see* **Note 10**). For example, if your calculation shows that you need a total of 40 μL Streptavidin-PE, add 4 μL Streptavidin-PE every 10 min for a total of 10 times.
4. Store the tetramerized pMHC away from light until use in 4 °C (*see* **Note 11**).

3.2 Preparation & Quantification of the Peptide Library

This protocol is written for a 96-well plate format, but it can be adapted to an individual tube format.

1. Each peptide library typically comes in the format of 1–3 mg lyophilized peptide per peptide in a rack or in individual tubes. Dissolve each peptide in 40 μL Dimethyl Sulfoxide (DMSO) (target <1.66% final DMSO concentration in water) (*see* **Note 12**). If you have more than 1–3 mg of each peptide, weight out about 1–3 mg peptide lyophilized powder per each peptide directly into a new Eppendorf tube before dissolving the peptide in 40 μL DMSO (*see* **Note 13**).
2. Vortex to dissolve the peptide in DMSO. Make sure that there is little to no solid left.

3. Split 40 μL peptide in DMSO into 2 × 20 μL/aliquot. One aliquot can be stored in the −20 °C freezer for later use. Alternatively, you can keep everything in 40 μL-DMSO aliquots; then add 800 μL RNAse-free water in the next step.
4. For each 20-μL aliquot of peptide in DMSO, dilute 21X by adding 400 μL RNAse-free water. Vortex to dissolve the peptide.
5. In new clean Eppendorf tubes or a flat-bottom clear 96-well plate, further dilute each peptide 3X by adding 100 μL of the peptide solution in the previous step to 200 μL RNAse-/DNAse−/protease-free water. The rest of the stock peptides from step 4 can be stored in 20 °C freezer for later use.
6. Use a pipette or multichannel pipette to mix 10 times to ensure that all peptides are dissolved and diluted properly. The peptide stocks quantification should be around or less than 1 mg/mL.
7. If the solution is clear, you can proceed to the next step (*see* **Note 14**).
8. Quantification of Peptide Concentrations: We suggest using the Pierce Quantitative Fluorescent Peptide Assay (e.g., Thermo Fisher) since it is a high-throughput quantification method that can tolerate additives such as DMSO (up to 50%). Equilibrate all kit components for the Pierce assay to room temperature prior to the assays (*see* **Note 15**).
9. If using the Pierce Quantitative Fluorescent Peptide Assays, prepare a 9-point standard curve (0–1000 μg/mL) in triplicate of the peptide standard in the appropriate diluent (similarly to the stock peptide diluent), according to the manufacturer's protocol such as Thermo Fisher (23290) [17].
10. Prepare the assay samples (our peptides of interest) according to the specific manufacturer's protocol (*see* **Note 16**). For example, to a black, fluorescent-compatible 96-well plate, pipette 10 μL of each standard or sample to each well, then add 70 μL Fluorometric Peptide Assay Buffer, then 20 μL Fluorometric Peptide Assay Reagent to each well [17].
11. Incubate at room temperature for 5 min or for the time specified by your assay kit.
12. Using the plate reader, use a single time-point fluorescent reading method. For Pierce assays, the fluorescence is typically measured at excitation 390 nm and emission 475 nm.
13. Using GraphPad Prism or other analysis software, plot the standard curve to determine the peptide concentration of each peptide sample. Use the curve-fitting, four-parameter (quadratic) fit in GraphPad Prism (*Sigmoidal, 4PL, X is the concentration*) to fit the standard curve, then use the

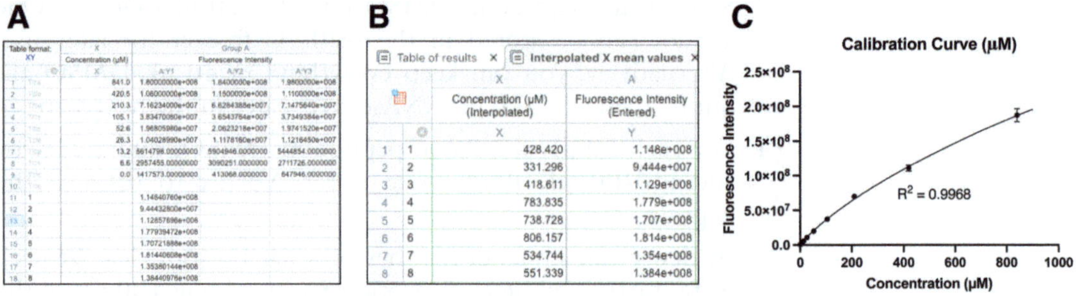

A

Table format: XY		X Concentration (μM)	Group A Fluorescence Intensity		
		X	A:Y1	A:Y2	A:Y3
1		841.0	1.80000000e+008	1.8400000e+008	1.9800000e+008
2		420.5	1.08000000e+008	1.1500000e+008	1.1100000e+008
3		210.3	7.16234000e+007	6.6284388e+007	7.1475640e+007
4		105.1	3.83470080e+007	3.6543784e+007	3.7349384e+007
5		52.6	1.96805980e+007	2.0623218e+007	1.9741520e+007
6		26.3	1.04028990e+007	1.1178160e+007	1.1216450e+007
7		13.2	8614798.00000000	5904946.0000000	5444854.0000000
8		6.6	2957455.00000000	3090251.0000000	2711726.0000000
9		0.0	1417573.00000000	413068.0000000	647946.0000000
10					
11	1		1.14840760e+008		
12	2		9.44432800e+007		
13	3		1.12857696e+008		
14	4		1.77939472e+008		
15	5		1.70721888e+008		
16	6		1.81440608e+008		
17	7		1.35380144e+008		
18	8		1.38440976e+008		

B

Table of results × | Interpolated X mean values ×

		X Concentration (μM) (Interpolated)	A Fluorescence Intensity (Entered)
		X	Y
1	1	428.420	1.148e+008
2	2	331.296	9.444e+007
3	3	418.611	1.129e+008
4	4	783.835	1.779e+008
5	5	738.728	1.707e+008
6	6	806.157	1.814e+008
7	7	534.744	1.354e+008
8	8	551.339	1.384e+008

Fig. 2 Peptide Quantification Assay. Example of the raw fluorescence data of the standards and samples (**a**), interpolation data to obtain the peptide concentrations (**b**), and curve-fitting of the standard curve (**c**)

interpolate function to find the peptide concentration of each sample (*see* **Note 17**). Examples of the Prism spreadsheet, result page, and calibration curve are shown in Fig. 2.

3.3 Peptide Exchange

1. Thaw the peptide stocks at room temperature.
2. Thaw the chaperone protein (TAPBPR) on ice.
3. Centrifuge the chaperone sample at 4 °C for 10 min at 14,000 rpm to get rid of any potential aggregates.
4. Calculate the pMHC concentration after tetramerization: pMHC is previously tetramerized using the protocol above **without the wash step**, so that you can calculate the pMHC concentration *by dilution factor*. We recommend using Excel to perform all the calculations prior to the peptide exchange reaction. Example calculations:

   Calculate pMHC concentration after tetramerization:

   $[pMHC]_{\text{before tetrameriztion}*} \frac{\text{Volume}_{\text{pMHC before tetramerization}}}{\text{Total Volume}_{pMHC+tetramer\ reagent}}$

   Calculate the volume of tetramerized pMHC to add to each exchange reaction:

   If your pMHC is at 20 μM after tetramerization with Streptavidin-PE, we should add.

   (X μL pMHC)*(20 μM pMHC) = 490 pmol pMHC.

   → X = 24.5 μL pMHC tetramer at 20 μM per reaction

5. Calculate the peptide volume for each exchange reaction: we use a 10:1 molar ratio of peptide:pMHC (4900 pmol peptide). Below is a sample calculation:

   If the peptide stock is at 500 μM

   (X μL peptide)*(500 μM peptide) = 4900 pmol peptide

   → X = 9.8 μL of 500 μM peptide

6. Calculate how much TAPBPR to use. TAPBPR should be at least 1X molar ratio to pMHC (490 pmol). You can use up to

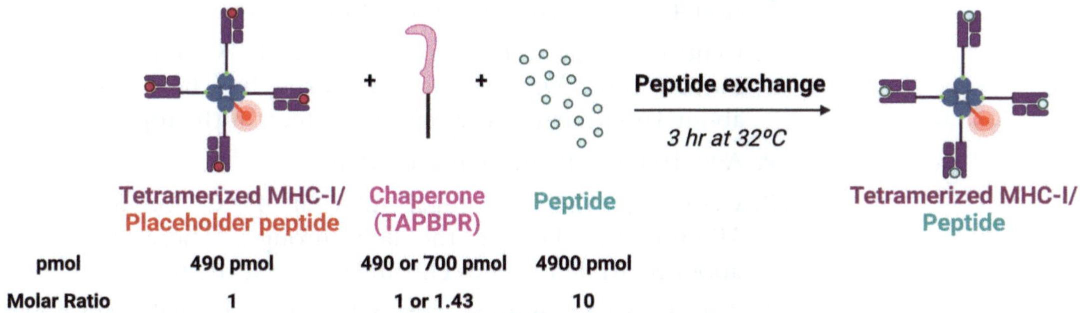

Fig. 3 A summary of the peptide exchange protocol. Figure created with BioRender.com

1.43X molar ratio to pMHC of TAPBPR here (700 pmol). Below is a sample calculation:

If TAPBPR is at 200 μM

(X μL TAPBPR)*(200 μM TAPBPR) = 490 pmol TAPBPR.

→ $X = 2.5$ μL of 200 μM TAPBPR.

7. For each peptide exchange reaction, combine the pMHC, peptide, and TAPBPR in an Eppendorf tube (*see* **Note 18**). *See* Fig. 3 for a graphical summary of the protocol.
8. Incubate the mixture above for 3 h at 32 °C, preferably in a heat block set to this precise temperature (*see* **Note 19**).

3.4 Tetramer Purification

Briefly, each exchanged pMHC-I tetramer will be washed 3 times to remove TAPBPR and peptide excess, using a 100 kDa centrifugal column pre-equilibrated with cold, sterile 1X PBS pH 7.2 (tetramer buffer). For optimal conditions, pre-chill sterile 1X PBS pH 7.2. tetramer buffer prior to the wash protocol. Below is the detailed protocol:

1. For each individual tetramer, place each Amicon Ultra Centrifuge Filter 100 kDa (centrifugal filter) in a new, clean Eppendorf tube.
2. Pre-equilibrate the Amicon Ultra Centrifuge Filter 100 kDa (centrifugal filter) by first adding 500 μL cold sterile 1X PBS pH 7.2 to the filter (*see* **Note 20**).
3. Centrifuge at 4 °C for 5 min at 14,000 rpm. Discard the flow-through solution.
4. Add 400 μL of cold sterile 1X PBS to the centrifugal filter.
5. For each individual tetramer, add the resulting reaction from the previous peptide exchange step to each individual pre-equilibrated Amicon centrifugal filter.
6. Centrifuge the centrifugal filter at 4 °C for 3 min at 10,000 rpm. Discard the flow-through. There should be about 100 μL left in the centrifugal filter at the top.

7. Add 400 μL of cold sterile 1X PBS.
8. Centrifuge the centrifugal filter at 4 °C for 3 min at 10,000 rpm. Discard the flow-through. There should be about 100 μL left in the centrifugal filter at the top.
9. Add 400 μL of cold sterile 1X PBS.
10. Centrifuge the centrifugal filter at 4 °C for 5 min at 10,000 rpm. Discard the flow-through. There should be about 50 μL left in the centrifugal filter at the top.
11. Transfer to a new tube, vortex lightly, and briefly spin down using a mini centrifuge.
12. If the volume is not at 50 μL, adjust the volume to 50 μL for each individual tetramer (*see* **Note 21**).
13. Store the tetramers at 4 °C away from light. Generally, pMHC-I tetramers generated via peptide exchange are stable at 4 °C for about 2–3 weeks.
14. Peptide exchange is typically validated by Differential Scanning Fluorometry. However, if the T_m is already high (> 50 °C) due to a higher-affinity placeholder peptide, peptide exchange can be validated using mass spectrometry (e.g., MALDI or LC/MS).
15. Tetramers can be further validated by Jurkat/T cell staining with a known target. A peptide of a known epitope should be included as a control.

4 Notes

1. The biotinylated pMHC-I protein can be from BirA-tagged either wild-type MHC-I or disulfide-stabilized, open MHC-I proteins refolded with a placeholder peptide, using previously published protocols and plasmid DNA encoding protein constructs for E.coli expression available on addgene for common HLA allotypes [10, 12]. For example, we previously tested this method with biotinylated open HLA-A*11:01 refolded with the HIV-1 gag peptide (AIFQSSMTK) to generate libraries with KRAS 10-mer and KRAS 9-mer peptides. For biotin-protein ligase, please follow the manufacturer's specific protocol for handling of your biotinylating reagents (e.g., Avidity, LLC).
2. Depending on your fluorophore of choice for your downstream application, other fluorescent conjugation moieties can be used, such as Streptavidin-Allophycocyanin (APC, Agilent). For the optimal concentrations of Streptavidin-APC for tetramerization, see the NIH Tetramer Core Facility protocol [16].

3. Use RNAse-/DNAse-/Protease-free water to prevent contamination and minimize unwarranted proteolysis of the peptide stocks.
4. A 96-well plate is recommended for production and storage of the peptide stocks for subsequent high-throughput quantification of the peptide concentrations, but alternative storage vessels can be used such as Eppendorf Tubes.
5. Pierce Quantitative Fluorometric Peptide Assay kit (Thermo Scientific) is recommended, but an alternative peptide concentration quantification method can be used. The recommended assay is a sensitive and fast method for quantification of synthetic peptides or peptide mixtures in a plate-reader format. This method utilizes an amine-reactive fluorescent detection reagent that labels the peptide N-terminus specifically. The fluorescently labeled peptides are then detected by a plate reader in fluorescence reading mode at excitation and emission wavelengths of 390 nm and 475 nm, respectively. More information about this peptide quantification assay can be found on the manufacturer's website [18].
6. If using the Pierce assay for quantification of the peptides, another fluorescent-compatible black plate can be used.
7. We have optimized this protocol using chicken TAPBPR. Human TAPBPR (hTAPBPR; wild-type) or engineered TAPBPR^{HIFI} can also be used. If you work with open pMHC-I refolded with a destabilized placeholder peptide (tm of approx. 50 °C), the exchange reaction can proceed in the absence of TAPBPR. Protocols for production chTAPBPR, hTAPBPR, or engineered TAPBPR^{HiFi} are outlined in previous publications [5, 13–15].
8. Amicon Ultra Centrifugal Filter, 50 kDa MCWO or equivalent centrifugal filters can also be used as a substitute for Amicon Ultra Centrifugal Filter, 100 kDa MWCO.
9. The amount of streptavidin needed depends on the molecular weight of the pMHC-I monomer and the actual percent biotinylation from your biotinylation reaction [16].
10. To ensure that we maximize tetramer formation, it is essential to add the streptavidin to pMHC over a 10-time interval instead of a 1-time addition. At the beginning, we will have excess biotinylated pMHC to saturate all the streptavidin binding sites. By the end of the process, most of the pMHC should be tetramerized, and any potential excess streptavidin will be minimized. If all streptavidin were added at the same time, the excess amount of streptavidin might result in a polydispersed mixture of monomers, and multimers of pMHC instead of a homogeneous mixture of tetramers [16].

11. We recommend proceeding directly to the peptide exchange reactions the day after the tetramerization step.
12. If possible, use sterile DMSO manufactured in a 5 mL-amber glass vial format for the initial solubilization of the lyophilized peptides. Besides, this is a recommendation to prepare a library in a high-throughput and consistent manner for subsequent high-throughput quantification. If you have only a few peptides, they can be dissolved in a different initial solvent (e.g., water, organic solvent), depending on the inherent characteristics of your peptides. Guidelines for dissolving peptides can be found in the instructions of the peptide production vendors [19, 20].
13. When weighing out a small amount of lyophilized peptide, be careful of static charge that can interfere with your weighing or impact your weighing results.
14. If the solution still looks cloudy, you can quickly centrifuge to pellet and get rid of any insoluble peptides or impurities, then transfer to a new plate/tube. If you observe that the insoluble peptides or impurities are floating on top of the peptide solution instead of forming a pellet at the bottom, you may need to filter using a Whatman GD/X 0.2 μm Polyethersulfone (PES) 13 mm diameter syringe filter (or equivalent filters) to avoid significant loss of volume.
15. Peptides in an aqueous solution free of DMSO can be quantified using absorbance and their calculated extinction coefficients at 205 nm using a spectrophotometer such as the NanoDrop. However, here, we suggest using the Pierce Quantitative Fluorescent Peptide Assay since it is a reliable, sensitive, and standardized peptide quantification method that can tolerate additives such as DMSO (up to 50%), allowing consistent and simultaneous measurements of multiple peptide concentrations in a high-throughput plate-reader format.
16. If we have a library of more than 100 peptides, the assay samples (for each peptide of interest) can be conducted in single replicates. Otherwise, the peptide samples can also be done in triplicates for the Pierce peptide quantification assays.
17. It is possible that a few values of the unknown samples may be slightly higher than the range; in that case, you can adjust the parameters in the nonlinear regression in Prism: under "Range," manually end the curve at a higher X value instead of the automatically chosen value.
18. We recommend doing 15 peptide exchange reactions per batch and 3 batches per day. That way, a library of 135 peptides can be achieved within 3 days.

19. Please note that the chaperone concentration used and incubation temperature for the peptide exchange reactions are higher here compared to our previously published protocols.
20. Addition of 0.02% sodium azide to the tetramer buffer is optional.
21. If the final volume here is more than 50 μL, do not reconcentrate and record the volume for calculation of the final concentration. If it is less than 50 μL, which is more likely to be the case, measure how much it is and add additional volume of the tetramer buffer to have a final volume of 50 μL. A tip to measure the volume here is to set the pipette to 50 μL, then draw all the volume of the tetramer up. If the volume in the pipette tip does not entirely fill up to 50 μL (i.e., there is empty space at the bottom of the tip), slowly turn the knob for the volume setting until the liquid in the tip hits the bottom of the tip and there is no space left. Then, the reading on the volume should be approximately the total volume of the liquid in the tip.

Acknowledgments

This work was supported by The Children's Hospital of Philadelphia Cell and Gene Therapy collaborative, and by NIH grant 5R01AI143997 (NIAID). We thank Dr. Yi Sun for her suggestions and feedback regarding the tetramer and peptide exchange protocols. This work was delivered as part of the MATCHMAKERS team, supported by the Cancer Grand Challenges partnership financed by CRUK (CGCATF-2023/100004), the National Cancer Institute (OT2CA297575), and The Mark Foundation for Cancer Research.

Conflict of Interest N.G.S is listed as co-inventor in provisional patent applications and published patents related to this work, as outlined in detail in previous publications [5, 10, 11, 14].

References

1. Altman JD, Moss PAH, Goulder PJR et al (1996) Phenotypic Analysis of Antigen-Specific T Lymphocytes. Science 274:94–96. https://doi.org/10.1126/science.274.5284.94
2. Davis MM, Altman JD, Newell EW (2011) Interrogating the repertoire: broadening the scope of peptide–MHC multimer analysis. Nat Rev Immunol 11:551–558. https://doi.org/10.1038/nri3020
3. Dash P, Fiore-Gartland AJ, Hertz T et al (2017) Quantifiable predictive features define epitope-specific T cell receptor repertoires. Nature 547:89–93. https://doi.org/10.1038/nature22383
4. Glanville J, Huang H, Nau A et al (2017) Identifying specificity groups in the T cell receptor repertoire. Nature 547:94–98. https://doi.org/10.1038/nature22976

5. Overall SA, Toor JS, Hao S et al (2020) High throughput pMHC-I tetramer library production using chaperone-mediated peptide exchange. Nat Commun 11:1909. https://doi.org/10.1038/s41467-020-15710-1
6. Du H, Mallik L, Hwang D et al (2024) Targeting peptide antigens using a multiallelic MHC I-binding system. Nat Biotechnol 1–11. https://doi.org/10.1038/s41587-024-02505-8
7. Toebes M, Coccoris M, Bins A et al (2006) Design and use of conditional MHC class I ligands. Nat Med 12:246–251. https://doi.org/10.1038/nm1360
8. Bakker AH, Hoppes R, Linnemann C et al (2008) Conditional MHC class I ligands and peptide exchange technology for the human MHC gene products HLA-A1, -A3, -A11, and -B7. Proc Natl Acad Sci 105:3825–3830. https://doi.org/10.1073/pnas.0709717105
9. Luimstra JJ, Garstka MA, Roex MCJ et al (2018) A flexible MHC class I multimer loading system for large-scale detection of antigen-specific T cells. J Exp Med 215:1493–1504. https://doi.org/10.1084/jem.20180156
10. Sun Y, Young MC, Woodward CH et al (2023) Universal open MHC-I molecules for rapid peptide loading and enhanced complex stability across HLA allotypes. Proc Natl Acad Sci 120:e2304055120. https://doi.org/10.1073/pnas.2304055120
11. Phan HAT, Hwang D, Young MC et al (2024) Ready-to-load MHC-I nanoparticles for high-throughput T cell screening studies. bioRxiv:2024.12.17.628918. https://doi.org/10.1101/2024.12.17.628918
12. Garboczi DN, Hung DT, Wiley DC (1992) HLA-A2-peptide complexes: refolding and crystallization of molecules expressed in Escherichia coli and complexed with single antigenic peptides. Proc Natl Acad Sci 89:3429–3433. https://doi.org/10.1073/pnas.89.8.3429
13. Jiang J, Natarajan K, Boyd LF et al (2017) Crystal structure of a TAPBPR–MHC I complex reveals the mechanism of peptide editing in antigen presentation. Science 358:1064–1068. https://doi.org/10.1126/science.aao5154
14. Sun Y, Papadaki GF, Devlin CA et al (2023) Xeno interactions between MHC-I proteins and molecular chaperones enable ligand exchange on a broad repertoire of HLA allotypes. Sci Adv 9:eade7151. https://doi.org/10.1126/sciadv.ade7151
15. Sun Y, Pumroy RA, Mallik L et al (2025) CryoEM structure of an MHC-I/TAPBPR peptide-bound intermediate reveals the mechanism of antigen proofreading. Proc Natl Acad Sci 122:e2416992122. https://doi.org/10.1073/pnas.2416992122
16. Production Protocols – Class I MHC Tetramer Preparation: Overview. In: NIH Tetramer Core Facil. https://tetramer.yerkes.emory.edu/support/protocols. Accessed 26 Feb 2025
17. Pierce™ Quantitative Peptide Assays & Standards. https://www.thermofisher.com/order/catalog/product/23290. Accessed 27 Feb 2025
18. Pierce™ Quantitative Peptide Assays & Standards. https://www.thermofisher.com/order/catalog/product/23290. Accessed 26 Feb 2025
19. Peptide solubility and stability. https://www.genscript.com/peptide_solubility_and_stablity.html. Accessed 27 Feb 2025
20. Solubility Guidelines for Peptides. https://www.sigmaaldrich.com/US/en/technical-documents/technical-article/research-and-disease-areas/cell-and-developmental-biology-research/solubility-guidelines?srsltid=AfmBOoqhTrTSZ3540s6eZxDwoBDaI2APUe_3plmQdQDw3JjXiI4H-Oky. Accessed 27 Feb 2025

Chapter 2

Empty MHC Class I Protein Production

Ankur Saikia, Yelyzaveta Makedon, Bianka Nagy, and Sebastian Springer

Abstract

This chapter contains a detailed protocol for the recombinant production and in vitro refolding of disulfide-stabilized peptide-empty MHC class I proteins. MHC class I proteins are heterotrimeric complexes consisting of a heavy chain, light chain (β_2m), and a peptide. This protocol covers **MHC class I heavy chain and light chain expression in *E. coli* as inclusion bodies, their purification, and subsequent refolding using a dilution method**. The protocol details the steps for expressing the proteins, purifying inclusion bodies, and refolding disulfide-stabilized MHC class I molecules with dipeptides. This approach ensures the production of correctly folded, functional, peptide-empty MHC class I proteins suitable for various biochemical and biophysical studies.

Key words MHC Class I, Disulfide-mediated stabilization, Inclusion bodies, *E. coli* expression, Protein refolding, Size exclusion chromatography

1 Introduction

Major Histocompatibility Complex class I proteins (MHC-I) present intracellular or endocytosed peptides to cytotoxic T cells (CTL), playing a key role in the adaptive immune response. Recombinant production of MHC-I is necessary to make MHC-I tetramers (and similar oligomers) for the antigen-specific detection of patient CTL [1, 2]. Recombinant peptide-MHC class I complexes (pMHC-Is) are also necessary for T cell receptor (TCR) safety screening, where candidate therapeutic TCRs (or TCR-like antibodies) are tested on pMHC-I arrays or other high-throughput approaches for cross-reactivity to self-antigens [3]. Such comprehensive analysis of TCRs or CTL typically requires tens to hundreds of pMHC-Is with different peptides, and the great polymorphism of the MHC-I heavy chain means that, for future personalized immunotherapy, every patient will require different pMHC-Is.

Such multitudes of pMHC-Is are difficult to achieve with classic production methods [2]. pMHC-Is are made by producing the heavy chain and the light chain (beta-2 microglobulin, β_2m)

Efstratios Stratikos (ed.), *Antigen Processing and Presentation: Methods and Protocols*, Methods in Molecular Biology, vol. 3007, https://doi.org/10.1007/978-1-0716-5092-9_2,

separately in *E.coli* as inclusion bodies and then folding them in vitro with the peptide of interest, followed by several rounds of purification. This procedure requires a separate production batch for each pMHC-I and is slow since the folding and purification procedure typically takes days to weeks. To accelerate the generation of multiple pMHC-I, several techniques for exchanging peptides on pMHC-I have therefore been developed, the most common being the exchange of UV-cleavable peptides, or dipeptide-mediated peptide exchange [4–8].

An alternative solution is peptide-empty MHC-I proteins that have an additional disulfide bond, which prevents the peptide-binding groove from unfolding in the empty state. Such disulfide-stabilized MHC class I proteins (dsMHC-I) rapidly bind exogenous peptide and provide excellent tetramer staining without influencing binding specificity and affinity [9, 10]. Instead of a full-length peptide, the in vitro folding reaction uses a dipeptide that is readily removed by size exclusion chromatography, yielding peptide-empty disulfide-stabilized MHC-I [9, 11].

This chapter provides a detailed protocol for the production of peptide-empty dsMHC-I by expressing MHC-I heavy chain and β_2m as inclusion bodies in *E. coli*, followed by in vitro refolding with dipeptide via dilution in a controlled redox environment, and purification of the correctly folded protein complex by size exclusion chromatography.

2 Materials

2.1 Chemicals

1. **Ampicillin**: PanReac/AppliChem A0839.
2. **L-Arginine**: PanReac/AppliChem A3709.
3. **Chloramphenicol**: PanReac/AppliChem A1806.
4. **Dipeptide GM** (glycyl-methionine): Bachem 4001203.
5. **DTT (dithiothreitol)**: PanReac/AppliChem A11001.
6. **EDTA (ethylene diamine tetraacetic acid)**: PanReac/AppliChem 131669.
7. **L-Glutathione reduced (MW 307.3)**: PanReac/AppliChem A2084.
8. **L-Glutathione oxidized (MW 612.6)**: PanReac/AppliChem A2243.
9. **Guanidinium hydrochloride**: Sigma Aldrich 50950.
10. **HEPES**: Roth 6763.
11. **IPTG (Isopropyl β-D-1-thiogalactopyranoside)**: PanReac/AppliChem A1008.
12. **β-mercaptoethanol**: Sigma A4338.

13. **PMSF (phenyl methyl sulfonyl fluoride)**: Sigma P7626.
14. **Sodium chloride (NaCl)**: PanReac/AppliChem 14884.
15. **Sucrose**: PanReac/AppliChem A3935.
16. **Tris base**: PanReac/AppliChem A1379.
17. **Triton X-100**: PanReac/AppliChem A4975.
18. **Tryptone**: PanReac/AppliChem 1,198,401.
19. **Urea**: PanReac/AppliChem A1361.
20. **Yeast Extract**: PanReac/AppliChem 1198400.

2.2 Buffers and Stock Solutions

Prepare all solutions using ultrapure water and analytical grade reagents (see Sect. 2.1). Prepare and store all reagents at room temperature, unless indicated otherwise. Make sure to follow local regulations regarding workplace safety, biosafety, and waste disposal.

1. **EDTA stock solution**: 500 mM in water. To make 1 L, add 186.1 g of EDTA (ethylenediaminetetraacetic acid) disodium salt 2-hydrate to 800 mL of water, dissolve completely, and adjust the volume to 1 L with water. Store at 4 °C.
2. **IPTG stock solution**: 1 M in water. To make 10 mL, add 2.38 g of IPTG (isopropyl β-D-1-thiogalactopyranoside) to 8 mL of water, dissolve completely, and adjust the volume to 10 mL with water. Store at −20 °C.
3. **DTT stock solution**: 1 M in water. To make 10 mL, add 1.54 g of DTT (dithiothreitol) to 8 mL of water, dissolve completely, and adjust the volume to 10 mL with water. Store at −20 °C.
4. **PMSF stock solution**: 0.5 M in DMSO (500×). To make 10 mL, add 870 mg of PMSF (phenylmethyl sulfonyl fluoride) to 8 mL of DMSO. Dissolve completely and adjust the volume to 10 mL with water. Store at −20 °C.
5. **NaCl stock solution**: 5 M in water. To make 500 mL, add 146.1 g of NaCl (sodium chloride) to 450 mL of water, dissolve completely (possibly with slight heating), and adjust the volume to 500 mL with water. Store at room temperature.
6. **Tris·Cl stock solutions:** 1 M in water. To make 500 mL, add 60.6 g of Tris (tris(hydroxymethyl)aminomethane) base to 450 mL of water pre-cooled to 4 °C, adjust pH to 8.0 or to 7.5 (both required), and adjust the volume to 500 mL with water. Store at 4 °C.
7. **HEPES stock solution:** 1 M in water. To make 500 mL, add 119.2 g of HEPES (4-(2-hydroxyethyl)-1-piperazineethane-sulfonic acid) to 450 mL of water pre-cooled to 4 °C, adjust

pH to 6.5, and adjust the volume to 500 mL with water. Store at 4 °C.

8. **Ampicillin stock solution**: 50 mg/mL in water. Store at −20 °C.
9. **Kanamycin stock solution**: 25 mg/mL in water. Store at −20 °C.
10. **Chloramphenicol stock solution**: 25 mg/mL in 70% Ethanol. Store at −20 °C.
11. **Dipeptide stock solution**: Varies depending on the solubility of the dipeptide; usually 1 M or 100 mM. Store at −20 °C.
12. **Refold buffer**:
    - Add 100 mL of water to a 1 L graduated cylinder or a glass beaker.
    - 100 mM Tris·Cl (pH 8.0): Add 100 mL of 1 M Tris·Cl pH 8.0 stock solution.
    - 500 mM L-Arginine hydrochloride: Add 500 mL of 1 M L-Arginine stock solution.
    - 2 mM EDTA: Add 4 mL of 500 mM EDTA.
    - Add water up to 900 mL and then adjust the buffer to pH 8.0 with HCl or NaOH. Make up the volume to 1 L with water. Store at 4 °C.
    - 5 mM reduced glutathione and 0.5 mM oxidized glutathione: Add 1.53 g of reduced glutathione and 0.306 g of oxidized glutathione to the chilled refolding buffer just before beginning the folding reaction.
    - 10 mM dipeptide: Add from the stock solution to the chilled refolding buffer just before beginning the folding reaction.
    - 0.1 mM PMSF: Add 200 μL of 0.5 M PMSF stock solution to the chilled refolding buffer just before beginning the folding reaction.
13. **SEC** (size exclusion chromatography) **buffer** (*see* **Note 1**):
    - Add 100 mL of water to a 1 L graduated cylinder or a glass beaker.
    - 150 mM NaCl: add 8.8 g NaCl.
    - 50 mM sodium phosphate: add 4.69 g $NaH_2PO_4 \cdot H_2O$ and 4.28 g $Na_2HPO_4 \cdot 7H_2O$ (*see* **Note 2**).
    - 10 mM dipeptide (*see* **Note 3**).
    - Add water up to 900 μl, adjust to pH 6.5 with HCl or NaOH, and make up the volume to 1 L with water.
    - Filter with a bottletop 45 μm filter to remove any particles before using for chromatography. Store at 4 °C.

14. **LB Medium:** 10 g tryptone, 5 g yeast extract, 10 g NaCl. Combine in 1 L of distilled water, dissolve completely, and autoclave.

15. **Buffer A**:
    - Add 80 mL of water to a graduated cylinder or glass beaker.
    - 50 mM Tris pH 8.0: Add 5 mL 1 M Tris·Cl pH 8.0.
    - 1 mM EDTA: Add 200 μl 500 mM EDTA.
    - 25% sucrose: Add 25 g sucrose.
    - Dissolve completely, then adjust the pH with HCl or NaOH to 8.0.
    - Adjust the volume to 100 mL with water.
    - Store at 4 °C for 1–2 months maximum.

16. **Buffer B**:
    - Add 160 mL of water to a graduated cylinder or glass beaker.
    - 50 mM Tris pH 8.0: Add 10 mL 1 M Tris·Cl pH 8.0.
    - 5 mM EDTA: Add 2 mL 500 mM EDTA.
    - 25% sucrose: Add 50 g sucrose.
    - 1% Triton X-100: Add 10 mL 20% (v/v) Triton X-100.
    - Dissolve completely, then adjust the pH with HCl or NaOH to 8.0.
    - Adjust the volume to 200 mL with water.
    - Store at 4 °C for 6 months maximum.

17. **Buffer C**:
    - Add 160 mL of water to a graduated cylinder or glass beaker.
    - 50 mM Tris pH 8.0: Add 10 mL 1 M Tris·Cl pH 8.0.
    - 2 M Urea: Add 24 g urea.
    - 2 M NaCl: Add 80 mL 5 M NaCl.
    - Dissolve completely, then adjust the pH with HCl or NaOH to 8.0.
    - Adjust the volume to 200 mL with water.
    - Store at 4 °C for 6 months maximum.

18. **Buffer D**:
    - Add 160 mL of water to a graduated cylinder or glass beaker.
    - 50 mM Tris pH 7.5: Add 10 mL 1 M Tris·Cl pH 7.5.
    - 150 mM NaCl: Add 6 mL 5 M NaCl.

- Dissolve completely, then adjust the pH with HCl or NaOH to 7.5.
- Adjust the volume to 200 mL with water.
- Store at 4 °C for 6 months maximum.

19. **Buffer E**:
    - Add 40 mL of water to a graduated cylinder or glass beaker.
    - 6 M guanidinium chloride: add 27.8 g guanidinium chloride.
    - 50 mM HEPES, pH 6.5: add 2.5 mL 1 M HEPES pH 6.5.
    - While dissolving the guanidinium chloride, adjust the pH with HCl or NaOH to 6.5.
    - When the solution is clear, adjust the pH again to 6.5.
    - Adjust the volume to 50 mL with water.
    - Store at 4 °C for 6 months maximum.

2.3 Cells and Plasmids

1. Bacterial expression strain (such as *E.coli BL21DE3(pLysS)*).
2. Expression plasmids for the MHC class I heavy chain and for b_2m. We usually use a vector of the pET system (such as pET3a) to express codon-optimized genes. For the disulfide-mediated stabilization, tyrosine-84 and alanine-139 are mutated to cysteines, allowing for the formation of a disulfide bond between the ends of the helices in the F pocket region (*see* **Note 4**).

3 Methods

3.1 Expression of MHC Class I Proteins in E. coli

The protocol describes the preparation of a single protein, heavy chain or β_2m, from 500 mL of bacterial culture. For the folding reaction (3.2.), one preparation of heavy chain and one of β_2m are required.

Preparatory Work

1. Make 600 mL LB medium and autoclave.
2. Transform the bacterial expression strain with the expression plasmid, streak out on plates with the appropriate antibiotic, and restreak to generate single colonies.
3. Inoculate one single colony into one flask with 50 mL LB medium with the appropriate antibiotics. Incubate at 37 °C, 200 rpm, overnight.

Protein Expression in *E. coli*

1. Centrifuge the starter culture in two 50 mL screw-cap centrifuge tubes ("Falcon" tubes) at 3000× *g* (rcf) for 10 minutes at room temperature (*see* **Note 5**).
2. Fill one 2 L Erlenmeyer flask with 500 mL LB medium containing appropriate antibiotics (*see* **Note 6**). Retrieve 15 mL of this medium and use it to resuspend the bacterial cell pellet. Transfer the resuspended pellet to the large flask to inoculate it with the starter culture.
3. Incubate the large flask in the shaker incubator: 37 °C, 180 rpm.
4. Take samples of the bacteria every hour and measure the OD_{600} with water as the blank.
5. Grow the 500 mL culture to an OD_{600} of about 0.6. Take a sample of the bacterial suspension (non-induced culture).
6. Add IPTG to induce expression (final concentration 0.5 mM).
7. Incubate in the 37 °C shaker incubator for another 4 h.
8. Take a sample of the bacterial suspension (induced culture).
9. Transfer into two 500 mL centrifuge bottles.
10. Harvest by centrifugation at 5000× *g* for 10 min at 4 °C.
11. Resuspend each of the two bacterial pellets in 20 mL Buffer A (sucrose buffer) (*see* **Note 7**). Transfer each resuspended pellet to one 50 mL Falcon tube.
12. Freeze the resuspended bacterial pellets by placing the tubes into the −80 or − 20 freezer (*see* **Note 8**).

Lysis of the Bacteria and Isolation and Purification of Inclusion Bodies

1. Lyse the cells by thawing the suspensions in a 30 °C water bath. Swirl the tubes to make sure the thawing is fast and uniform. Place the samples on ice.
2. Sonicate using a tip sonicator (*see* **Note 9**) until the solution is no longer viscous. Keep on ice between the bursts of sonication.
3. Take a 50 μl sample from the crude lysate and label it #1.
4. Transfer the cell lysates to 25 mL screw-cap high-speed centrifuge tubes ("Oak Ridge" tubes). Spin at 40,000× *g*, 4 °C, for 15 min. Take a sample from the supernatant and label it #2. Discard the supernatant.
5. Add 10 mL Buffer B (detergent buffer) to each Oak Ridge tube. Resuspend using a pipette. Transfer to a 50 mL Falcon tube. Sonicate as above. Take a sample from the suspension and label it #3.

6. Spin as above and discard the supernatant.
7. Resuspend in Buffer C (urea buffer), sonicate, take sample #4, and centrifuge as above.
8. Resuspend in Buffer D (tris-buffered saline), sonicate, take sample #5, and centrifuge as above.
9. Resuspend the pellet by sonication in 5 mL Buffer E. Take sample #6.
10. Leave the suspension to solubilize by gentle shaking at 4 °C for 24–48 hours.
11. Centrifuge the solubilized protein suspension at 40,000× *g*, 4 ° C, for 15 minutes. Spin the supernatant a second time to make sure you get rid of the insoluble pellet. Pool the supernatants from the two tubes of the same preparation.
12. Determine the concentration by measuring the OD_{280} against water. The extinction coefficients of human β_2m and of HLA-A*02:01 (extracellular domain) are approximately 19,300 and 82,400 M^{-1} cm^{-1}, respectively (exact numbers depend on the expression construct), similar for other MHC-I heavy chains. The expected concentration is 8–15 mg/mL for MHC-I heavy chain (yield from a 500 mL expression culture, 80–150 mg) and about 7.5–10 mg/mL for β_2m (yield, 75–100 mg).
13. Store the solubilized inclusion bodies at 4 °C (for several days) or freeze at −20 °C (for up to a year).
14. Check samples #1–6 on a 12% SDS-PAGE gel. The recombinant protein should be visible in the lysate (sample #1) but not in the supernatant (sample #2), and then in increasing purity in samples #3–6 (Fig. 1).

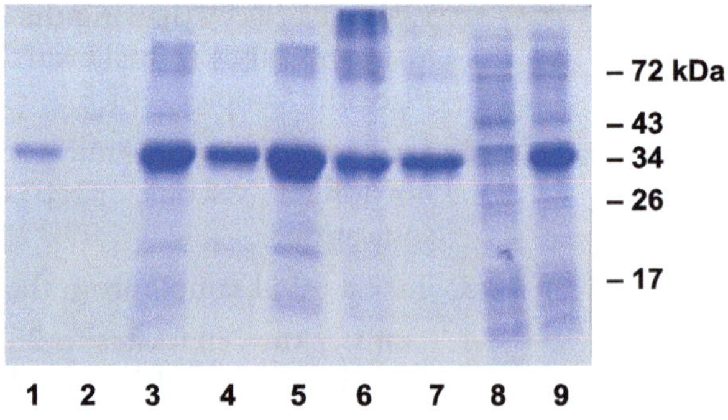

Fig. 1 Inclusion body preparation. Lanes: 8, uninduced culture; 9, induced culture; 1, buffer A lysate; 2, buffer A supernatant after centrifugation; 3, buffer B suspension; 4, buffer C suspension; 5, buffer D suspension; 6, buffer E suspension; 7, solubilized inclusion bodies

3.2 In Vitro Refolding of Peptide-Empty dsMHC-I

The refolding reaction is performed in the cold room.

1. The day before the folding reaction, prepare 1 L of refolding buffer (without glutathione, PMSF, peptide) as described above.
2. Filter the refold buffer through a 0.45 μm bottle-top filter.
3. Put the refold buffer into the cold room overnight.
4. On the morning of the folding reaction, thaw out the frozen vials of MHC-I heavy chain and β_2m solubilized inclusion bodies (or retrieve from the 4 °C storage). Mix well by inversion and keep them on ice.
5. To the cold refold buffer, add reduced and oxidized glutathione (as powders), the dipeptide (from a stock solution or as powder), and the PMSF (from a stock solution). Stir for 5 minutes to dissolve.
6. To begin with refolding, add the light chain (β_2m) dropwise and slowly (over 5 minutes) into the cold refold buffer with constant stirring, and then allow it to refold on its own for 1 hour, with continued stirring. The final concentration of β_2m in the refolding reaction should be 2 μM (about 22 mg/L, *see* **Note 10**).
7. Then start injecting the MHC-I heavy chain into the refold buffer in 8–10 portions added dropwise and slowly (over 5 minutes) at 1 hour intervals (*see* **Note 11**). The final concentration of the heavy chain in the refolding reaction should be 1 μM (about 30 mg/mL).
8. Then allow the refolding reaction to proceed for 24 to 48 hours with constant slow stirring in the cold room (*see* **Note 12**). The completed folding reaction is referred to as the "refold" in the following.
9. Take a 1 mL sample of the crude refold for analysis, and store at −20 °C.

3.3 Concentration of Refolded Protein

1. Filter the refold by using a nylon filter disk in the bottle-top filter (*see* **Note 14**), no vacuum needed.
2. Take a 1 mL sample of the filtered refold for analysis, and store at −20 °C.
3. Concentrate the filtered refold with an ultrafiltration cross-flow device with a cut-off range (MWCO) of 30,000 Da to about 25 mL (*See* **Notes 15** and **16**).
4. Take a 1 mL sample of the ultrafiltrate, and store at −20 °C.
5. Take a 100 μL sample of the concentrated refold, and store at −20 °C.
6. Further concentrate the sample with centrifugal ultrafiltration devices with a cut-off range (MWCO) of 30,000 Da to about 2 mL (*see* **Notes 17** and **18**).

3.4 Size Exclusion Chromatography (SEC

1. Prepare the SEC buffer as described above.
2. Prepare the Superdex-200 (26/600) column by washing with 2 column volumes (CV) of water, then equilibrating with 1 CV of SEC buffer (*See* **Note 19**).
3. Centrifuge the concentrated refold at 20,000× *g*, 15 min, 4 °C, in a tube appropriate to its volume.
4. Filter the concentrated refold with a 0.2 μm syringe filter.
5. Inject 1–2 mL of the concentrated refold into the SEC column and elute according to the procedure of your chromatography equipment.
6. Three peaks are expected (Fig. 2): an aggregate peak at the dead volume, the MHC-I heavy chain/β_2m complex peak at approximately 40–45 kDa, and a β_2m peak at approximately 10 kDa.
7. Analyze the SEC-purified protein fractions on a 12% SDS-PAGE gel with standard protein markers (Fig. 3). The aggregate peak usually contains only the heavy chain, the MHC-I heavy chain/β_2m complex peak contains both the heavy and the light chains, and the β_2m peak contains only the light chain.
8. Pool the desired protein fractions.

3.5 Storage

1. Determine the protein concentration.
2. Dilute the protein solution to a concentration of 10 mg/mL in SEC buffer.
3. Add dipeptide to 1 mM, or more if appropriate (*see* **Note 3**).

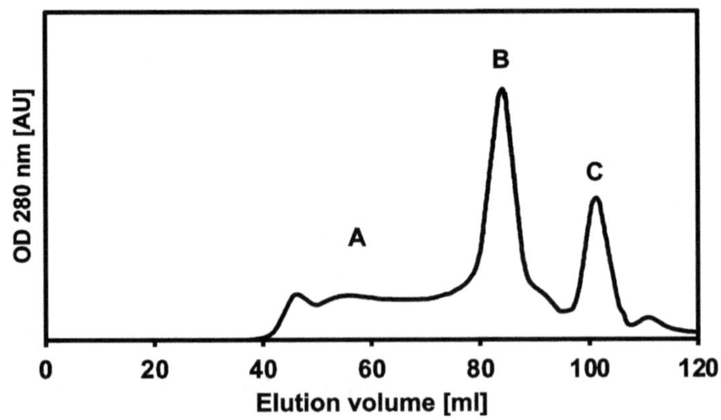

Fig. 2 Size exclusion chromatography of a dsMHC-I refold. Disulfide-stabilized A*02:01 was refolded with β_2m as described in the text and separated on a Cytiva Superdex 200 Increase 10/300 GL column. The volume on the x-axis counts from the point of injection. A, aggregates; B, dsMHC-I heavy chain/β_2m complex; C, free β_2m. See Fig. 3 for an SDS gel of the fractions

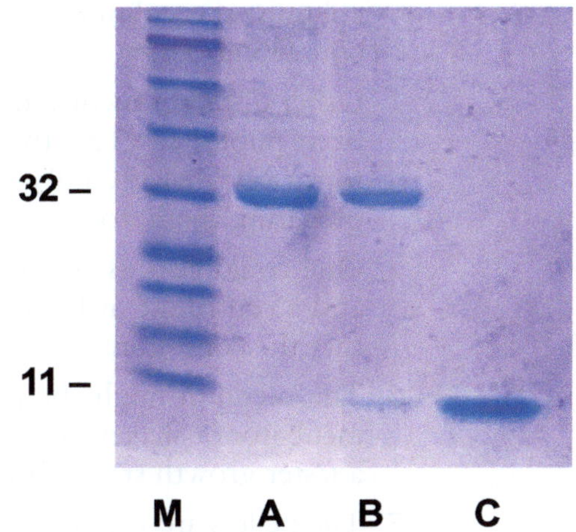

Fig. 3 Analysis of the fractions from size exclusion chromatography. Pooled samples corresponding to the A region and the B and C peaks in Fig. 2 were separated by SDS-PAGE. M, molecular weight marker (molecular weights in kDa). The A region contains high-molecular-weight aggregates of the heavy chain mostly; the B peak contains the dsMHC-I heavy chain/β_2m complex (the β_2m band is weaker because of the 3:1 mass ratio of the subunits), and the C peak contains free β_2m

4. Aliquot in 1 mL aliquots.
5. Store at 4 °C for days, or snap-freeze in liquid nitrogen (*see* **Note 20**) and keep at −80 °C for years.

4 Notes

1. The pH of the SEC buffer should be adjusted to the maximum stability of the peptide-empty dsMHC, which may vary between allotypes. We find pH 7.0 to be a relatively safe bet, but for some allotypes, such as HLA-A*02:01, pH 6.5 provides better results. There is no stringent reason that this should be a phosphate buffer, so Good's buffers are also adequate, especially when proceeding with crystallization (instead of making pMHC tetramers).
2. Mixing basic and acidic sodium phosphates in the right proportion (instead of just using one sodium phosphate) allows for better control of the ionic strength of the buffer. For changing the pH of the buffer, these amounts must be recalculated.
3. Some MHC-I allotypes, even if disulfide-stabilized, require conformational support by dipeptides at all times prior to the addition of peptides. For the SEC purification of such MHC-I allotypes, the respective dipeptide must be present in the SEC buffer at 10 mM. Importantly, in such cases, since during SEC, the protein migrates into the buffer that is present on the

column, the column must be equilibrated with a dipeptide-containing buffer.

4. Other disulfide-mediated stabilization efforts for MHC-I have been published [12]; this protocol, while originally designed for (Y84C/A139C) disulfide stabilization, might be useful in preparing them, too.
5. One could also use screw-cap high-speed "Oak Ridge" tubes and a centrifuge at higher g forces (20,000 x g), but we find that the conditions given are sufficient.
6. The medium might be pre-warmed by placing it in a 37 °C incubator or shaker for an hour before inoculating; this leads to a faster growth start of the expression culture.
7. The pellets are voluminous and sticky, and we find it useful to break them up and dislodge them from the wall of the tube with a Teflon stir bar retriever, or with a plastic pipette. Resuspension then continues by pipetting up and down until a smooth suspension is achieved.
8. The expression strain, *BL21(DE3)pLysS*, produces intracellular lysozyme. When the bacteria are frozen and thawed, the lysozyme gains access to the cell wall, and the bacteria autolyze. For this reason, it is better to freeze the bacterial suspension overnight at this step and then proceed with lysis on the next day.
9. We use an EpiShear probe sonicator (Active Motif, Carlsbad, CA, USA) with a medium tip, running a program with 2 seconds on and 4 seconds off, amplitude 80%.
10. If the concentration of b_2m in the folding reaction is too high, then covalent dimers and higher-order oligomers form that interfere with MHC-I folding. This might be worth optimizing.
11. Slow addition of the heavy chain is important for maximizing the yield. Folding and aggregation/denaturation are competing reactions, and since the rate of aggregation is proportional to the square of the concentration, it makes sense to keep the concentration of unfolded MHC-I heavy chain as low as reasonably possible. The addition intervals may be somewhat customized, depending on allotype, peptide, and other considerations. We usually add 5–6 aliquots with 1-hour breaks in between, then leave to fold overnight, and add another 3–4 aliquots at 1-hour intervals.
12. Optimal refolding time may vary with allotype and peptide; this is worth optimizing.
13. For wild-type MHC class I proteins folded with dipeptide or with short peptides, maintain the dipeptide or the short peptide in the SEC buffer throughout the entire purification.

14. The purpose of this step is to remove any larger aggregates and perhaps contamination that would interfere with the ultrafiltration. We use Merck Millipore NY4104700 nylon disk filters with a Thermo Scientific Nalgene DS0320–5045 reusable bottle-top filter.
15. We use a Sartorius (Göttingen, Germany) Vivaflow 200 Laboratory Cross Flow Cassette with a 30 kDa MWCO.
16. The filtrate contains the refold buffer and the dipeptide and may be used to immediately start another round of refolding by adding new β_2m and heavy chain inclusion bodies. This can even be repeated for a third round. This reuse saves on the cost of dipeptides.
17. We use Vivaspin 20 Centrifugal Concentrators from Sartorius with a 30 kDa MWCO.
18. Depending on the size exclusion column used, this centrifugation step may not be required. Omitting it (and using a larger size exclusion column that will allow 10–20 mL sample volume) will increase the yield since (in our experience) MHC-I aggregates and precipitate over time at concentrations of >10 mg/mL.
19. Before starting size exclusion chromatography, it is essential to ensure that the dsMHC-I will be stable in the absence of the dipeptide. This is, for example, true for HLA-A*02:01 and for H-2K^b. Newly designed allotypes must be tested for stability in the absence of dipeptide. On the size exclusion column, the dipeptide will be stripped away from the MHC-I, which might lead to precipitation (and loss of the column) if the dsMHC-I is unstable. If unsure, add 10 mM of the respective dipeptide to the SEC buffer used to equilibrate the column. This will make sure the dipeptide is present throughout the column run. The SEC buffer that is applied following injection of the sample into the column does not need to contain dipeptide.
20. Do not put the purified proteins into the −80 °C or the −20 °C freezer to freeze them. Slow freezing leads to local overconcentration, precipitation, and loss of the protein. This is especially pronounced with empty MHC-I.

5 Troubleshooting

1. **No protein expression**
   - Confirm that the right glycerol stock was used for expression.
   - Prepare a fresh plasmid and transform it into the expression strain.

- Use freshly prepared LB media.
- Confirm the IPTG concentration.

2. **Low expression yield**
   - Check the protein expression level by SDS-PAGE before purification.
   - Check proper cell lysis.
   - Check for over-sonication.
   - Use only freshly prepared buffer with the exact pH.
3. **Precipitation during refolding**
   - Include enough protease inhibitors.
   - Use only clean bottles for refolding.
   - Do not over-speed the magnetic stirrer.
   - Use the exact heavy chain:light chain ratio.
   - Check pH of buffer.
   - Check peptide concentration.
   - Use a different MHC-I allotype or construct.
   - Use lower MHC-I heavy chain concentration in the refolding reaction.
4. **Precipitation during concentration**
   - Skip the last concentration step, leave the refold at 10–20 mL volume and inject into a larger SEC column, and/or several times.
   - Try adjusting the refold to lower pH values (down to 6.5) to stabilize the MHC-I [13].

Conflict of Interest

The authors have no conflicts of interest to declare that are relevant to the content of this chapter.

References

1. Tiwari R, Singh VK, Rajneesh et al (2024) MHC tetramer technology: exploring T cell biology in health and disease. Adv Protein Chem Struct Biol 140:327–345. https://doi.org/10.1016/bs.apcsb.2024.02.002
2. Altman JD, Moss PA, Goulder PJ et al (1996) Phenotypic analysis of antigen-specific T lymphocytes. Science 274:94–96
3. Bentzen AK, Hadrup SR (2019) T-cell-receptor cross-recognition and strategies to select safe T-cell receptors for clinical translation. Immuno-Oncol Technol 2:1–10. https://doi.org/10.1016/j.iotech.2019.06.003
4. Toebes M, Rodenko B, Ovaa H, Schumacher TN (2009) Generation of peptide MHC class I monomers and multimers through ligand exchange. Curr Protoc Immunol Chapter 18: Unit 18:16
5. Luimstra JJ, Garstka MA, Roex MCJ et al (2018) A flexible MHC class I multimer loading system for large-scale detection of antigen-

specific T cells. J Exp Med 215:1493–1504. https://doi.org/10.1084/jem.20180156

6. Saini SK, Schuster H, Ramnarayan VR et al (2015) Dipeptides catalyze rapid peptide exchange on MHC class I molecules. Proc Natl Acad Sci USA 112:202–207. https://doi.org/10.1073/pnas.1418690112
7. Overall SA, Toor JS, Hao S et al (2020) High throughput pMHC-I tetramer library production using chaperone-mediated peptide exchange. Nat Commun 11:1909. https://doi.org/10.1038/s41467-020-15710-1
8. Hadeler A, Saikia A, Zacharias M, Springer S (2022) Rapid peptide exchange on MHC class I by small molecules elucidates dynamics of bound peptide. Curr Res Immunol 3:167–174. https://doi.org/10.1016/j.crimmu.2022.08.002
9. Saini SK, Tamhane T, Anjanappa R et al (2019) Empty peptide-receptive MHC class I molecules for efficient detection of antigen-specific T cells. Sci Immunol 4:eaau9039. https://doi.org/10.1126/sciimmunol.aau9039
10. Anjanappa R, Garcia-Alai M, Kopicki J-D et al (2020) Structures of peptide-free and partially loaded MHC class I molecules reveal mechanisms of peptide selection. Nat Commun 11:1314. https://doi.org/10.1038/s41467-020-14862-4
11. Saini SK, Ostermeir K, Ramnarayan VR et al (2013) Dipeptides promote folding and peptide binding of MHC class I molecules. Proc Natl Acad Sci USA 110:15383–15388. https://doi.org/10.1073/pnas.1308672110
12. Sun Y, Young MC, Woodward CH et al (2023) Universal open MHC-I molecules for rapid peptide loading and enhanced complex stability across HLA allotypes. Proc Natl Acad Sci U S A 120:e2304055120. https://doi.org/10.1073/pnas.2304055120
13. Saikia A, Hadeler A, Prasad P et al (2022) Fast peptide exchange on major histocompatibility complex class I molecules by acidic stabilization of a peptide-empty intermediate. Protein Sci 31:e4478. https://doi.org/10.1002/pro.4478

Chapter 3

Determining the Half-Life of MHC Class I Molecules by Blocking Protein Synthesis Using Cycloheximide

A. W. Peshala Amarajeewa, Alexander Fish, and Malgorzata A. Garstka

Abstract

Major histocompatibility complex class I (MHC I) molecules play a crucial role in activating adaptive immune responses by presenting viral and tumor antigens to cytotoxic CD8$^+$ T cells. Additionally, MHC I molecules have been implicated in autoimmunity through the presentation of self-peptides. The effective presentation of antigenic peptides depends on the stability of MHC I allotypes. Evaluating MHC I stability involves analyzing its degradation kinetics and determining its half-life. In this chapter, we present an assay to evaluate the degradation kinetics of MHC I molecules (both endogenous and overexpressed) in cell lines by inhibiting protein synthesis with cycloheximide. Additionally, we incorporated MG132, a proteasomal degradation inhibitor, to examine the impact of the proteasome on MHC I degradation. Furthermore, we detail a method for calculating the half-life of MHC I molecules by fitting the degradation data into a one-phase decay model.

Key words Major histocompatibility complex class I molecules, HLA-B*39:06, HLA-B*38:01, Cycloheximide, MG132, Degradation kinetics, One-phase decay, Half-life

1 Introduction

Major Histocompatibility Complex class I (MHC I) is a transmembrane glycoprotein composed of a highly polymorphic heavy chain (HC), a non-covalently associated invariant β_2-microglobulin (β_2m) light chain, and a peptide. MHC I molecules play a crucial role in adaptive immune responses by presenting antigens to cytotoxic CD8$^+$ T cells, thereby creating a favorable environment for these cells to combat pathogenic viral infections and tumors [1–3]. Additionally, MHC I is also known to elicit autoimmune responses by presenting self-antigens, which can lead to autoimmunity [4]. The biosynthesis and folding of MHC I take place in the

Supplementary Information The online version contains supplementary material available at https://doi.org/10.1007/978-1-0716-5092-9_3.

Efstratios Stratikos (ed.), *Antigen Processing and Presentation: Methods and Protocols*, Methods in Molecular Biology, vol. 3007, https://doi.org/10.1007/978-1-0716-5092-9_3,

endoplasmic reticulum (ER). The peptides generated by the proteasome are transported to the lumen of the ER by the transporter associated with antigen processing (TAP) [5, 6]. In the ER, these peptides are further trimmed by endoplasmic reticulum-associated aminopeptidases [7, 8]. MHC I can bind a diverse array of peptides, ranging from optimal (high-affinity) to suboptimal (low-affinity) ones. ER-resident chaperones, such as calnexin, calreticulin, ERp57, and tapasin, assemble together with MHC I to form a peptide-loading complex that supports folding, peptide loading, and peptide optimization on MHC I. Once loaded with peptides, MHC I molecules are transported through the secretory pathway to the cell surface, where they present peptides to cytotoxic $CD8^+$ T cells for immunosurveillance.

MHC I loaded with sub-optimal peptides results in the formation of unstable peptide-MHC I complexes [9], consequently leading to altered immune responses. Rather than the presentation of peptides at the cell surface, unstable peptide-MHC I complexes tend to be retained in the secretory pathway and recycling pathways subjected to cellular quality control [10–12]. Typically, these forms of peptide-MHC I complexes are degraded and removed from the antigen presentation pathway mainly by various quality control pathways, including the endoplasmic reticulum-associated degradation (ERAD); supported by ubiquitin-associated 20S proteasomal degradation [13], which plays a vital role in maintaining cellular homeostasis.

The affected stability and degradation of MHC class I molecules have been observed in viral infections [14, 15], tumors [16], and autoimmunity [17, 18]. Stability has been proposed as a factor influencing the predisposition of MHC I to autoimmunity [4, 19]. Studying the degradation kinetics of MHC I could reveal their stability and the mechanisms behind the altered immune responses in various diseases. Thus, examining MHC I degradation patterns is crucial as it provides valuable insights into immune responses in tumor, pathogen-related and autoimmune diseases [20].

Many commercial antibodies for the detection of MHC class I molecules by Western blot are available. Moreover, small epitope tags located at the cytosolic tail at the C terminus of MHC class I heavy chain are successfully used by us and others to detect specific MHC I allotypes [4, 21]. The inclusion of small tags does not affect the maturation, surface expression, or degradation of a variety of MHC class I allotypes [22, 23].

Cycloheximide is a well-known chemical compound originally isolated from *Streptomyces griseus*. It inhibits protein synthesis in eukaryotic cells by blocking translation elongation, subsequently creating ER stress due to the accumulation of misfolded/unfolded proteins in the ER [24]. Therefore, cycloheximide can be effectively used to study the degradation kinetics of proteins, including

MHC I. MG132 is a peptide aldehyde derived from a Chinese medical plant that blocks the proteasomal degradation pathway. It can covalently bind to the active site of the β sub-units of the 20S proteasome, resulting in blocking the proteolytic activity of the 26S proteasomal complex [25]. Therefore, MG132 can be effectively used to study the proteasomal degradation of MHC I by blocking the activity of proteasomes.

In this chapter, we describe a method to study the degradation kinetics of MHC I molecules by blocking protein synthesis using cycloheximide. This approach allows for the observation of MHC I degradation over time. Furthermore, we modified the cycloheximide assay to determine the effect of proteasomal degradation on MHC I by treating the cells with the proteasomal inhibitor MG132. Additionally, we describe the method for fitting degradation kinetics data to a one-phase decay model to determine the half-life ($t_{1/2}$) of MHC I.

2 Materials

2.1 General Reagents

1. Cell lines of interest. In this chapter, we use a human cervical cancer cell line (HeLa) and HeLa cells stably expressing HLA-B*39:06-HA and HLA-B*38:01-HA.
2. Cell culture medium, Dulbecco's modified Eagle's medium (DMEM) with 8% Fetal Bovine Serum (FBS) and 1% penicillin/streptomycin (referred to as a full medium).
3. 1x Phosphate Buffered Saline (PBS).
4. Eppendorf tubes and tissue culture grade 6 well plates or 3.5 cm dishes.
5. Sterile pipettes, disposable tips, and cell culture flasks.

2.2 Studying the Degradation of MHC Class I

1. Dimethyl Sulfoxide (DMSO).
2. 10 mg/mL cycloheximide (protein synthesis inhibitor) stock solution in DMSO, aliquot and store at −20 °C.
3. 20 mM MG132 (proteasomal inhibitor) stock solution in DMSO, aliquot, and store at −20 °C.

2.3 Cell Harvesting, Lysis, and Sample Preparation

1. Lysis buffer containing detergent (0.5–1% CHAPS, 1X M-PER, 1X T-PER, or any other detergent that does not absorb light at 280 nm (*see* **Note 1**), supplemented with complete protease cocktail (e.g., Roche, cOmplete protease inhibitor cocktail EDTA-free, cat.11873580001) which targets serine and cysteine proteases, 10 mM NEM, an inhibitor of deubiquitinating enzymes, in 1X PBS (Prepare freshly before use and store at 4 °C).

2. Cell scraper.
3. NanoDrop (Thermofisher) to measure protein concentration.

2.4 Sodium Dodecyl Sulfate Polyacrylamide Gel Electrophoresis (SDS-PAGE)

1. 30% acrylamide/bis-acrylamide (37.5:1).
2. 10% ammonium persulfate (APS): Weigh 5 g of APS. Dissolve in 50 mL of ultra-pure water. Aliquot and store at −20 °C.
3. 10% SDS: Weigh 5 g of SDS. Dissolve in 50 mL of ultra-pure water. Aliquot and store at RT.
4. Separation gel buffer: 1.5 M Tris-HCl, pH = 8.8. Weigh 181.7 g of Tris-HCl. Dissolve in 750 mL of ultra-pure water. Adjust the pH with concentrated HCl, and bring the final volume to 1 L. Store at RT.
5. Stacking gel buffer: 0.5 M Tris-HCl, pH = 6.8. Weigh 60.6 g of Tris-HCl. Dissolve in 750 mL of ultra-pure water. Adjust the pH with concentrated HCl and bring the final volume to 1 L. Store at RT.
6. Tetramethylethylenediamine (TEMED): Store at 4 °C.
7. Prestained molecular weight standards.
8. SDS-PAGE running buffer: 25 mM Tris–HCl, pH 8.3. 192 mM glycine, 0.1% SDS. Dissolve in ultra-pure water. Store at RT(*see* **Note 2**).
9. 5X SDS-PAGE sample buffer: 10% SDS, 30% glycerol, 5% β-mercaptoethanol, 0.04% bromophenol blue, 250 mM Tris-HCl (pH = 6.8). Dissolve in ultra-pure water. Aliquot and store at −20 °C.
10. 0.1% Bromophenol blue solution: Weigh 0.1 g of bromophenol blue. Dissolve in 100 mL ultra-pure water or ethanol. Store the solution in a dark, cool place to maintain its stability.
11. Gel casting and running apparatus.

2.5 Western Blot

1. 10X Western blot transfer buffer: Weigh 30 g of Tris base and 144 g of glycine. Dissolve in 1 L of ultra-pure water and store at RT (to use, dilute in ultra-pure water to 1X, and add methanol to 20%).
2. Components for western blot assembly: PVDF membranes, filter papers and sponge pads.
3. Transfer apparatus: Typically, an electrotransfer system (e.g., wet or semi-dry transfer).
4. Blocking solution: e.g., 5% milk in PBS. Weigh 2.5 g of skimmed milk powder and dissolve in 50 mL of PBS (*see* **Note 3**).

5. Antibody incubation buffer: e.g., 2% milk in PBST. Weigh 1 g of skimmed milk powder and dissolve in 50 mL of PBST (*see* **Note 3**).
6. PBST: 0.1% Tween-20 in PBS.
7. Appropriate primary antibodies: anti-MHC class I and loading control (here, GAPDH) antibodies.
8. Appropriate secondary antibodies conjugated with horseradish peroxidase (HRP).
9. Luminescent HRP detection reagent.
10. Chemiluminescence detection system.

3 Methods

3.1 Treating Cells with Cycloheximide

We use the HeLa cell line to study endogenous MHC I (Fig. 1) and HeLa cells that stably express HLA-B with a C-terminal HA tag to

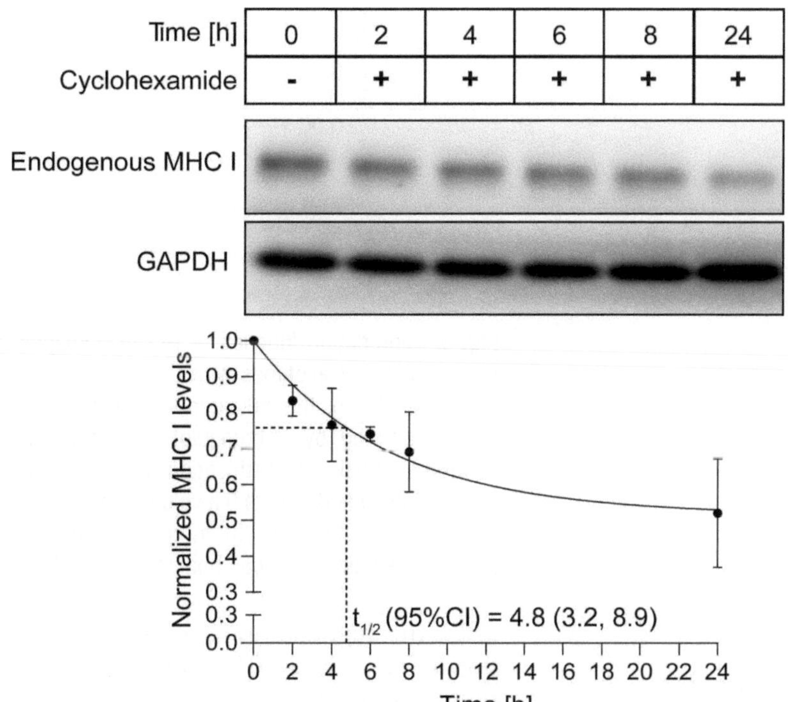

Fig. 1 Degradation kinetics of endogenous MHC class I molecules in HeLa cells. HeLa cells were treated with 25 μg/mL cycloheximide for the indicated time points and lysed. Protein samples (10 μg) were examined by immunoblot with anti-HLA I and GAPDH (loading control) antibodies. Quantification: The signal for MHC I bands was first normalized to GAPDH and then to the 0 h value. Data are represented as mean ± SD, $n = 5$, and fitted by nonlinear fit to one phase decay to calculate half-life ($t_{1/2}$, h)

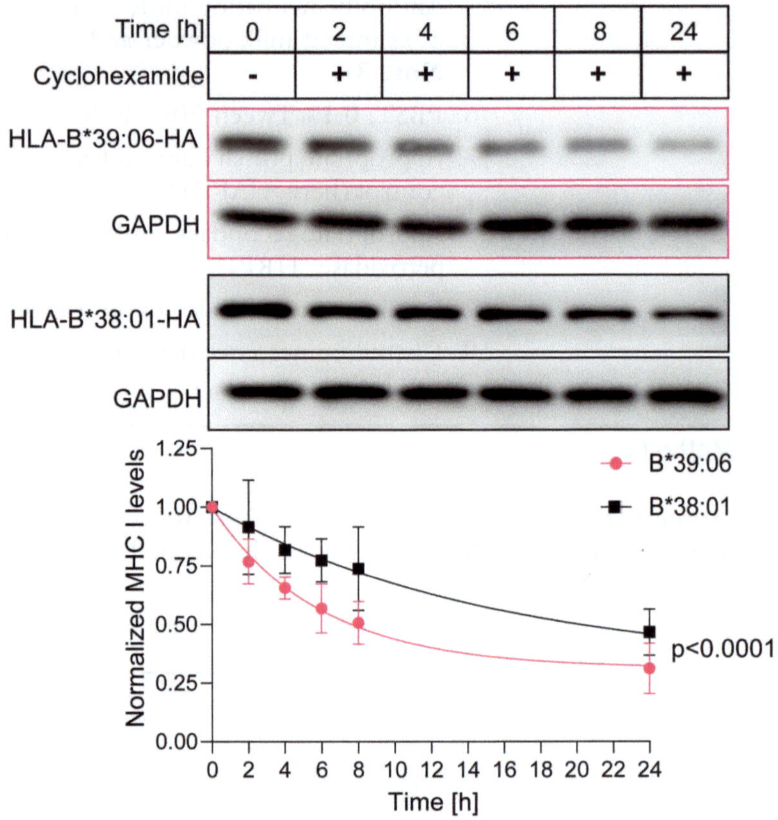

	B*39:06	B*38:01
$t_{1/2}$ (95% CI)	4.1 (3.2, 5.7)	10.7 (8.4, 14.9)

Fig. 2 Degradation kinetics of HLA-B allotypes. HeLa cells stably expressing indicated HLA-B allotypes with C-terminal HA tag were treated with 25 μg/mL cycloheximide for the indicated time points and lysed. Protein samples (10 μg) were examined by immunoblot with anti-HA (12CA5) and GAPDH (loading control) antibodies. Quantification: The signal for MHC I bands was first normalized to GAPDH and then to 0 h value. Data are represented as mean ± SD, $n = 7$, and fitted by nonlinear fit to one phase decay to calculate half-life ($t_{1/2}$). Statistical significance was analyzed using the two-way ANOVA

assess the degradation of overexpressed MHC I (here, HLA-B*39:06 and HLA-B*38:01 allotypes) (Fig. 2). The cells are cultured in tissue culture flasks with the complete medium until they reach 70–80% confluency. The growth medium and incubation conditions remain consistent across all experiments unless specified otherwise.

1. Seed 0.25–0.30 × 10^6 cells in 2 mL of complete medium per 3.5 cm dish, using one plate for each time point tested (here, six dishes for 0, 2, 4, 6, 8, and 24 h time points) and culture

them under standard conditions until they reach approximately 80% confluency (*see* **Note 4**).

2. Prepare a cycloheximide solution in a complete medium. Cycloheximide concentration and incubation time may vary depending on the cell line used. For HeLa cells, the cycloheximide concentration ranges from 5 to 50 μg/mL, with an incubation time of 12–24 h (*see* **Note 5**).
3. Discard the old culture medium from the plates and add 2 mL of culture medium containing cycloheximide to each plate, except for the 0 h time point (*see* **Note 6**).
4. Immediately harvest and lyse the 0 h time point sample (without adding cycloheximide), following the steps outlined in Sect. 3.3 (*see* **Note** 7).
5. Process the samples collected at other time points in the same way by following the steps outlined in Sect. 3.3 (*see* **Note** 7).

3.2 Treating Cells with Cycloheximide in the Presence of MG132 (see Note 8)

Here, we use HeLa cells stably expressing HLA-B*39:06 with the C-terminal HA tag (Fig. 3).

1. Seed 0.25–0.30 $\times 10^6$ cells in 2 mL of complete medium per 3.5 cm dish, using two sets of plates (one for treatment with MG132, the other as vehicle control, DMSO) and culture them under standard conditions until they reach approximately 80% confluency (*see* **Notes 4** and **9**).
2. Prepare MG132 solution in a complete medium. The concentration of MG132 and incubation time may vary depending on the cell line used. For HeLa cells, MG132 concentration ranges from 1 to 10 μM, with an incubation time of 6–26 h (*see* **Notes 9** and **10**).
3. Pre-treat the cells with MG132. Discard the old medium, add medium with MG132 or DMSO to each well, and incubate for at least 2 h (*see* **Notes 10** and **11**).
4. Prepare cycloheximide solution in a complete medium (*see* **Note 12**).
5. Add medium with cycloheximide to all plates in the experimental setup except for the 0 h time point (*see* **Note 13**).
6. Immediately harvest and lyse the 0 min time point sample (without adding cycloheximide), following the steps outlined in Sect. 3.3 (*see* **Note** 7).
7. Process the samples collected at other time points in the same way by following the steps outlined in Sect. 3.3 (*see* **Note** 7).

3.3 Harvest and Lysis of Cells (see Note 14)

1. At each time point, take the corresponding plate with cells from the incubator and place it on ice (*see* **Note** 7).
2. Discard the medium and wash the cells with 1 mL of 1X PBS.

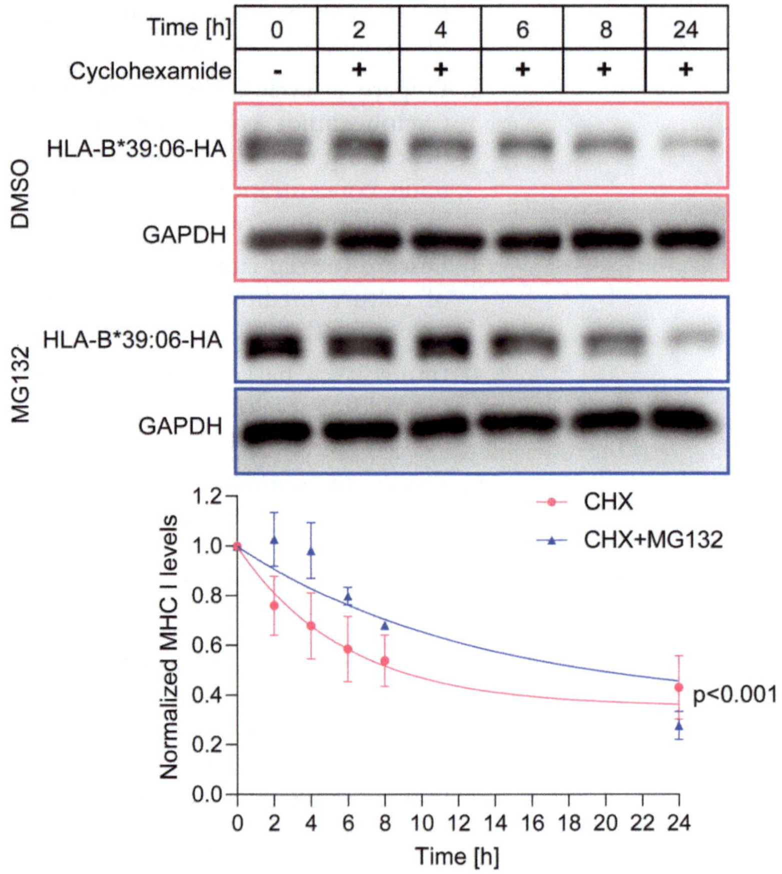

	CHX	CHX+MG132
$t_{1/2}$ (95% CI)	4.1 (2.8, 7.3)	9.1 (6.2, 17.5)

Fig. 3 Degradation kinetics of HLA-B*39:06 under inhibition of proteasome. HeLa cells stably expressing HLA-B*39:06 with C-terminus HA-tag were pre-treated with 5 μM MG132 for 2 h, after which time 25 μg/mL cycloheximide was added, cells were cultured for the indicated time points and lysed. Protein samples (10 μg) were examined by immunoblot with anti-HA (12CA5) and GAPDH (loading control) antibodies. Quantification: The signal for MHC I bands was first normalized to GAPDH and then to the 0 h value. Data are represented as mean ± SD, $n = 3$, and fitted by nonlinear fit to one phase decay to calculate protein half-life ($t_{1/2}$, h)

3. Add 120 μL of lysis buffer and scrape the cells with a cell scraper.
4. Transfer the cell lysate to a 1.5 mL Eppendorf tube.
5. Allow the cells to lyse in a rotator set at 20 rpm for 30 min at 4 °C.
6. Spin the lysed cell suspension at 16,000× g for 15 min, 4 °C.

7. Transfer the supernatant to a 1.5 mL Eppendorf tube and store it at −80 °C until further use.

3.4 SDS-Page

3.4.1 Preparation of 10% SDS-PAGE Gels

1. Clean and set up the gel casting apparatus (*see* **Note 15**).
2. Prepare 8 mL of 10% separation gel by mixing 3.2 mL of ultra-pure water, 2.7 mL of 30% acrylamide/bis-acrylamide, 2 mL of 1.5 M Tris HCl (pH = 8.8), 75 μL of 10% SDS, 10 μL of TEMED and 75 μL of 10% APS (7.5 mL is enough for one gel, *see* **Note 16**). Avoid forming air bubbles while mixing.
3. After mixing, immediately pour the separation gel into the gel casting tray, then add 1 mL of isopropanol on top to ensure an even surface formation of the gel (*see* **Note 17**).
4. Let the gel solidify for at least 30 min before pouring the stacking gel.
5. Discard the isopropanol once the gel has completely solidified, then rinse at least three times with ultra-pure water.
6. Prepare 3 mL of 4% stacking gel by mixing 2.1 mL of ultra-pure water, 500 μL of 30% acrylamide/bis-acrylamide, 380 μL of 0.5 M Tris HCL (pH = 6.8), 30 μL of 10% SDS, 30 μL of 10% APS and 3 μL of TEMED. (Avoid forming air bubbles while mixing).
7. Carefully pour the stacking gel over the separation gel, quickly insert the comb, and allow the gel to solidify for at least 20 min. Gels can be stored for up to one week in ultra-pure water or wrapped in damp tissue paper at 4 °C.

3.4.2 Preparing Samples for SDS-PAGE

1. Determine the protein concentration in the cell lysates using Nanodrop.
2. Resuspend 10 μg of protein in 5X SDS-PAGE sample buffer to achieve a final concentration of 1X SDS-PAGE sample buffer.
3. Denature the samples at 95 °C for 10 min. and let them cool down on ice.
4. Store the samples at −80 °C until further use.

3.4.3 Running SDS-PAGE

1. Assemble the gel in the SDS-PAGE running apparatus and fill both the inner and outer chambers with 1x SDS-PAGE running buffer.
2. Carefully remove the comb without damaging the wells.
3. Load prestained molecular weight standards and samples into the wells of 10% polyacrylamide gel.
4. Run the gel at 80 V for 30 min, until the dye front enters the separation gel.
5. Increase the voltage to 100 V and continue running the gel for approximately 1.5 h until the dye front exits the gel.

3.5 Western Blot

1. Transfer the proteins from the gel to 45 μm PVDF membrane (Millipore) at 100 V for 2 h.
2. Incubate the membrane in a blocking solution for 1 h at RT.
3. Prepare primary antibodies in an antibody incubation buffer. The concentration/dilution of antibodies, along with the time and temperature of incubation, will vary and need to be determined experimentally.
4. Wash the membrane 6 times for 5 min with 1x PBST.
5. Incubate the membrane with HRP-conjugated secondary antibodies for 30–60 min. at RT.
6. Wash the membrane 3 times for 5 min. With 1x PBST, and 3 times for 5 min. With 1x PBS (*see* **Note 18**).
7. Develop the signal by using a chemiluminescent reagent and visualize the proteins on the membrane.

3.6 Quantification of the Signal and Data Normalization

1. Quantify the signal from the immunoblotted membrane using appropriate software (we use ImageJ software).
2. First, normalize the MHC I signal to the GAPDH signal at each time point (*see* **Note 19**).
3. Second, normalize all data to the 0 min time point for each degradation assay (*see* **Note 20**).

3.7 Fitting the Data to a One-Phase Decay Model

Conduct a nonlinear regression analysis and fit the data to a one-phase decay curve represented by the equation below, as outlined in the Supplementary Information.

$$Y(t) = (Y_0 - \text{Plateau}) \times e^{-K \times t} + \text{Plateau}$$

where

Y(t) = dependent variable; here, normalized MHC I levels.

Y_0 = Initial value of Y at $t = 0$.

Plateau = value of Y as t approaches infinity.

K = rate constant, which indicates the speed of decay.

t = independent variable; here, time.

4 Notes

1. The most common method for protein quantification involves measuring direct absorbance at 280 nm with Nanodrop. CHAPS, M-PER and T-PER do not absorb at 280 nm and do not affect the accuracy of A280 quantification. TritonX-100 and NP-40 detergents absorb at a wavelength of 280 nm and are unsuitable for determining protein concentration using the

Nanodrop, as the use of buffers with significant absorbance at 280 nm may lead to quantification errors when making direct measurements at this wavelength.

2. We prepare 10X SDS-PAGE running buffer and dilute it to 1X to run the gel. For the 10X SDS-PAGE running buffer, weigh 30.3 g of Tris base, 144 g of glycine, and 10 g of SDS, dissolve in 1 L of ultra-pure water, and store at RT (to use, dilute to 1X in ultra-pure water).
3. Consult the supplier's information to determine the best blocking reagent for your antibody. Some antibodies may need different blocking solutions, like BSA.
4. When using cells overexpressing MHC I, you may need to culture them with the appropriate antibiotic. For the cycloheximide assay, refrain from including the antibiotic in the medium to prevent antibiotic-induced effects on cell proliferation and viability during the experiment, thereby ensuring the accuracy of the results.
5. Determine the timeline for the cycloheximide chase beforehand. There can be variations between cells, and cycloheximide can be toxic to some cells when exposed to high concentrations and prolonged incubation times.
6. When preparing media with inhibitors, it is advisable to prepare enough solution for at least one extra plate to ensure that the same volume of media can be added to each cell culture plate. This guarantees that the cells in every plate are exposed to the same amount of MG132 or cycloheximide, resulting in consistent results.
7. After being removed from the incubator, the cells should be placed on ice to stop their metabolic activities and harvested immediately to ensure data accuracy.
8. Other degradation inhibitors (e.g., chloroquine or wortmannin) can be used by following the same procedure.
9. The experimental setup should include two sets of plates (12 plates in total). The first set (6 plates) is to check the effect of MG132 (cells are treated with both MG132 and cycloheximide), while the second set (6 plates) serves as the control (cells are treated with DMSO and cycloheximide).
10. Determine the timeline for the MG132 treatment beforehand. There can be variations between cells, and MG132 can be toxic to some cells when exposed to high concentrations and prolonged incubation times.
11. Prepare MG132 solution in a complete medium at a concentration 1.3X the working concentration. For example, if the working concentration is 5 μM, prepare the solution at 6.7 μM. Add 1.5 mL of this solution to the cells in each well. After

adding the cycloheximide to the cells, the desired working concentration of MG132 will be 5 μM.

12. Add cycloheximide solution to cells cultured already in a complete medium with MG132. We prepare cycloheximide solution in a complete medium at a concentration 4X the working concentration. For example, if the working concentration is 25 μg/mL, prepare the solution at 100 μg/mL. Add 0.5 mL of this solution to 1.5 mL of medium with MG132. This will dilute cycloheximide to the desired 25 μg/mL working concentration.
13. Do not discard the medium with MG132/DMSO at this stage. Add medium with cycloheximide to cells incubated in medium with MG132. Ensure that the cells remain in the combined medium with both MG132 and cycloheximide throughout the entire cycloheximide chase.
14. All the steps must be done at 4 °C to prevent protein degradation during sample harvesting and processing.
15. Clean the glasses and combs of the gel-casting trays with ultrapure water to remove any dust or impurities and let them dry.
16. The volume needed to cast the gel depends on the gel casting apparatus used. If you are using the BIORAD Mini-PROTEAN Tetra cell system, 7.5 mL of separation gel is sufficient.
17. Leave enough space to pour the stacking gel over the separation gel (which should be at least 3 cm high).
18. Tween-20 may interfere with luminescent HRP detection reagents, causing high background signals. To minimize this, perform the final wash with PBS without Tween-20 or briefly rinse the membrane with ultra-pure water before signal detection.
19. Normalizing the data to the GAPDH signal eliminates variability caused by unequal protein loading during SDS-PAGE.
20. To check the degradation kinetics of MHC I molecules (using either cycloheximide or cyclohexmide+MG132) normalize the data at all time points to the 0 min time point for each experiment.

Acknowledgments

This study was funded by the National Natural Science Foundation of China grant (81870536 to MAG).

Conflict of Interest This study was funded by the National Natural Science Foundation of China (grant number 81870536 to MAG).

References

1. Neefjes J, Jongsma ML, Paul P, Bakke O (2011) Towards a systems understanding of MHC class I and MHC class II antigen presentation. Nat Rev Immunol 11(12):823–836
2. Vyas JM, Van der Veen AG, Ploegh HL (2008) The known unknowns of antigen processing and presentation. Nat Rev Immunol 8(8): 607–618
3. York IA, Goldberg AL, Mo X, Rock KL (1999) Proteolysis and class I major histocompatibility complex antigen presentation. Immunol Rev 172(1):49–66
4. Amarajeewa AP, Özcan A, Mukhtiar A, Ren X, Wang Q, Ozbek P, Garstka MA, Serçinoğlu O (2024) Polymorphism in F pocket affects peptide selection and stability of type 1 diabetes-associated HLA-B39 allotypes. Eur J Immunol:2350683
5. Goldberg AL, Rock KL (1992) Proteolysis, proteasomes and antigen presentation. Nature 357(6377):375–379
6. Uebel S, Tampé R (1999) Specificity of the proteasome and the TAP transporter. Curr Opin Immunol 11(2):203–208
7. Brouwenstijn N, Serwold T, Shastri N (2001) MHC class I molecules can direct proteolytic cleavage of antigenic precursors in the endoplasmic reticulum. Immunity 15(1):95–104
8. Fruci D, Niedermann G, Butler RH, van Endert PM (2001) Efficient MHC class I-independent amino-terminal trimming of epitope precursor peptides in the endoplasmic reticulum. Immunity 15(3):467–476
9. Harndahl M, Rasmussen M, Roder G, Dalgaard Pedersen I, Sørensen M, Nielsen M, Buus S (2012) Peptide-MHC class I stability is a better predictor than peptide affinity of CTL immunogenicity. Eur J Immunol 42(6): 1405–1416
10. Domnick A, Winter C, Sušac L, Hennecke L, Hensen M, Zitzmann N, Trowitzsch S, Thomas C, Tampé R (2022) Molecular basis of MHC I quality control in the peptide loading complex. Nat Commun 13(1):4701
11. Owen BA, Pease LR (2001) Thermal stability of MHC class I-β2-microglobulin peptide complexes in the endoplasmic reticulum is determined by the peptide occupancy of the transporter associated with antigen processing complex. J Immunol 166(3):1740–1747
12. Springer S (2015) Transport and quality control of MHC class I molecules in the early secretory pathway. Curr Opin Immunol 34: 83–90
13. Hughes EA, Hammond C, Cresswell P (1997) Misfolded major histocompatibility complex class I heavy chains are translocated into the cytoplasm and degraded by the proteasome. Proc Natl Acad Sci 94(5):1896–1901
14. Wiertz EJ, Jones TR, Sun L, Bogyo M, Geuze HJ, Ploegh HL (1996) The human cytomegalovirus US11 gene product dislocates MHC class I heavy chains from the endoplasmic reticulum to the cytosol. Cell 84(5):769–779
15. Jones TR, Sun L (1997) Human cytomegalovirus US2 destabilizes major histocompatibility complex class I heavy chains. J Virol 71(4): 2970–2979
16. Yamamoto K, Venida A, Yano J, Biancur DE, Kakiuchi M, Gupta S, Sohn AS, Mukhopadhyay S, Lin EY, Parker SJ (2020) Autophagy promotes immune evasion of pancreatic cancer by degrading MHC-I. Nature 581(7806):100–105
17. Abualrous ET, Fritzsche S, Hein Z, Al-Balushi MS, Reinink P, Boyle LH, Wellbrock U, Antoniou AN, Springer S (2015) F pocket flexibility influences the tapasin dependence of two differentially disease-associated MHC class I proteins. Eur J Immunol 45(4):1248–1257
18. Austin MC, Muralidharan C, Roy S, Crowder JJ, Piganelli JD, Linnemann AK (2025) Dysfunctional β-cell autophagy induces β-cell stress and enhances islet immunogenicity. Front Immunol 16:1504583
19. Ren X, Amarajeewa AP, Jayasinghe MT, Garstka MA (2024) Differences in F pocket impact on HLA I genetic associations with autoimmune diabetes. Front Immunol 15: 1342335
20. Guiliano DB, Antoniou AN (2013) Measuring synthesis and degradation of MHC class I molecules. In: Antigen Processing: Methods and Protocols, pp 93–108
21. Lenart I, Truong L-H, Nguyen DD, Rasiukienė O, Tsao E, Armstrong J, Kumar P, McHugh K, Pereira BI, Maan BS (2022)

Intrinsic folding properties of the hla-B27 heavy chain revealed by single chain trimer versions of peptide-loaded class I major histocompatibility complex molecules. Front Immunol 13:902135
22. Dirscherl C, Hein Z, Ramnarayan VR, Jacob-Dolan C, Springer S (2018) A two-hybrid antibody micropattern assay reveals specific in cis interactions of MHC I heavy chains at the cell surface. Elife 7:e34150
23. Roeth JF, Williams M, Kasper MR, Filzen TM, Collins KL (2004) HIV-1 Nef disrupts MHC-I trafficking by recruiting AP-1 to the MHC-I cytoplasmic tail. J Cell Biol 167(5):903–913
24. Schneider-Poetsch T, Ju J, Eyler DE, Dang Y, Bhat S, Merrick WC, Green R, Shen B, Liu JO (2010) Inhibition of eukaryotic translation elongation by cycloheximide and lactimidomycin. Nat Chem Biol 6(3):209–217
25. Guo N, Peng Z (2013) MG132, a proteasome inhibitor, induces apoptosis in tumor cells. Asia Pac J Clin Oncol 9(1):6–11

Chapter 4

Probing Thermal Transitions of Peptide-Major Histocompatibility Complexes by Differential Scanning Fluorimetry

Holly Anne Martin and Lance M. Hellman

Abstract

Differential scanning fluorimetry (DSF) is a versatile and accessible technique for probing the thermal stability of peptide-major histocompatibility complexes (pMHCs). Understanding the thermal transition midpoint (T_m) between the folded and unfolded states of pMHCs provides critical insights into their structural stability, which is essential for studying antigen presentation and immune responses. pMHC stability is influenced by the inherent structural differences between MHC class I and class II molecules. Notably, class I MHCs require bound peptides for stability, while class II MHCs remain stable without them. Class I MHCs typically present peptides of 8–12 amino acids anchored at specific residues, whereas class II MHCs accommodate longer peptides with distinct anchoring positions. This manuscript outlines a comprehensive DSF protocol optimized for pMHCs, highlighting the method's advantages over other thermal stability techniques, such as differential scanning calorimetry (DSC) and circular dichroism (CD). DSF offers a simpler, cost-effective alternative, utilizing minimal sample volume and readily available real-time PCR (qPCR) instrumentation. We detail critical steps for sample preparation, including optimal buffer selection, dye addition, and degassing procedures, along with specific instrument setup guidelines for both qPCR-based systems and the NanoTemper Prometheus. Data analysis strategies using Microsoft Excel and Origin software are also discussed, including normalization, derivative calculation, and T_m determination. By providing a standardized DSF protocol tailored to pMHC analysis, this manuscript aims to support researchers in efficiently measuring thermal stability, thereby facilitating investigations into pMHC dynamics and immune function.

Key words Differential scanning fluorimetry (DSF), Peptide-major histocompatibility complex (pMHC), Thermal transition (T_m)

1 Introduction

The interaction between T-cells and antigen-presenting cells is the hallmark of cellular immunity. The determination of a healthy cell versus an abhorrent or infected cell is mediated through the contact between the T-cell receptor (TCR) and the peptide-major

Efstratios Stratikos (ed.), *Antigen Processing and Presentation: Methods and Protocols*, Methods in Molecular Biology, vol. 3007, https://doi.org/10.1007/978-1-0716-5092-9_4,

histocompatibility complex (pMHC). MHCs differ based on their recognition by either $CD8^+$ or $CD4^+$ T-cells, MHC class I and II, respectively. There are particular differences between class I and II MHC. Class I MHCs are comprised of a human leukocyte antigen (HLA) and a β_2-microglobulin (β_2m), while class II is a heterodimer with an α- and β-chain [1]. Of particular interest to this discussion is that class I MHCs are not stable without the presence of the peptide, while class II are. Class I MHC present peptides ranging 8–10 amino acids in length [2], while class II MHC tend to express peptides 13–25 amino acids long [1]. Peptides are anchored to the MHCs at positions P2 and C-terminal end for class I, and at positions P1, P4, P6, and P9 for class II [1, 3]. The amino acids at these positions are critical to the overall stability of the complexes. Depending on the allelic variant of MHC protein, the anchoring amino acids vary [4, 5].

Methods to determine the thermal transition (T_m), which is the midpoint between the folded and unfolded states of pMHCs, center around three technologies. Each has its own particular strengths and weaknesses. These include differential scanning calorimetry (DSC), circular dichroism (CD), and differential scanning fluorimetry (DSF); this paper will focus on the latter. Differential scanning calorimetry directly measures the change in temperature (ΔT) between a reference and the sample. DSC looks at the changes in heat capacity (ΔC_p) of a macromolecule over a temperature range [6]. This technique is the gold standard for probing the biothermodynamics of this folded/unfolded transition and whether it is an endothermic or exothermic process [7]. Circular dichroism looks at the differential absorption of circularly polarized left- and right-handed light [8]. To determine macromolecular stability, samples can be heated with a Peltier water-controlled device [9]. It is recommended to use a stirrer with CD to guarantee thermal equilibrium evenly throughout the sample. A significant limitation to both DSC and CD is the volume and concentration of sample needed compared with DSF. In addition, CD heats the sample through a water bath, which is less accurate. Nitrogen gas is also a requirement for circular dichroism.

DSC and CD require specialized instrumentation; and although specialized instruments exist for DSF (one which we will discuss below) it can be performed on a real-time PCR (qPCR) instrument, something most research- or education-based labs have. DSF instruments available, one that we will mention below. DSF typically uses the least amount of sample, with possibly the simplest method for determining T_m of the macromolecule. Sample volumes for an qPCR is 20 μL and the NanoTemper Prometheus™ only requires 10 μL. DSF instruments use a thermal element instead of a water-controlled device, which leads to greater temperature accuracy. Like CD, qPCR instruments will collect a data set at every temperature increment programmed, while the Prometheus™ collects data every three seconds. Some companies, like NanoTemper and ThermoFisher Scientific, either directly provide the T_m or have separate software packages to aid in its determination. All data can be analyzed using any graphing software package, Microsoft Excel®, or Google Sheets®. We will discuss using Origin® and Excel® software later. If using an RT-PCR instrument, dye must be added. Typical dyes are SYPRO-Orange® (ThermoFisher), GloMelt™ (Biotium), or Protein Thermal Shift™ dye kit (ThermoFisher). These dyes fluoresce once bound to hydrophobic surfaces. The NanoTemper Prometheus™ is a dye-less method that utilizes tryptophan/tyrosine fluorescence. The number of samples that can be melted at a time varies depending on the instrument, from a single sample up to a 384-well plate, while some instruments can be automated. Some widely available PCR instruments include the QuantStudio™ (ThermoFisher) series, CFX Opus (BioRad), and AriaMix (Agilent), to name a few. This discussion will primarily focus on typical qPCR and Prometheus instruments. Similarly, there are many different graphing programs that can be used for peak integration. We will discuss how to use OriginPro® and compare that value to the same data analyzed by Excel®.

2 Materials

Use a buffer that is adequate for your sample. However, buffers like TRIS, phosphate, and bicine are typically not optimal buffers to use due to their change in pH over a temperature range, whereas the pH of HEPES is not affected by temperature. Additives like glycerol and sugars may have an impact on the obtained measurements. Two useful buffers are listed below.

2.1 HEPES Buffered Saline (HBS) Buffer (1x)

1. 10 mM HEPES (4-[2-hydroxyethyl]-1-piperazineeethanesulfonic acid), 150 mM NaCl, pH 7.4. Add 2.38 g of HEPES acid-free to approximately 900 mL ddH_2O while stirring. Then add 8.77 g NaCl. pH between 6.8 and 8.2, then add more ddH_2O to bring the final volume to 1 L. Store at 4 °C. Typical HBS buffer is pH 7.4.

2.2 HBS-EP Buffer (1x)

1. 10 mM HEPES, 150 mM NaCl, 3 mM EDTA, 0.005% surfactant P20, pH 7.4. Add 2.38 g of HEPES acid-free to approximately 900 mL ddH_2O while stirring. Add 8.77 g NaCl and 1.12 g EDTA disodium salt dihydrate. Add 500 μL of Tween 20. pH between 6.8 and 8.2, then add more ddH_2O to bring the final volume to 1 L. Store at 4 °C. Typical HBS buffer is pH 7.4.

2.3 Sample Devices

1. 96-well plates (VWR, Radnor, PA, USA, cat. no. 82006–664) and adhesive film (VWR Radnor, PA, USA, cat. no. 89024–752).
2. PCR tubes with flat cap (VWR, Radnor, PA, USA, cat. no. 20170–012).
3. NanoTemper Prometheus™ Series Capillaries (Munich, Germany, PR-C002).

3 Methods

Lab purification protocols can vary, so we will not discuss those here.

3.1 Sample Preparation

1. Purify and concentrate pMHC according to individual lab protocols.
2. Dilute the sample to the desired concentration with a final volume of 20 μL (or appropriate volume for instrument) into a 1.5 mL tube. A good target concentration for DSF is 10 μM, although lower concentrations (<2 μM) can be used.
3. Add dye to a final concentration of 1x to the appropriate volume. NOTE: SYPRO™ Orange comes as 5000x concentration. It is highly recommended to make dye dilutions at the time of use and not to store these dilutions over time. NOTE: No dye is needed if using the NanoTemper Prometheus™. Mix the sample and dye well, then pulse centrifuge.
4. OPTIONAL: Degas samples under house vacuum while on ice for 15 min prior to loading.

3.2 qPCR Instrument Setup

1. Place the sample with dye into a 96-well plate or individual PCR tubes that are appropriate for the instrument. If using a 96-well plate, cover with appropriate adhesive film. If using PCR tubes, do not use markers on the side of the tube as they may interfere with detection. Ensure no air bubbles are at the bottom of the wells or tubes. Tap tubes gently or centrifuge to remove bubbles if necessary.
2. When running samples with a reporter dye, it is good practice to run a buffer and dye-only control well, which can be used to subtract background noise if necessary.
3. Open the instrument and place the 96-well plate or tubes into the instrument. Close.
4. Select "ROX" as the reporter and "None" for passive reference. NOTE: if your instrument contains multiple lasers, make sure the appropriate one is selected.

5. Select and label the wells where the samples are in the instrument.
6. Set the temperature to ramp at the desired rate; most use 1 °C/min. NOTE: The newer QuantStudio™ operating software does not allow for 1 °C/min temperature ramping. Either choose 0.016 °C/sec (0.96 °C/min) or 0.017 °C/sec (1.02 °C/min).
7. Data can be downloaded for analysis or to generate a higher resolution image with graphing or other software.

3.3 Prometheus Instrument Setup

1. Open the Prometheus, remove the magnetic lid and clean the mirrored surface with ethanol and lint-free cloth. Load the sample into the capillary tube by just placing the capillary into the sample.
2. Dry the exterior of the capillary tube with a lint-free wipe.
3. Place the tube into Prometheus, ensuring capillary tubes are centered on the capillary tray, then cover with the magnetic lid.
4. Perform a "Discovery" scan. This will determine if your sample is concentrated enough or too concentrated to get adequate results from the thermal melt.
5. If the sample is adequate, label the samples in the program and then "Run" the sample.
6. Results will be presented as "Melting Scan," which displays the 350/330 nm ratio and also the first derivative curve. To see numeric values, select the tab at the top for "Annotations and Results." This will display the integrated peak value for each peak it determines. However, only the T_m is given, not standard deviations.
7. Data can be downloaded for analysis or to generate a higher resolution image with graphing or other software.

3.4 Data Analysis with Microsoft Excel®

1. Create and label two or more columns, one for temperature and the other(s) for sample measurements.
2. It is recommended to normalize data if comparing the melting profiles of multiple samples visually on the same plot. Note that this step is not essential to determine the T_m.

$$\text{Normalized Data} = (D - D_{\min})/(D_{\max} - D_{\min})$$

Whereas, D represents the data point, D_{min} = the minimum data value, and D_{max} = the maximum data value. Be sure to lock D_{min} and D_{max} in place with "$" in front of both the column letter and the row number (Excel shortcut is F4). If values are properly selected, then the minimum value will be 0 and the maximum will be 1. If a negative value is obtained, then the wrong D_{min} was selected. If a number larger than 1 was obtained, then the wrong D_{max} was used.

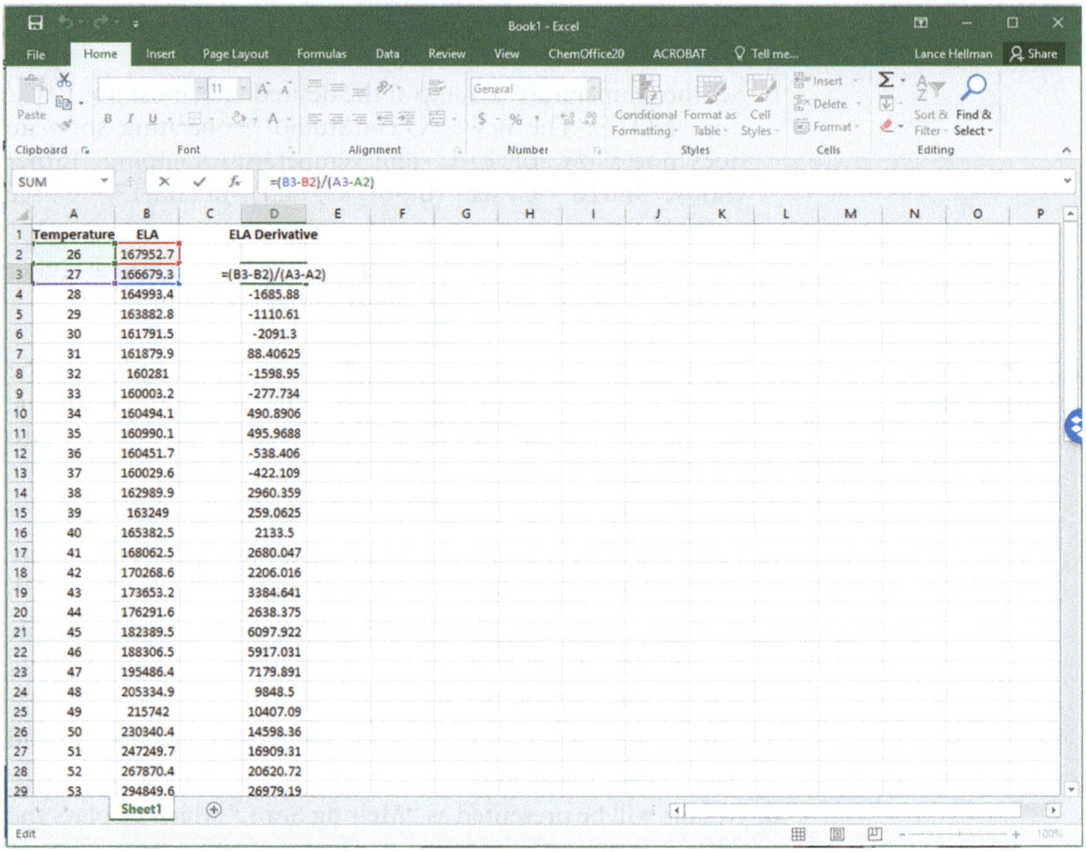

	Temperature	ELA		ELA Derivative
2	26	167952.7		
3	27	166679.3		=(B3-B2)/(A3-A2)
4	28	164993.4		-1685.88
5	29	163882.8		-1110.61
6	30	161791.5		-2091.3
7	31	161879.9		88.40625
8	32	160281		-1598.95
9	33	160003.2		-277.734
10	34	160494.1		490.8906
11	35	160990.1		495.9688
12	36	160451.7		-538.406
13	37	160029.6		-422.109
14	38	162989.9		2960.359
15	39	163249		259.0625
16	40	165382.5		2133.5
17	41	168062.5		2680.047
18	42	170268.6		2206.016
19	43	173653.2		3384.641
20	44	176291.6		2638.375
21	45	182389.5		6097.922
22	46	188306.5		5917.031
23	47	195486.4		7179.891
24	48	205334.9		9848.5
25	49	215742		10407.09
26	50	230340.4		14598.36
27	51	247249.7		16909.31
28	52	267870.4		20620.72
29	53	294849.6		26979.19

Fig. 1 A screen capture that shows how to calculate the T_m using the derivative of a data set in Excel®. The T_m will be the greatest value after taking the derivative

3. Take the derivative of the data set(s).

$$\text{Derivative} = (D_2 - D_1)/(T_2 - T_1)$$

Whereas D_2-D_1 represents the change in data points and T_2-T_1 the corresponding change in temperature (Fig. 1). The derivative calculation should start on the second data point; the first will be blank.

4. The T_m is the greatest numeric value of the derivative. If plotting the derivative, the peak corresponds to the T_m.

3.5 Data Analysis with Origin/OriginPro® Software

1. Open a New Workbook and label two or more columns, one for temperature and the other(s) for sample measurements. Fill in the appropriate columns with experimental data.
2. If comparing the melting profiles of multiple samples visually on the same plot. Normalization of data is not required for the determination of the T_m. Most graphing software packages have a normalization function. To normalize data, select column, right-click, then select "Normalize."

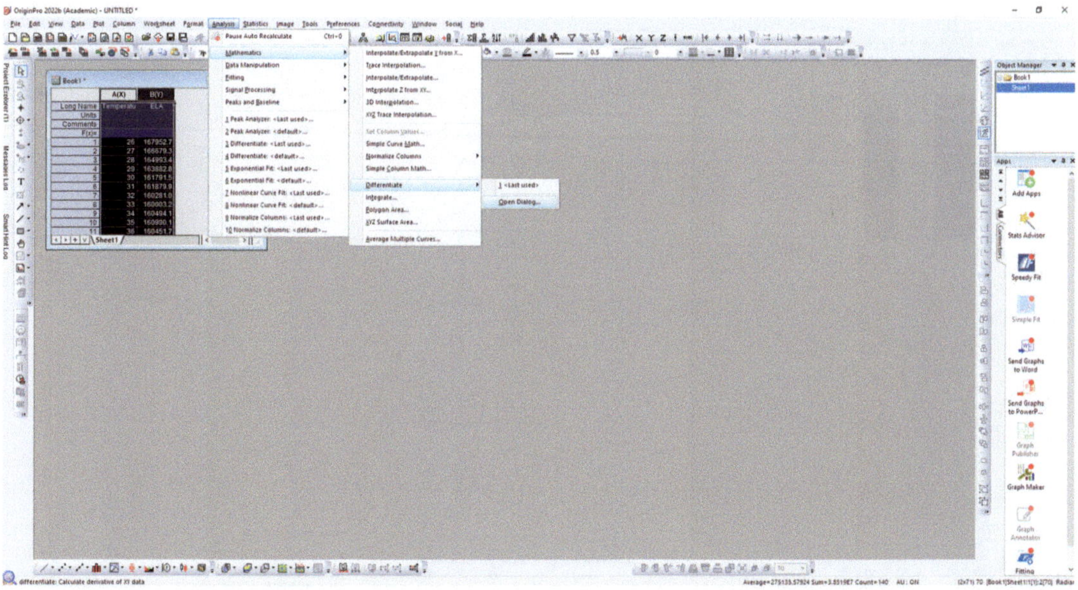

Fig. 2 A screen capture showing how to take the derivative of a data set using OriginPro®

3. To analyze data, first highlight Temperature and Sample columns, then under "Analysis" on the top, select "Mathematics," then "Differentiate," then "Open Dialog…," then "OK" (Fig. 2). This will generate the derivative.
4. To determine the T_m value, select the temperature and Derivative column together. Under "Analysis," select "Peaks and Baseline," then "Peak Analyzer," then "Open Dialog…" (Fig. 3).
5. This should open a window called "pa_goal," select "Fit Peaks (Pro)," then "Next."
6. This shows "pa_baseline," select "XPS," then "Next."
7. Now, under "pa_xpsbase" select "Shirley," then "Next."
8. Do not select anything under "pa_basetreat," select "Next."
9. This opens "pa_peaks," select "Find" (you can manually add/subtract peaks), then "Next."
10. This window will be "pa_fit," at the bottom select "Fit Control," then select "Bigaussian," then keep selecting "1 Iteration" until the fit converges; then select "OK."
11. Then "Finish." The T_m will be displayed as the "xc" value with Standard Error also reported.
12. To generate an image with both Derivative and Fit, select the "FitPeakCurve1" tab. Copy both the "Independent Variable" and "Fit Peak 1." Paste into a "New Workbook"; then add

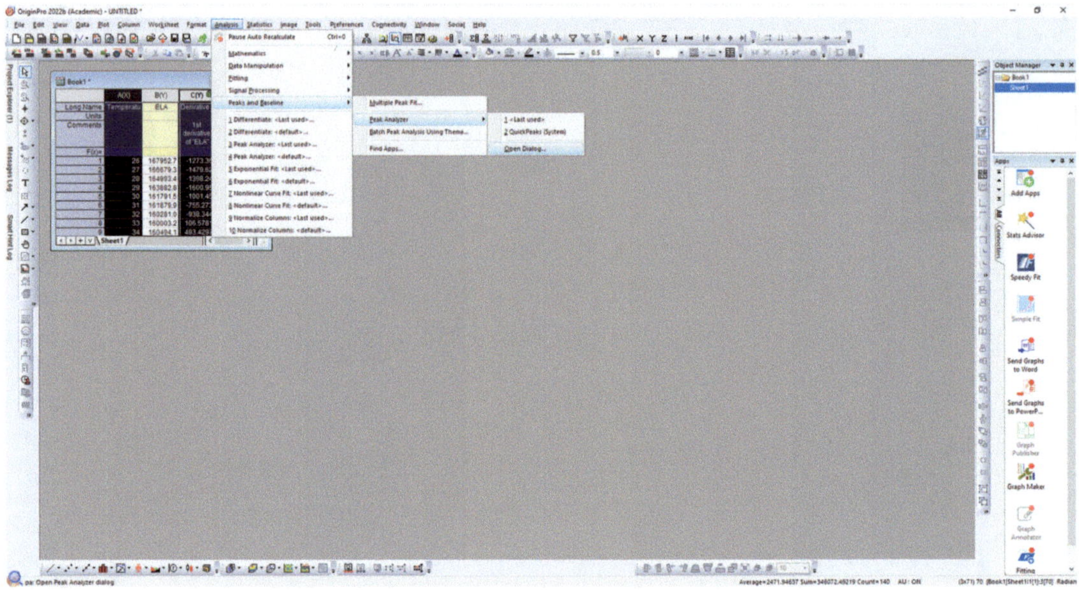

Fig. 3 A screen capture showing how to select Peak Analyzer in OriginPro®

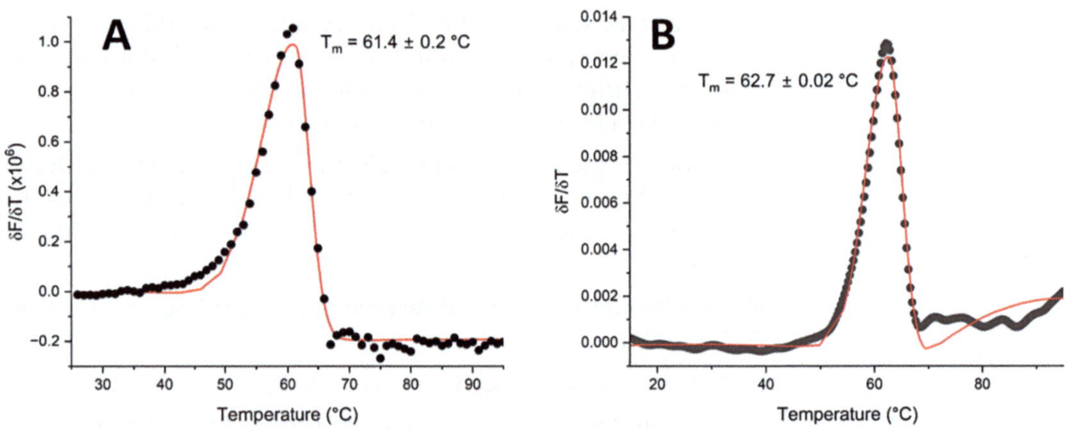

Fig. 4 The derivative and fit of HLA-A2 with the ELA (ELAGIGILTV) peptide in HBS-EP buffer using OriginPro to determine T_m. Derivative data points are black circles, while the red line is the fit of that data using OriginPro®. Samples used in Panels A and B were refolded from different protein stocks on different dates. (**a**) Calculated $T_m = 61.4 \pm 0.2$ °C, the same data set determined the $T_m = 60$ °C when using Excel® as described in Sect. 3.4. Data was collected on StepOnePlus™ qPCR (Applied BioSystems). (**b**) Calculated $T_m = 62.7 \pm 0.02$ °C. Data was collected on NanoTemper Prometeus™. Panel B is only showing every 20th data point, although the Fit included all points

both the original temperature and Derivative values from the first workbook. Make sure to make both temperature columns the X-axis. Highlight all columns and generate a scatter plot with lines. Edit image as desired (Fig. 4).

4 Notes

1. An alternate experiment using an qPCR, is to look at the dissociation rates of the pMHC complex [10, 11]. Set up the qPCR as a normal PCR run but only have one stage. Keep the temperature set to the desired temperature and adjust the time it stays at that cycle before scanning. Adjust the number of cycles to equal the number of hours you wish to cover, so 5 min per cycle for 80 cycles will cover 6.7 hrs. Once done, fit the data to a biexponential decay curve on graphing software.
2. NanoTemper recommends using Capillary Sealing Paste (PR-P001) and Capillary Sealing Applicators (PR-P002) for experiments that exceed 95 °C and/or last 3 hrs.
3. Programs, like Excel or Google Sheets, are less accurate in T_m determination compared to graphing software like OriginPro®. They will not readily provide standard deviations as well. However, most individuals have the Microsoft Office® package and Google Sheets® is free. At the same time, graphing software can be expensive.

Conflict of Interest

The authors have no conflicts of interest to declare that are relevant to the content of this chapter.

References

1. Wieczorek M, Abualrous ET, Sticht J et al (2017) Major histocompatibility complex (MHC) class I and MHC class II proteins: conformational plasticity in antigen presentation. Front Immunol 8:292. https://doi.org/10.3389/fimmu.2017.00292
2. Neefjes J, Jongsma MLM, Paul P et al (2011) Towards a systems understanding of MHC class I and MHC class II antigen presentation. Nat Rev Immunol 11:823–836. https://doi.org/10.1038/nri3084
3. Engelhard VH (1994) Structure of peptides associated with class I and class II MHC molecules. Annu Rev Immunol 12:181–207. https://doi.org/10.1146/annurev.iy.12.040194.001145
4. Collins EJ, Garboczi DN, Wiley DC (1994) Three-dimensional structure of a peptide extending from one end of a class I MHC binding site. Nature 371:626–629. https://doi.org/10.1038/371626a0
5. Rammensee HG, Bachmann J, Emmerich NPN et al (1999) SYFPEITHI: database for MHC ligands and peptide motifs. Immunogenetics 50:213–219. https://doi.org/10.1007/s002510050595
6. Morikis D, Lambris JD (2004) Physical methods for structure, dynamics and binding in immunological research. Trends Immunol 25:700–707. https://doi.org/10.1016/j.it.2004.09.009
7. Gill P, Moghadam TT, Ranjbar B et al (2010) Differential scanning calorimetry techniques: applications in biology and nanoscience. J Biomol Tech 21:167–193
8. Ranjbar B, Gill P (2009) Circular dichroism techniques: biomolecular and nanostructural analyses—a review. Chem Biol Drug Des 74:101–120. https://doi.org/10.1111/j.1747-0285.2009.00847.x
9. Morgan CS, Holton JM, Olafson BD et al (2008) Circular dichroism determination of

class I MHC-peptide equilibrium dissociation constants. Protein Sci 6:1771–1773. https://doi.org/10.1002/pro.5560060819

10. Miles KM, Miles JJ, Madura F et al (2011) Real time detection of peptide-MHC dissociation reveals that improvement of primary MHC-binding residues can have a minimal, or no, effect on stability. Mol Immunol 48:728–732. https://doi.org/10.1016/j.molimm.2010.11.004

11. Hellman LM, Yin L, Wang Y et al (2016) Differential scanning fluorimetry based assessments of the thermal and kinetic stability of peptide-MHC complexes. J Immunol Methods 432:95–101. https://doi.org/10.1016/j.jim.2016.02.016

Chapter 5

Method for Measuring the Kinetic Stability of Peptides Bound onto MHC Class I Using MALDI-TOF Mass Spectrometry

George Mavridis, Manousos Makridakis, Jerome Zoidakis, and Efstratios Stratikos

Abstract

The kinetic stability of peptides bound onto MHC class I molecules is a critical parameter that helps shape their interaction with immune receptors and consequently their antigenic potential. We present here a method for measuring the kinetic stability and sensitivity to proteolytic degradation of peptides bound onto MHC class I molecules after in vitro refolding, by analyzing the time-resolved MALDI-TOF Mass Spec signal from the peptide, using the whole MHC-I/peptide complex in situ. This approach can provide important information on the dynamic nature of the MHC-peptide interaction, the kinetic half-life of binding and the sensitivity of the peptide to external proteolytic digestion or other modifications.

Key words MHC, Peptide, Enzyme, Mass spectrometry, MALDI-TOF, ERAP1, Kinetics

1 Introduction

Major Histocompatibility Complex I molecules (MHC-I) bind small peptides originating from the proteolytic degradation of intracellular proteins and present them on the cell surface to T lymphocytes [1]. The sequence of the presented peptides is critical to the fate of the cell. Presentation of peptides derived from pathogens can activate CD4+ T cells, which will proceed to eliminate the presenting cell. This activation relies on the productive interaction between the MHC-I molecule/peptide complex and the T cell receptor [2]. This interaction is largely determined by the sequence and structure of the bound peptide, but also by the dynamic nature of the MHC-I/peptide complex. Indeed, MHC-I molecules are "kinetics traps," limiting the off-rate of the bound peptide [3]. This is important because on the cell surface the external concentration of the peptide is practically null and quick release could result in

Efstratios Stratikos (ed.), *Antigen Processing and Presentation: Methods and Protocols*, Methods in Molecular Biology, vol. 3007, https://doi.org/10.1007/978-1-0716-5092-9_5,

binding of random extracellular peptides. The dynamics of MHC-I/peptide interaction are also very important during intracellular loading and they are controlled by the Peptide Loading Complex [4]. Although most peptides form stable complexes with the MHC-I, this may not be true for all and this can determine immunogenicity [5]. Thus, being able to analyze the stability of peptide binding onto an MHC-I is very useful for understanding the immunogenic potential of each peptide. In addition, MHC-I binding protects the peptide from proteolytic degradation and thus highly kinetically labile peptides may be prone to degradation [6, 7]. This may be especially important inside the ER, where ER-resident aminopeptidases specialized for antigenic peptide trimming (ERAP1 and ERAP2) may act on labile peptides as they are loaded onto MHC-I. In addition, ERAP1 and ERAP2 have been proposed to be able to trim some extended peptides while bound onto MHC-I [8]. Here, we present a detailed protocol that allows the detection of a peptide bound onto an MHC-I molecule by MALDI-TOF MS. This method is suitable to follow peptide dissociation and degradation by peptidases [6–9] but could be adapted to include other factors that affect the kinetics of MHC-I/peptide interaction. The protocol is presented for a specific pair of MHC-I allele/peptide and peptidase but can be easily adapted for any allele/co-factor combination.

2 Materials

2.1 Molecular Biology Reagents

1. Tryptone.
2. Yeast extract for molecular biology.
3. Sodium chloride for molecular biology.
4. Agar for molecular biology.
5. D(+)-Glucose.
6. Potassium chloride.
7. Magnesium chloride.
8. Ampicillin.
9. Isopropyl-β-D-thiogalactopyranoside (IPTG).
10. Magnesium sulfate.
11. Calcium chloride.

2.2 Preparation of MHC-I/Peptide Complex

1. Sucrose Solution: 25% w/v sucrose, 50 mM Tris-HCl pH 8.0, 1 mM EDTA, and 0.1% w/v sodium azide (NaN_3).
2. Deoxycholic Acid–Triton X-100 Solution: 1% w/v deoxycholic acid, 1% v/v Triton X-100, 20 mM Tris-HCl pH 7.5, 100 mM NaCl, and 0.1% w/v NaN_3.

3. Triton X-100 Solution: 0.5% v/v Triton X-100, 50 mM Tris-HCl pH 8.0, 100 mM NaCl, 1 mM EDTA, and 0.1% w/v NaN_3.
4. Tris-HCl Solution: 50 mM Tris-HCl pH 8.0, 1 mM EDTA, and 0.1% w/v NaN_3.
5. Urea Solution: 8 M urea, 25 mM [2-(N-morpholino) ethane-sulfonic acid] MES pH 6.0, and 10 mM EDTA.
6. DNase Solution: 75 mM NaCl, 50% v/v glycerol, and 2 mg/mL deoxyribonuclease (DNase).
7. Lysozyme (1 mg/ml suspension in Tris-HCl pH 6, 150 mM NaCl).
8. Arginine 0.4 M in ddH_2O.
9. GSH 5 mM in ddH_2O.
10. GSSG O.5 Mm in ddH_2O.
11. 5% v/v glycerol solution in ddH_2O.
12. EDTA 2 mM in ddH_2O.
13. PMSF 0.1 mM in ethanol.
14. DTT 5 mM in ddH_2O.
15. Peptide ELRSRYWAI 40 μM dissolved in DMSO.

2.3 Mass Spectrometry Analysis

1. Matrix solution: 7 mg/mL α-cyano-4-hydroxycinnamic acid (CHCA), 50% v/v acetonitrile and 0.1% v/v TFA (see Note 1).
2. Peptide calibration standard mix. Any commercially available peptide standard is sufficient provided it covers the mass range between 500 and 3500 Da. Resuspend the lyophilized standards in 125 μL of 50% v/v acetonitrile, 0.1% v/v TFA and prepare aliquots of 20 μL, which can be stored at −20 °C.

3 Methods

This section provides a detailed, step-by-step guide for generating MHC class I / peptide complexes using as a working example the human MHC-I allele HLA-B*08:01 and the peptide ELRSRY-WAI, an antigenic epitope from the influenza A virus nucleoprotein [10]. The complex is then incubated with an aminopeptidase that trims antigenic peptides to probe peptide kinetic stability or onto-MHC peptide trimming [6, 8]. Finally, the detection of the MHC-bound peptide is performed in situ using MALDI-TOF MS.

3.1 Preparation Growth Media

1. Dissolve 10 g of tryptone, 5 g of yeast extract, and 5 g of NaCl in 1 L of double-distilled water (ddH_2O).
2. Add 17 g of agar and sterilize the mixture in an autoclave at 121 °C for 30 minutes. After autoclaving, transfer the

container to a water bath set at 45 °C. Once the temperature of the medium has dropped to 45 °C, add ampicillin to a final concentration of 100 μg/Ml in double-distilled water (ddH_2O).

3. Aseptically dispense the medium into Petri dishes and allow it to solidify. Store the plats at 4 °C until use.
4. To prepare LB growth medium, dissolve 10 g of tryptone, 5 g of yeast extract, and 5 g of NaCl in 1 L of ddH_2O. Sterilize the solution in an autoclave at 121 °C for 30 minutes.
5. To prepare SOC growth medium, dissolve 10 g of tryptone, 5 g of yeast extract, and 5 g of NaCl in 1 L of ddH_2O. Sterilize the solution in an autoclave at 121 °C for 30 minutes. Once cooled, aseptically add 10 mM $MgCl_2$, 10 mM $MgSO_4$, 25 mM KCl, and 20 mM glucose.

3.2 Preparation of Competent E. coli

1. A single colony of *E. coli* BL21 DE3, grown on solid LB agar medium in the absence of antibiotic, is selected and inoculated into a 15 mL Falcon tube containing 2 mL of LB broth without antibiotic.
2. The following day, the 2 mL of bacterial culture is used to inoculate 200 mL of LB medium in an Erlenmeyer flask. The culture is incubated with shaking (200 rpm, 37 °C) until the optical density at 600 nm (OD_{600}) reaches 0.4.
3. The culture is then aliquoted into four 50 mL Falcon tubes and centrifuged at 4000 rpm for 20 minutes at 4 °C. The supernatant is discarded, and each pellet is resuspended in 25 mL of cold (4 °C) 100 mM $MgCl_2$ solution.
4. The suspension is centrifuged again under the same conditions, the supernatant is discarded, and the pellet is resuspended in 25 mL of cold 100 mM $CaCl_2$ solution. A third centrifugation step is performed (4000 rpm, 4 °C, 20 minutes), the supernatant is removed, and each pellet is resuspended in 2 mL of a cold solution containing 100 mM $CaCl_2$ and 14% (v/v) glycerol.
5. Aliquots of 50 μL are distributed into Eppendorf tubes and flash-frozen in liquid nitrogen. Competent cells are stored at −80 °C.

3.3 Transformation

Competent *E. coli* BL21 DE3 cells are transformed with a plasmid carrying an ampicillin resistance gene as a selection marker, and a gene encoding the heavy chain of the HLA-B*08:01 allele of the human MHC-I.

1. Add 20 ng of plasmid DNA (in a volume of ~2 μL) to 50 μL of competent cells in an Eppendorf tube. Incubate the mixture on ice for 30 minutes.

2. Perform a heat shock at 42 °C for 90 seconds. The cells are then returned to ice for 2 min, after which 150 μL of SOC medium is added.
3. Incubate the cells at 37 °C with shaking (200 rpm) for 2 h.
4. After incubation, 200 μL of the bacterial culture is aseptically spread onto an LB agar Petri dish containing 100 μg/mL ampicillin.
5. Plates are incubated at 37 °C without shaking for 12–14 h to allow colony formation.

3.4 Expression and Isolation of Recombinant HLA-B*08:01

1. Select a single colony from step 3.3 and inoculate it into a 50 mL Falcon tube containing 5 mL of LB medium supplemented with 100 μg/mL ampicillin. Incubate the culture at 37 °C with shaking at 200 rpm for 8 h.
2. Use 4 mL of this starter culture to inoculate a sterile 2 Lt Erlenmeyer flask containing 400 mL of LB medium with 100 μg/mL ampicillin (1:100 dilution).
3. When the optical density (OD) of the bacterial culture at 600 nm exceeds 0.8, add isopropyl β-D-1-thiogalactopyranoside (IPTG) to the Erlenmeyer flask to a final concentration of 1 mM to induce protein expression. Incubate for 4 hours at 37 °C.
4. Following incubation, centrifuge the bacterial culture (15 minutes, 4000 rpm, 15 °C), discard the supernatant and store the bacterial pellet at −80 °C.
5. To each Falcon tube containing bacterial pellet harvested from a 50 mL culture, add 5 mL of sucrose solution and thoroughly resuspend the pellet.
6. Add 5 mg of lyophilized lysozyme (prepared as a 1 mg/mL suspension) and incubate with gentle agitation for 20 minutes.
7. Add 10 mL of the deoxycholic acid–Triton X 100 solution, 5 mM $MgCl_2$ (final concentration), and 100 μL of the DNase solution and incubate the samples with continuous agitation until complete bacterial lysis is achieved, as indicated by the solution reaching water-like viscosity. The Falcon tubes are then stored at −20 °C.
8. On the following day, thaw samples in a 37 °C water bath until completely liquified, centrifuge at 8000× *g* for 20 min at 4 °C, and then discard the supernatant.
9. Homogenize the pellet mechanically, resuspend in 15 mL of Triton X-100 solution, and centrifuge again under the same conditions. Repeat this step four times.

10. Resuspend the pellet in 15 mL of Tris-HCl solution and centrifuge again (8000× *g*, 20 min, 4 °C). Repeat this washing step four times.
11. After the final wash, discard the supernatant and add 10 mL of the urea solution to each tube. Perform a final centrifugation step at 10,000 × g for 1 hour at 20 °C to solubilize the cellular inclusions.
12. Following centrifugation, filter the solution through a sterile 0.2 μm pore-size disposable syringe filter (such as Whatman® Puradisc 13, Merck).
13. Determine protein concentration by measuring absorbance at 280 nm using a standard spectrophotometer. The molar extinction coefficient of the protein subunits can be calculated using the Expasy ProtParam tool based on their amino acid sequence.
14. The concentration of the denatured protein subunits is calculated using the Beer-Lambert Law, and the resulting urea-solubilized proteins were stored at −80 °C.
15. The same procedure can be followed to isolate b2 microglobulin (b2m) as inclusion bodies (using a b2m containing expression plasmid).

3.5 Refolding of HLA B*08:01-b$_2$m-ELRSRYWAI

1. Prepare 50 mL aqueous refolding buffer containing 0.4 M arginine, 100 mM Tris-HCl pH 8, 5 mM reduced glutathione, 0.5 mM oxidized glutathione, 5% v/v glycerol, 2 mM EDTA, and 0.1 mM PMSF. The buffer is stored at 4 °C.
2. Combine in a 15 mL Falcon tube the following components: 1.6 mg of HLA heavy chain (final concentration: 1 μM in 50 mL), 1.2 mg of β2-microglobulin (final concentration: 2 μM in 50 mL), and 40 μM peptide ELRSRYWAI dissolved in dimethyl sulfoxide (DMSO).
3. Add dithiothreitol (DTT) to a final concentration of 5 mM and incubate the mixture at 37 °C for 30 minutes.
4. Transfer the protein mixture into a falcon tube containing the refolding buffer and incubate at 4 °C for 2 hours.
5. After incubation, transfer the solution into a dialysis membrane with a molecular weight cut-off of 10 kDa, place it in a 2 Lt graduated cylinder containing 10 mM Tris-HCl pH 8, maintained at 4 °C.
6. Dialyze the sample at 4 °C for 48 hours, with one exchange of the external buffer after the first 24 hours.

3.6 Isolation of Complex by Size-Exclusion Chromatography

1. Prepare an aqueous buffer containing 10 mM Tris-HCl (pH 8.0) and 150 mM NaCl and use it to equilibrate a Sephacryl S-200 chromatography column using two column volumes (240 mL) of this buffer.
2. Transfer the protein complex solution from the dialysis membrane to a 50 mL Falcon tube and filter through a sterile, single-use 0.2 μm pore-size filter.
3. Concentrate the filtrate to a final volume (cut-off value of the concentrator: 10 kDa) of 1 mL and inject into the size-exclusion column using a standard FPLC system at 4 °C.
4. Collect elution fractions and analyze by SDS-PAGE (Fig. 1).
5. Pool the fractions corresponding to the refolded complex and concentrate them to a final protein concentration of 8–11 μM.
6. Add sterile glycerol to a final concentration of 10% v/v, and store the samples at −80 °C until needed (*see* **Note 2**).

3.7 MHC-I Complex Treatment with Protease

1. Incubate 50 μL of the refolded complex with a protease such as ERAP1 or ERAP2 at various concentrations (ranging from 2 to 100 nM) at 37 °C for 30–60 min.
2. Perform a control reaction with just peptide. Dissolve peptide in a buffer containing 10 mM Tris-HCl (pH 8.0) and 150 mM NaCl, at the same concentration used for the equivalent peptide in the protein complex.

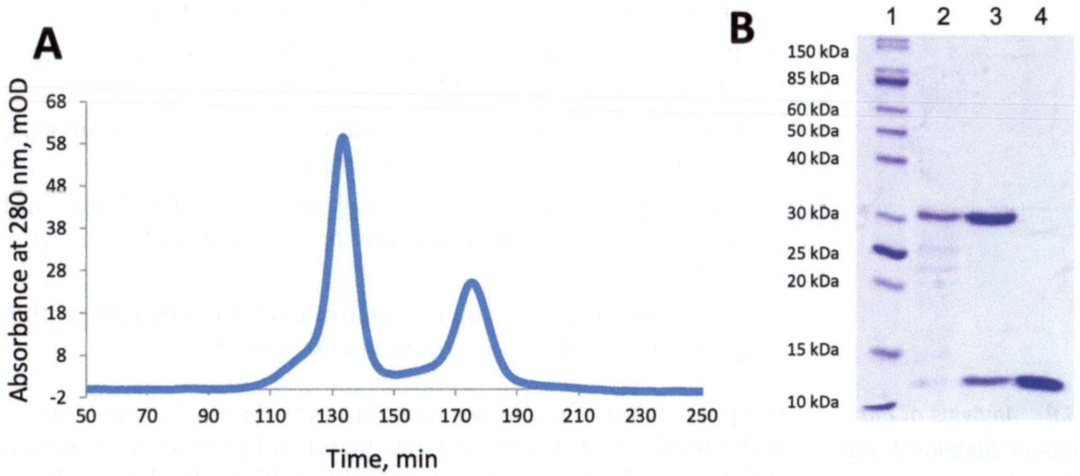

Fig. 1 Panel A, representative chromatogram from the size exclusion column analysis of samples after refolding HLA B*08:01 with peptide ELRSRYWAI. Panel B, representative SDS-PAGE from samples from chromatogram of panel A. Lane 1, MW standards; Lane 2, elution at 85 min, consisting of primarily heavy chain aggregates; Lane 3, elution at 130 min, consisting of correctly folded HLA B*08:01-ELRSRYWAI complex; Lane 4, elution at 175 min, consisting of excess β2m

3. At defined time points, collect 8 μL aliquots, and add trifluoroacetic acid (TFA) to a final concentration of 1% v/v to stop the reaction. Store the samples at −80 °C until analysis.
4. Prior to analysis, thaw the samples and dilute 1:10 in matrix solution in a final volume of 20 μL (*see* **Note 3**). Spot the diluted samples (2 μL for each sample) onto a stainless-steel MALDI target plate (in different positions) using the dried-droplet method. In addition, spot 0.5 μL of the peptide calibration standard mix and wait for 5 min until air-dry. In the same position where the calibration mix was spotted, add 0.5 μL matrix solution and wait for another 5 min until air-dry. When all samples are dried, proceed with the MALDI-TOF MS analysis.

3.8 MALDI-TOF MS Analysis

Peptide molecular weights are determined using a MALDI-TOF/TOF mass spectrometer (for example, the Ultraflex TOF/TOF, Bruker Daltonics, Bremen, Germany).

1. Insert the stainless-steel target plate into the instrument.
2. Perform calibration in positive ion mode using a peptide calibration standard mix covering a range of 1000–3500 Da (*see* **Note 4**).
3. After calibration, analyze samples in positive ion mode.
4. Acquire spectra (m/z range: 750–3500) by accumulating 10 × 1000 laser shots per sample at a laser frequency of 500 Hz. Spectra acquisition should be performed from different areas of each sample spotted into the target plate.
5. Perform spectral smoothing using the Savitzky-Golay algorithm (smoothing width 0.2 m/z, 1 cycle).
6. Use a signal-to-noise ratio threshold of 2.5 and generate peak lists using FlexAnalysis v3.3 or similar.
7. Export peak lists from FlexAnalysis v3.3 in Excel files and spectra in .emf. The latter can be accessed with paint (see Note 5).
8. For each time point, sum the intensities of the substrate peptide peaks and degradation product peaks (Fig. 2).

3.9 Analysis of the Kinetic Stability of the Complex

The proportion of the intact substrate peptide is calculated relative to the total signal (substrate + products) and plotted as a function of time (Fig. 3). To calculate the peptide half-life the data can be fit to a one-phase decay model (Y = (Y0 - Plateau)*exp.(-K*X) + Plateau, where Y is the % peptide remaining, X is the time, and K is the calculated rate) using any modeling software such as Graphpad Prism (Fig. 4). Alternatively use R with a sample script as the one

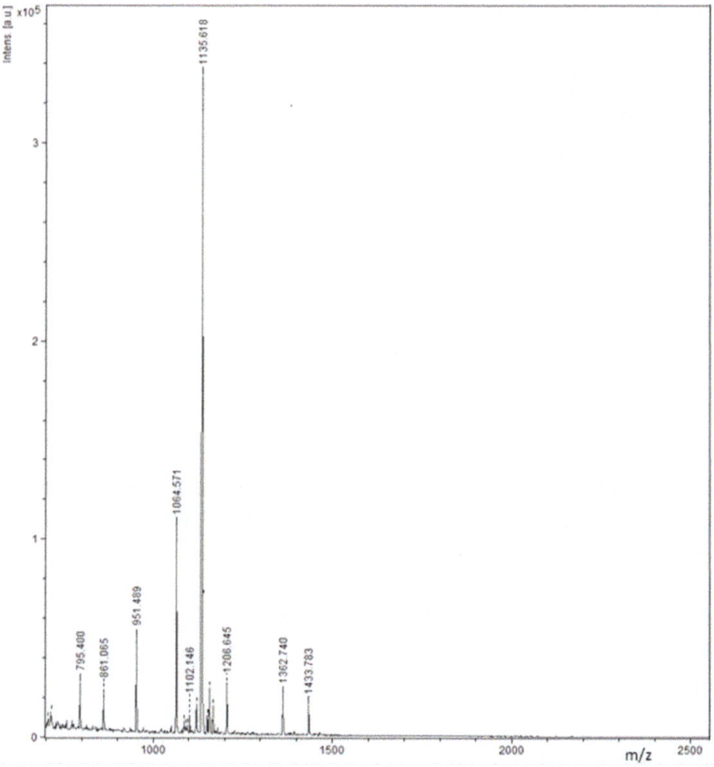

m/z	Intens.	Area	Rel. Intens.	% Abundance	Peptides
701,361	6150,952	8417,345	0,031	2,6	
708,363	9704,822	2541,594	0,049	0,8	RYWAI (5mer)
767,356	4480,582	1303,855	0,022	0,4	
795,367	96787,635	32948,039	0,486	10,0	SRYWAI (6mer)
817,347	8922,742	2399,510	0,045	0,7	
827,359	3859,089	778,145	0,019	0,2	
838,366	3329,063	1250,913	0,017	0,4	
861,023	6052,752	2777,471	0,030	0,8	
876,990	3494,991	1625,432	0,018	0,5	
880,416	4355,478	1146,081	0,022	0,3	
937,794	5708,844	4190,484	0,029	1,3	
951,446	145412,711	55350,255	0,730	16,8	RSRYWAI (7mer)
967,437	3149,790	879,132	0,016	0,3	
973,424	8828,331	2971,324	0,044	0,9	
983,418	4013,941	1598,993	0,020	0,5	
1022,475	4799,471	1592,953	0,024	0,5	
1031,083	3279,921	2143,184	0,016	0,7	
1050,849	7784,956	5381,862	0,039	1,6	
1064,524	167951,857	71396,954	0,843	21,7	LRSRYWAI (8mer)
1080,521	2862,139	955,745	0,014	0,3	
1086,506	10378,867	3950,833	0,052	1,2	
1096,518	3917,815	1266,384	0,020	0,4	
1102,076	4052,487	3084,890	0,020	0,9	
1108,495	2686,851	880,747	0,013	0,3	
1121,876	9945,156	7391,289	0,050	2,2	
1135,560	199216,954	92130,040	1,000	28,0	ALRSRYWAI (9mer)
1151,562	3071,937	1115,010	0,015	0,3	
1157,546	12912,885	5007,738	0,065	1,5	
1167,559	4226,547	1501,725	0,021	0,5	
1179,529	3279,964	1203,401	0,016	0,4	
1206,593	3699,571	1620,943	0,019	0,5	AALRSRYWAI (10mer)
1362,681	2137,204	861,540	0,011	0,3	RAALRSRYWAI (11mer)
1433,716	10368,765	5107,479	0,052	1,6	ARAALRSRYWAI (12mer)
1476,714	3516,589	1789,575	0,018	0,5	
2394,960	851,472	811,213	0,004	0,2	

Fig. 2 Representative mass spectrum from the analysis of an MHC-peptide complex. Peptide sequences that correspond to the measured masses are indicated in the rightmost column. The signal area and relative abundance of each peptide can be used to follow the stability of each peptide over time

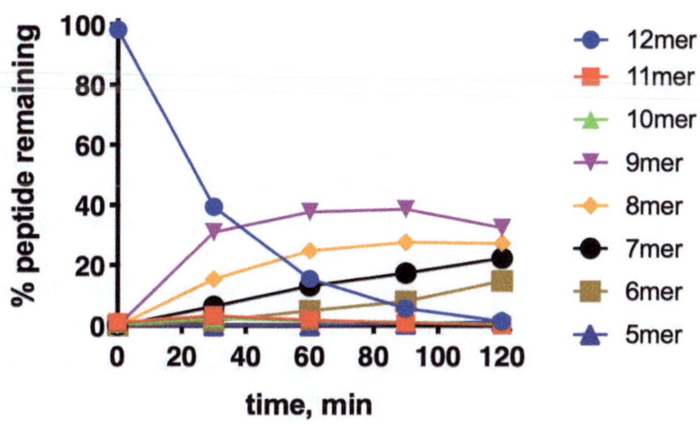

Fig. 3 Representative results from the analysis of 8 different length peptides bound onto HLA B*08:0, after exposure to the proteolytic action of ERAP1 for different time lengths. Notice the gradual disappearance of the 12mer parent peptide and accumulation of the 9mer product (data from [6])

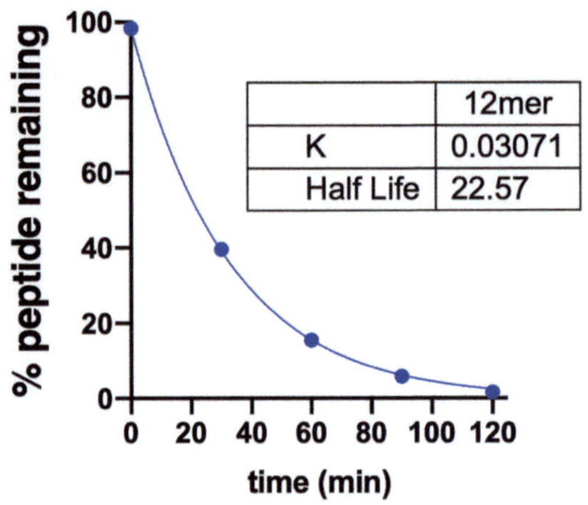

Fig. 4 Fitting the kinetics of 12mer peptide degradation to a one-phase decay model using GraphPad prism. The calculated rate and half-life are shown in the inset

provided below (data should be contained in file named "data.txt" formatted in two columns, each with a header "Time" and "Signal"):

```
# Load required libraries
library(ggplot2)
library(minpack.lm) # for robust nonlinear fitting
# Step 1: Read the data
data <- read.table("data.txt", header = TRUE, sep = "\t")
# Step 2: Check headers
if (!all(c("Time", "Signal") %in% names(data))) {
stop("The file must contain headers 'Time' and 'Signal'")
}
# Step 3: Set initial parameter guesses
start_Y0 <- data$Signal[1]
start_Plateau <- tail(data$Signal, 1)
start_K <- 0.1
# Step 4: Fit the one-phase decay model
fit <- nlsLM(
Signal ~ (Y0 - Plateau) * exp(-K * Time) + Plateau,
data = data,
 start = list(Y0 = start_Y0, Plateau = start_Plateau, K =
start_K),
control = list(maxiter = 500)
)
# Step 5: Extract rate constant (K) and format
K_value <- coef(fit)["K"]
K_formatted <- signif(K_value, 3)
subtitle_text <- paste("Rate constant (K) =", K_formatted)
```

```
# Step 6: Predict fitted curve over smooth time range
time_grid <- data.frame(Time = seq(min(data$Time), max(data
$Time), length.out = 200))
fit_values <- predict(fit, newdata = time_grid)
fit_df <- data.frame(Time = time_grid$Time, Signal = fit_va-
lues)

# Step 7: Plot data and fit, with K value
ggplot() +
geom_point(data = data, aes(x = Time, y = Signal), color =
"blue", size = 2) +
geom_line(data = fit_df, aes(x = Time, y = Signal), color =
"red", linewidth = 1) +
theme_minimal() +
labs(
title = "One-Phase Decay Fit",
subtitle = subtitle_text,
x = "Time",
y = "Signal"
)
```

4 Notes

1. Solutions used in MS should be prepared with ultra-pure water (18 MΩ·cm at 25 °C) and analytical grade reagents (99% purity). Always prepare fresh.
2. The concentration of the protein complex is calculated based on the sum of the molar extinction coefficients of the two subunits and the peptide, provided the latter contains one or more aromatic amino acids. The molecular weight (MW) of the heavy chain is 32 kDa, while the MW of b2m is 12 kDa and of the peptide ELRSRYWAI 1 kDa; therefore, the MW of the refolded complex is 45 kDa.
3. Different dilutions of the sample in matrix solution were tested (undiluted, 1:2, 1:5, 1:10, 1:20) to define optimal signal intensity and better ionization efficiency.
4. Peptide calibration standard mix allows calibrations and testing of MALDI-TOF mass spectrometers in a mass range between ~1000 and 3500 Da. The mixture contains seven standard peptides. It is highly recommended to perform calibration in order to assess the performance and condition of the instrument prior to any analysis.
5. Always observe the quality of the spectra obtained, the signal intensity and peak area of each identified mass.

Conflict of Interest

The authors have no conflicts of interest to declare that are relevant to the content of this chapter.

References

1. Rock KL, Goldberg AL (1999) Degradation of cell proteins and the generation of MHC class I-presented peptides. Annu Rev Immunol 17: 739–779
2. Wilson IA, Garcia KC (1997) T-cell receptor structure and TCR complexes. Curr Opin Struct Biol 7:839–848. https://doi.org/10.1016/s0959-440x(97)80156-x
3. Hermann C, van Hateren A, Trautwein N et al (2015) TAPBPR alters MHC class I peptide presentation by functioning as a peptide exchange catalyst. eLife 4. https://doi.org/10.7554/eLife.09617
4. Trowitzsch S, Tampe R (2020) Multifunctional chaperone and quality control complexes in adaptive immunity. Annu Rev Biophys. https://doi.org/10.1146/annurev-biophys-121219-081643
5. Wieczorek M, Abualrous ET, Sticht J et al (2017) Major histocompatibility complex (MHC) class I and MHC class II proteins: conformational plasticity in antigen presentation. Front Immunol 8:292. https://doi.org/10.3389/fimmu.2017.00292
6. Mavridis G, Arya R, Domnick A et al (2020) A systematic re-examination of processing of MHCI-bound antigenic peptide precursors by endoplasmic reticulum aminopeptidase 1. J Biol Chem 295:7193–7210. https://doi.org/10.1074/jbc.RA120.012976
7. Mavridis G, Mpakali A, Zoidakis J et al (2021) The ERAP1 active site cannot productively access the N-terminus of antigenic peptide precursors stably bound onto MHC class I. Sci Rep 11:16475. https://doi.org/10.1038/s41598-021-95786-x
8. Li L, Batliwala M, Bouvier M (2019) ERAP1 enzyme-mediated trimming and structural analyses of MHC I-bound precursor peptides yield novel insights into antigen processing and presentation. J Biol Chem 294:18534–18544. https://doi.org/10.1074/jbc.RA119.010102
9. Chen H, Li L, Weimershaus M et al (2016) ERAP1-ERAP2 dimers trim MHC I-bound precursor peptides; implications for understanding peptide editing. Sci Rep 6:28902. https://doi.org/10.1038/srep28902
10. Rowntree LC, Nguyen THO, Halim H et al (2018) Inability to detect cross-reactive memory T cells challenges the frequency of heterologous immunity among common viruses. J Immunol 200:3993–4003. https://doi.org/10.4049/jimmunol.1800010

Chapter 6

Methods for Isolating and Analyzing the HLA Class I Immunopeptidome

Martha Nikopaschou, Georgia Rouni, and Martina Samiotaki

Abstract

Immunopeptidomics is a mass spectrometry-based technique, crucial for characterizing the peptide repertoire displayed by Class I Human Leukocyte Antigens (HLA-I). This repertoire, known as the immunopeptidome, is central to adaptive immunity, as it facilitates T cell recognition of infected or malignant cells that present abnormal peptides. Understanding the immunopeptidome and the rules behind its generation is therefore essential for deciphering antigen presentation mechanisms and uncovering new diagnostic markers and therapeutic targets in infectious diseases, cancer and autoimmunity.

This chapter outlines a simple immunopeptidomics protocol, adapted from established workflows with the vision to be accessible and to rely on equipment and reagents commonly available in most biochemistry laboratories, as well as open-source software. Crucially, it includes instructions for the in-house production and purification of the monoclonal antibody needed for HLA-I immunopeptidomics. Practical considerations and tips are also incorporated to empower less experienced researchers, allowing them to successfully apply immunopeptidomics to their own research questions.

Key words Immunopeptidomics, Immunoaffinity purification, LC-MS/MS

1 Introduction

Class I antigen presentation is a cornerstone of adaptive immunity, as it enables $CD8^+$ T lymphocytes to detect and eliminate infected or cancerous cells. This vital process is mediated by Major Histocompatibility Class I (MHC-I) molecules, known as Human Leukocyte Antigen (HLA)-I in humans. These molecules present peptide fragments—collectively termed the immunopeptidome—derived from intracellular proteins, including host-derived self-peptides, pathogen-derived peptides and peptides generated during abnormal oncogenic processes [1, 2].

Under normal conditions, self-peptides presented by HLA-I do not elicit T cell responses due to the elimination of self-reactive T cells during thymic selection [3, 4]. Conversely, during infection or cancer, foreign or aberrant peptides can be recognized by $CD8^+$ T

Efstratios Stratikos (ed.), *Antigen Processing and Presentation: Methods and Protocols*, Methods in Molecular Biology, vol. 3007,
https://doi.org/10.1007/978-1-0716-5092-9_6,

cells, triggering crucial immune responses to clear problematic cells [5]. While this mechanism is highly beneficial and even lifesaving, cancer evasion remains a significant challenge. Moreover, dysregulation in T cell selection can lead to inappropriate immune activation against self-peptides and contribute to autoimmune inflammatory diseases [3, 6].

Collectively, these insights highlight the central role of the HLA-I immunopeptidome in health and disease and underscore the need for its comprehensive investigation [7, 8]. This need is now widely recognized [9], leading an increasing number of laboratories to incorporate immunopeptidome analysis into their research for mechanistic insights and for the identification of clinically relevant antigens.

Immunopeptidomics is the method of choice for the characterization of the immunopeptidome. It typically involves immunoaffinity purification of HLA-I complexes from cells or tissues using the pan-HLA W6/32 antibody, followed by peptide elution and sequence identification using liquid chromatography coupled to tandem mass spectrometry (LC-MS/MS) [10, 11].

The immunopeptidomics protocol described here and summarized in Fig. 1 represents the most recent version, building on previously published protocols [12, 13] and earlier applications in our lab [14–17] and has been applied in its current form in [18]. It offers a straightforward and robust solution, relying on equipment, reagents and techniques commonly available in most biochemistry laboratories. Crucially, we have included instructions for the in-house production and purification of the pan HLA-I monoclonal antibody W6/32, which we consider essential, given the antibody quantities required per experiment (mg range). Finally, we have included several practical considerations and tips to help less experienced researchers apply a robust immunopeptidomics workflow to their own research questions.

2 Materials

2.1 Isolation of the W6/32 Monoclonal Antibody from the HB-95 Hybridoma Cell Line

For a successful immunopeptidomics experiment, according to our workflow, antibody amounts in the range of milligrams are required. For this reason, in-house production and purification of the antibody is preferred.

1. W6/32 hybridoma (HB-95, ATCC).
2. Complete medium: High glucose Dulbecco's Modified Eagle Medium (DMEM) with 1% L-glutamine, 1% sodium pyruvate, 1% penicillin-streptomycin and 10% fetal bovine serum (FBS).

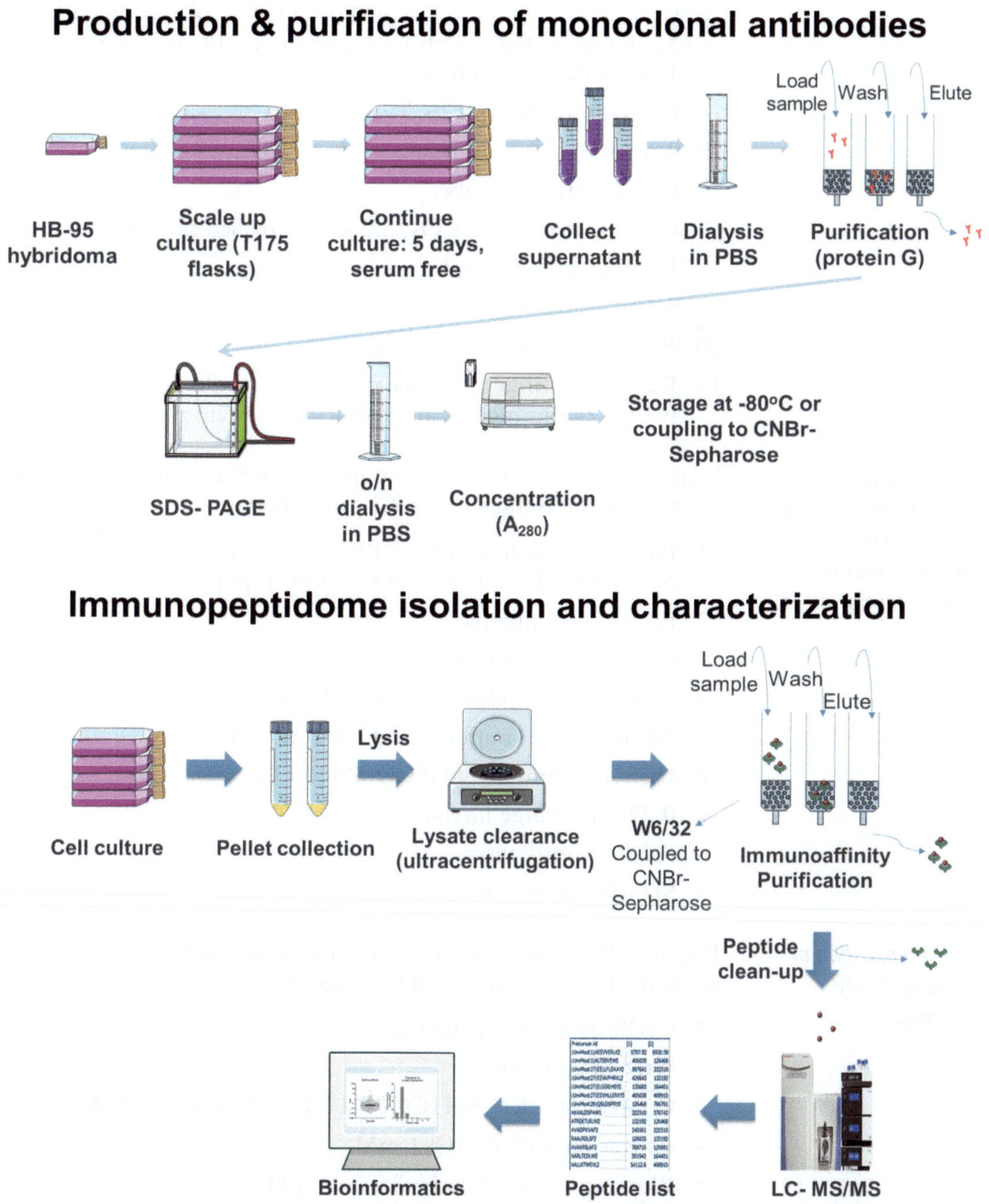

Fig. 1 General overview of the necessary steps for the characterization of Class I immunopeptidome from cells. This figure was drawn using images from Servier Medical Art, which is licensed under a Creative Commons Attribution 4.0 Unported License (CC BY) (https:/smart.servier.com/citation-sharing/) (https:/creativecommons.org/licenses/by/4.0/)

3. Collection medium: High glucose Dulbecco's Modified Eagle Medium (DMEM) with 1% L-glutamine, 1% sodium pyruvate, 1%penicillin-streptomycin.
4. Cell culture flasks (T25, T75, T175).
5. Trypan blue.
6. Cell counting chamber.
7. CO_2 incubator (5% CO_2, 37 °C, controlled humidity).
8. Laminar airflow hood.
9. 50 mL falcon tubes.
10. Refrigerated centrifuge (falcon).
11. 0.22 um syringe filters or disposable sterile bottle-top filters with 0.22 um membrane if a vacuum system is available.

2.2 Affinity Purification of the W6/32 Monoclonal Antibody from the Supernatant of HB-95 Cells

Prepare all solutions using ultrapure water and analytical grade reagents. Filter buffers using 0.22 um filters.

1. Phosphate-buffered saline (PBS): NaCl 137 mM, KCl 2.7 mM, Na_2HPO_4 10 mM, KH_2PO_4 1.8 mM, pH 7.4.
2. Protein G Sepharose.
3. Glass chromatography columns.
4. Elution buffer: Citric acid 0.1 M, pH 3.0.
5. Neutralization buffer: Tris-HCl 1 M, pH 9.0.
6. Dialysis membrane MWCO: 30000 Da.
7. 0.22 um syringe filters.
8. Spectrophotometer.
9. SDS-PAGE system.

2.3 Preparation of Immunoaffinity Columns

Prepare all solutions using ultrapure water and analytical grade reagents. Filter buffers using 0.22 um filters.

1. CNBr-activated Sepharose.
2. W6/32 antibody.
3. Coupling buffer: $NaHCO_3$ 0.1 M, NaCl 0.5 M, pH 8.3.
4. Activation buffer: HCl 1 mM.
5. Blocking buffer: Tris-HCl 0.1 M, pH 8.0.
6. Acidic buffer: CH_3COOH/CH_3COONa 0.1 M, NaCl 0.5 M, pH 4.0.
7. Basic buffer: Tris-HCl 0.1 M, NaCl 0.5 M, pH 8.0.
8. Lysis buffer without detergent: Tris-HCl 20 mM, NaCl 150 mM, pH 7.5.
9. Dialysis membrane MWCO: 30,000 Da.

10. Laboratory tube roller.
11. 50 mL centrifuge tubes.
12. 50 mL Lo-bind centrifuge tubes.
13. Reusable filter unit 250 ml.
14. 0.22 um PES membrane filters.
15. Rotary Vane Vacuum Pump.
16. Spectrophotometer.

2.4 Immunopeptidome Isolation

Prepare all solutions using LC-MS/MS grade water and reagents. Buffers should be made fresh on the day of the isolation. Use ultraclean glassware, free of detergent residues and protein lo-bind plasticware.

1. Cell pellet (*see* **Note 1**).
2. Protease inhibitors.
3. Lysis buffer: Tris-HCl 20 mM pH 7.5, NaCl 150 mM, 0.5% Igepal, 0.25% sodium deoxycholate.
4. EDTA 0.5 M, pH 8.0.
5. Lysis buffer without detergent: Tris-HCl 20 mM, NaCl 150 mM, pH 7.5.
6. Wash buffer 1: Tris-HCl 20 mM, NaCl 150 mM, pH 8.0.
7. Wash buffer 2: Tris-HCl 20 mM, NaCl 400 mM, pH 8.0.
8. Wash buffer 3: Tris-HCl 20 mM, pH 8.0.
9. Elution buffer: 1% trifluoroacetic acid (TFA).
10. Laboratory tube roller.
11. Refrigerated centrifuge.
12. Refrigerated ultracentrifuge.
13. 50 ml centrifuge tubes.
14. Lo-bind centrifuge tubes 50 mL.
15. Ultracentrifuge tubes.
16. Lo-bind microcentrifuge tubes 1.5 mL.
17. Glass chromatography columns.
18. SDS-PAGE and western blot analysis systems.
19. Primary antibody against human beta-2-microglobulin.
20. Secondary antibody recognizing the primary antibody, suitable for Western-blot detection.

2.5 Immunopeptidome Purification with C-18 Disposable Columns

Prepare all solutions using LC-MS/MS grade reagents.

1. Fractions from immunopeptidome isolation, in which beta-2-microglobulin was detected.

2. Disposable C18 Columns.
3. Lo-bind microcentrifuge tubes 1.5 mL.
4. Microcentrifuge tubes 1.5 mL.
5. Centrifuge (1.5 mL tubes).
6. SpeedVac.
7. Sample adjustment solution: 10% Acetonitrile (ACN).
8. Activation Solution: 50% ACN.
9. Wash Solution 1/Equilibration Solution: 5% ACN, 0.5% TFA.
10. Wash Solution 2: 5% ACN, 0.1% formic acid (FA).
11. Peptide Elution Solution: 30% ACN, 0.1% FA.
12. Protein Elution Solution: 80% ACN, 0.1% FA.

2.6 Peptide Clean-Up with the Sp3 Protocol

Prepare all solutions using LC-MS/MS grade reagents.

1. Carboxylate functionalized beads with a hydrophilic surface.
2. Carboxylate functionalized beads with a hydrophobic surface.
3. ACN 100%.
4. Water.
5. Buffer A: 2% (v/v) ACN and 0.1% (v/v) formic acid.
6. Buffer B: 80% (v/v) ACN and 0.1% (v/v) formic acid.
7. Magnetic rack for 1.5 mL tubes.
8. Lo-bind microcentrifuge tubes 1.5 mL.
9. Clear polypropylene vial with 300 ul insert and with 11 mm Snap caps.
10. Refrigerated centrifuge.
11. Thermomixer for microcentrifuge tubes.
12. Sonicating water bath.
13. Speedvac.

2.7 Liquid Chromatography/Mass Spectrometry

1. LC-MS/MS system (*see* **Note 2**).
2. HPLC configured in a peptide direct injection mode.
3. Analytical column (25 cm × 75 um ID, particle size 1.9 um, 120 Å).
4. Buffer A: 2% (v/v) Acetonitrile (ACN) and 0.1% (v/v) formic acid.
5. Buffer B: 80% (v/v) Acetonitrile (ACN) in 0.1% (v/v) formic acid.

3 Methods

3.1 Isolation of the W6/32 Monoclonal Antibody from the HB-95 Hybridoma Cell Line

1. Culture HB-95 hybridoma cells in complete DMEM at concentrations between $2*10^5$ and $1*10^6$ cells/ml in T25 culture flasks. Renew the medium every 2–3 days by either adding or partially replacing and adding new medium (*see* **Note 3**).
2. Once the viability is above 80%, start upscaling the cell culture to T75 and then T175 flasks.
3. Detach cells from confluent T175 flasks, count them (and assess viability) and collect them in 50 mL falcon tubes.
4. Centrifuge at 300 *g* for 5 min.
5. Discard supernatant, redistribute in 10 mL collection medium and centrifuge again under the same conditions.
6. Discard the medium and redistribute in fresh collection medium at a density of $0.8–1*10^6$ cells/ml.
7. Transfer cell suspension to a T175 flask and culture for 5 days (*see* **Note 4**).
8. On day 5, collect the content of the T175 flasks and centrifuge at 1200 g at 4 °C for 20–30 min.
9. Filter supernatant with 0.2 um filters and collect in new falcon tubes.
10. Proceed with antibody purification or store at –80 °C.

3.2 Affinity Purification of the W6/32 Monoclonal Antibody from the Supernatant of HB-95 Cells

1. Dialyze supernatant with pre-chilled PBS for 48 h at 4 °C. During the dialysis, change the PBS at least three times (*see* **Note 5**).
2. Pack 1 ml Protein G Sepharose beads in a glass chromatography column (*see* **Note 6**).
3. Wash with 20 bed volumes of ultrapure water and equilibrate with 20 bed volumes of PBS (*see* **Note 7**).
4. Pass the dialysate through the column with a flow rate of 1 mL/min at 4 °C (*see* **Note 8**).
5. Wash beads with 20 bed volumes of PBS.
6. Elute antibody with 10 bed volumes of citric acid. Collect the antibody in 1.5 mL microcentrifuge tubes containing 300 ul neutralization buffer in 1 mL fractions.
7. Analyze cell supernatant, flow-through, and the eluates with SDS-PAGE (12% polyacrylamide) and Coomassie staining (*see* **Note 9**) (Fig. 2).
8. Dialyze the highly concentrated samples (usually fractions 2–6) overnight with PBS at 4 °C.

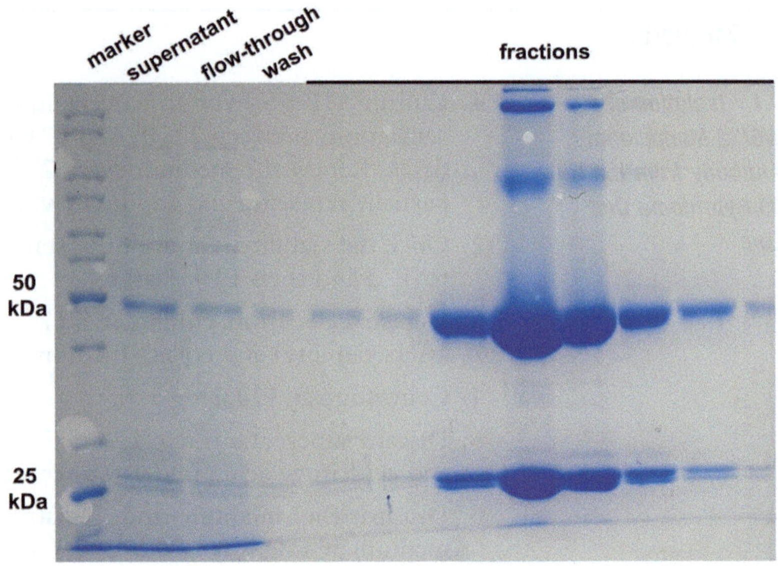

Fig. 2 Representative picture of an SDS-PAGE gel stained with Coomassie brilliant blue. The presence of the antibody is confirmed by a band at 50 kDa, corresponding to the antibody heavy chain, as well as a band at 25 kDa, corresponding to the light chain

9. Determine antibody concentration by measuring absorbance at 280 nm. Concentration is calculated based on the Lambert-Beer formula: A = ε*d*l, where A = absorbance, ε = molar absorption coefficient (210,000 M^{-1} cm^{-1} for IgG), C = concentration (M), and l = path length.
10. Store dialyzed antibody in 1 ml fractions at a concentration ≥ 1 mg/ml at -80 °C (*see* **Note 10**).

3.3 Preparation of Immunoaffinity Columns

This procedure requires 3 days.

Day 1

1. Thaw the necessary amount of antibody (*see* **Note 11**).
2. Dialyze the antibody in coupling buffer overnight at 4 °C.

Day 2

3. Measure absorbance at 280 nm to calculate antibody concentration and place at 4 °C in a lo-bind falcon tube until needed.
4. For each column, weigh 0.285 g of dry cyanogen bromide-activated Sepharose 4B beads in a lo-bind 50 ml falcon (*see* **Note 12**).
5. Activate the beads by adding 1 mM HCl in the lo-bind falcon and rotating for 30 min at room temperature.

6. Assemble the filtering device and connect it to the vacuum system. Place the beads in a reusable filter unit with a 0.22 um PES membrane filter, filter to remove the HCl and let the beads dry completely.
7. Wash twice with HCl and let the beads dry completely.
8. Wash twice with coupling buffer and let the beads dry completely.
9. Combine the activated beads with the purified antibody in coupling buffer at a ratio of 2 mg antibody per ml of slurry (*see* **Note 13**).
10. Rotate overnight at 4 °C.

Day 3

11. Start with the pre-columns: Perform steps 4–7 of day 2.
12. Wash three times with blocking buffer and let the beads dry completely.
13. Transfer beads to a 50 ml falcon and add blocking buffer.
14. Rotate for 3 h at room temperature.

After putting the pre-column beads on rotation, wait 30 min and continue with columns.

15. Let the column beads settle by placing the falcon in a stand and take a sample of the supernatant to measure absorbance at A280. Absorbance should be close to baseline if the coupling has been performed efficiently.
16. Assemble the filtering device and connect it to the vacuum system. Place the beads in a reusable filter unit with a 0.22 um PES membrane filter and filter to remove the coupling buffer. Let the beads dry completely.
17. Wash three times with blocking buffer and let the beads dry completely.
18. Transfer beads to a 50 ml falcon and add blocking buffer.
19. Rotate for 3 h at room temperature.

After 3 h blocking, continue with the pre-columns.

20. Assemble the filtering device and connect it to the vacuum system. Place the beads in a reusable filter unit with a 0.22 um PES membrane filter and filter to remove the blocking buffer. Let the beads dry completely.
21. Wash the beads with three cycles of acidic and basic buffer, always ending with the basic. Let the beads dry completely.

22. Wash beads twice with lysis buffer without detergent, transfer to a 50 ml falcon, add buffer and store at 4 °C until needed (*see* **Note 14**).

When blocking of columns is over, continue with columns.

23. Perform steps 20–22 for the coupled beads (*see* **Note 14**).

3.4 Immuno-peptidome Isolation

1. Freshly prepare lysis buffer and supplement it with 1 mM EDTA and protease inhibitors.
2. Thaw cells on ice and mix them with 2 mL lysis buffer per 10^8 cells in a lo-bind 50 mL falcon tube.
3. Lyse cells for 1 h while rotating at 4 °C.
4. During lysis, prepare columns and pre-columns by adding 1 mL of slurry per column. Let the preservation solution drain by gravity and wash with 10 bed volumes of lysis buffer without detergents. At the end of the wash, place the caps on the columns to avoid drying of the slurry.
5. Centrifuge lysates at 400 g for 15 min at 4 °C and collect the supernatant.
6. Further clear lysates by ultracentrifugation at 100,000 g for 1 h at 4 °C.

The following procedures must take place at 4 °C.

7. Pass the cleared lysates sequentially through the pre-columns and then the antibody-containing columns (*see* **Note 15**).
8. Pass the flow-through three more times through the antibody-containing columns to increase the percentage of bound HLA-I.
9. Wash the columns with 20 bed volumes of wash buffer 1.
10. Wash the columns with 20 bed volumes of wash buffer 2.
11. Wash the columns with 20 bed volumes of wash buffer 1.
12. Wash the columns with 40 bed volumes of wash buffer 3.
13. Transfer columns to room temperature and elute with 5 ml elution buffer. Collect 5 1 mL fractions in lo-bind microcentrifuge tubes.
14. Take 20 ul samples for Western-blot and store elution fractions at –80 °C.
15. Perform a Western-blot against beta-2-microglobulin to assess the efficiency of the purification. A beta-2-microglobulin band at ~12 kDa should be visible in the elution fractions but not in the flow-through (Fig. 3).

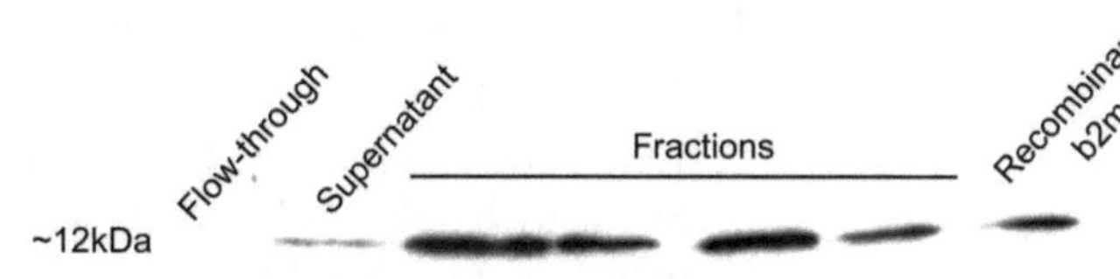

Fig. 3 Example of Western-blot against beta-2-microglobulin (b2m)

3.5 Immuno-peptidome Purification with C-18 Disposable Columns

1. Dry beta-2-microglobulin-containing samples using SpeedVac until the total volume per sample is approximately 100 ul.
2. Adjust the sample composition to 5% ACN, 0.5% TFA by mixing samples with the sample adjustment solution at a ratio 1:1 (*see* **Note 16**).
3. Place the C-18 columns in the receiver microcentrifuge tubes.
4. Activate the columns with 200 ul activation solution and centrifuge at 1500 g for 1 min (*see* **Note 17**).
5. Discard flow-through and repeat centrifugation.
6. Equilibrate columns by washing twice with 200 ul equilibration solution, with intermediate centrifugation and flow-through disposal, as above.
7. Gradually load samples, 150 ul at a time and centrifuge at 1500 g for 1 min. Collect flow-through in a new lo-bind tube.
8. After loading the samples once, repeat the procedure once more to ensure sample binding.
9. Wash columns with wash solution 1 and centrifuge as above.
10. Wash columns with wash solution 2 and centrifuge as above.
11. Elute peptides with the addition of 30 ul peptide elution solution and centrifuge as above.
12. Repeat step 11 and merge the two fractions.
13. Elute proteins with 25 ul protein elution solution and centrifuge as above.
14. Repeat the procedure with the flow-through using new columns and merge the peptide elution fractions (120 ul in total).
15. Keep 10 ul of the flow-through and 2–3 ul per sample for a Western-blot analysis and store the samples at -80 °C.
16. Check for the presence of beta-2-microglobulin in the fractions with Western-blot. If beta-2-microglobulin is detected, thaw the peptide fractions, adjust the composition to 5% ACN, 0.5% TFA and repeat the purification with new spin columns.

3.6 Peptide Clean-Up with the Sp3 Protocol [19]

Start with drying the peptide fractions and preparation of the beads:

1. Place the peptide fractions in SpeedVac and let them dry completely.
2. While the samples are being dried, mix 50 ul hydrophilic and 50 ul hydrophobic beads stock solution and place them on a magnetic rack for 2 min.
3. Carefully remove the supernatant and rinse the beads off-rack with 200 ul water. Gently pipette up and down and place the tubes back on the magnetic rack.
4. Carefully remove supernatant and repeat the washing two more times.
5. Resuspend beads in 500 ul water at a final concentration of 10 ug beads/ul of water. Place them at 4 °C until needed.
6. Once the peptide fractions are dried, add 10 ul water to dissolve the peptides.
7. Sonicate for 3 min.
8. Add 1 mL 100% ACN per tube.
9. Add 20 ul magnetic bead slurry and incubate for 30 min at room temperature with mild mixing every 10 min.
10. Place the samples onto the magnetic stand for 2 min.
11. Carefully remove the supernatants.
12. Add 200 ul 100% ACN off-rack, gently mix and place back on the magnetic rack.
13. Carefully remove the supernatants and repeat the procedure once more.
14. Remove the supernatant and add 50 ul Buffer A to elute the peptides.
15. Incubate for 5 min.
16. Transfer the eluates to the autosampler vials.
17. Transfer vials to SpeedVac and dry completely.

3.7 Liquid Chromatography/Mass Spectrometry.

This part of the protocol describes the LC-MS/MS analysis of HLA-I peptides using an Ultimate 3000 RSLCnano HPLC coupled to a Q Exactive HF-X Mass Spectrometer. The analytical column has a C-18 stationary phase with 1.9 um particle size, 120 Å pore size, length of 25 cm and internal diameter of 75 um. The column operates at room temperature and is mounted on the pepsprayer device in close proximity to the stainless-steel emitter (zero dead volume). Alternative systems can be used. In such cases, methods might need to be adjusted to the specific instrument requirements. Prior to analysis, ensure the cleanliness of the column and inject a blank sample to exclude sample carryover from previous runs. When injecting samples, aim for 1 ul per injection and try to include at least 2 technical replicates.

1. Resuspend the dried peptides in 5 ul Buffer A.
2. Sonicate for 3 min.
3. Place the vials in the cold autosampler inserts and place them in the autosampler.
4. Inject 1ul of the first sample.
5. Separate peptides using a 30 min linear gradient going from 7% to 35% Buffer B over 16.5 min with a flow rate of 250–300 nL/min. Then Buffer B reaches 45% in 1.5 min and 99% in 0.5 min. It is ultimately maintained constant for 1 min before returning to 7% over 10.5 min (*see* **Note 18**).
6. For DDA acquisition, used for building the hybrid library in FragPipe (see § 3.8), we recommend one technical replicate. Use a scanning range between 275 and 1100 m/z in the full MS, with 120 K resolving power, automatic gain control (AGC) of $3*10^6$ and max injection time (IT) of 100 ms. For the MS/MS, select the 5 most abundant ions and set resolving power to 30 K, AGC to $1*10^6$, max IT to 50 ms and normalized collision energy (NCE) to 30 (for full DDA method specifications in Q-Exactive HF-X see Tables 1 and 2, *see* **Note 19**).
7. For the DIA acquisition, we recommend two technical replicates per sample. For the full MS, scan a range between 375 and 1100 m/z in profile mode with 120 K resolving power, AGC of $3*10^6$ and max IT of 60 ms. For the MS/MS using high collision MS/MS fragmentation (HCD), set NCE to 26, the resolving power to 15 K, AGC $3*10^6$ and max IT to 22 ms (for full DIA method specifications in Q-Exactive HF-X see Table 3, *see* **Note 20**).

Table 1
Instrument settings for full MS acquisition using a Q Exactive HF-X Orbitrap mass spectrometer in DDA mode

General		Full MS	
Runtime	0 to 30 min	Microscans	1
Polarity	positive	Resolution	120,000
In-source CID	0.0 eV	AGC target	$3*10^6$
Default charge state	2	Maximum IT	100 ms
Inclusion	–	Number of scan ranges	1
Exclusion	–	Scan range	275 and 1100 m/z
Tags	–	Spectrum data type	Profile

Table 2
Instrument settings for dd-MS^2/dd-SIM acquisition using a Q Exactive HF-X Orbitrap mass spectrometer in DDA mode

dd-MS^2 / dd-SIM		dd settings	
Microscans	1	Minimum AGC target	$8.00*10^3$
Resolution	30,000	Intensity threshold	$1.6*10^5$
AGC target	$1*10^5$	Apex trigger	–
Maximum IT	50 ms	Charge exclusion	–
Loop count	12	Multiple charge states	all
MSX count	1	Peptide match	preferred
TopN	12	Exclude isotopes	on
Isolation window	4.0 m/z	Dynamic exclusion	10 s
		Isolation offset	0.0 m/z
		Scan range	275 to 1100 m/z
		Fixed first mass	—
		(N)CE / stepped (N)CE	30
		Spectrum data type	Profile
Mass Spectrometer Tuning			
Spray voltage: 1.8–2.20 kV (calibrating depending on the emitter) Capillary Temperature: 250 °C Funnel RF Level: 40			

3.8 Data Analysis

The Nonspecific-HLA-DIA workflow offered by Fragpipe [20], an open source and easily applicable platform with a Graphical User Interface (GUI), is a suitable and affordable solution for the identification immunopeptidome-derived peptides. It supports a hybrid spectral library building from both DDA and DIA data and eventually allows quantification from DIA files using DIA-NN [21].

To derive peptide lists, the following steps are needed, after installing the software:

1. Select the Nonspecific-HLA-DIA workflow on the workflow tab, add the raw DDA and DIA MS files and fill in the required experiment information.
2. In the Database tab add the required fasta file (*see* **Note 21).**
3. Leave the rest of the parameters to default, unless a more custom setting is required for your experiment.

Table 3
Instrument settings for a DIA acquisition using a Q Exactive HF-X Orbitrap mass spectrometer

Full MS—SIM general		**DIA general**	
Runtime	0 to 30 min	Runtime	0 to 30 min
Polarity	positive	Polarity	positive
In-source CID	0.0 eV	In-source CID	0.0 eV
Microscans	1	Default charge state	3
Resolution	120,000	Microscans	1
AGC target	$3*10^6$	Resolution	15,000
Maximum IT	60 ms	AGC target	$3*10^6$
Number of scan ranges	1	Maximum IT	22 ms
Scan range	275 to 1100 m/z	Loop count	39
Spectrum data type	Profile	MSX count	1
		MSX isochronous its	on
		Isolation window	8.0 m/z
		Isolation offset	0.0 m/z
		Fixed first mass	200.0 m/z
		(N)CE / stepped (N)CE	26
		Spectrum data type	Profile
Mass Spectrometer Tuning			
The tune file for the operation of the nano-flow mode is: Spray voltage: 1.8–2.20 kV (calibrating depending on the emitter) Capillary Temperature: 250 °C Funnel RF Level: 40			

4. Go to the run tab, choose your outcome directory and press RUN.

   The peptide list can then be found in your output directory, in the "diann-output" folder (file name: report.pr_matrix.tsv). For differential expression analysis, you can then use the FragPipe-Analyst [22] tool, available online: http://fragpipe-analyst.nesvilab.org/.

5. Go to Analysis on the top left side of the screen and press the arrow.
6. In the "Data type" dropdown menu, select "DIA (peptide)."
7. Then upload the "report.pr_matrix.tsv" and the "experiment_annotation.tsv" files—both present in your directory—in the designated sections.

8. By default, an adjusted p-value cutoff of 0.05, a Log2 fold change cutoff of 1 and a Benjamini Hochberg FDR correction are applied. These parameters can be found and changed in the "Advanced Options" section. In this section, you can further customize your analysis by adding normalization and imputation.
9. Press "Run."
10. In the output page you will find a downloadable "DE results. csv" file, which contains the results of the differential expression analysis. Typical graphs, such as volcano plots, heatmaps and PCA are also included. A PDF report containing the analysis parameters, as well as the graphs, can also be downloaded.

3.9 Result Evaluation-Quality Control

For results evaluation, several critical parameters need to be assessed. Example graphs are provided in Figs. 4 and 5 below.

1. Length distribution: Given the known preference of HLA-I molecules for the presentation of 9-mer peptides, a distribution of peptide lengths needs to be plotted. For successful immunopurification, the majority of the peptides should be 9 amino acids long (Fig. 4a).
2. Binding affinity prediction: Servers like HLAthena [23] and NetMHCpan 4.1 [24] can predict the binding affinity of the isolated peptides for the HLA alleles carried by the tested samples. For successful immunopurification, most of the peptides should have scores below 2 (Fig. 4b).
3. Motif analysis: Tools like MHCMotifDecon 1.0 [25] GibbsCluster 2.0 [26] and Weblogo [27] can generate motifs

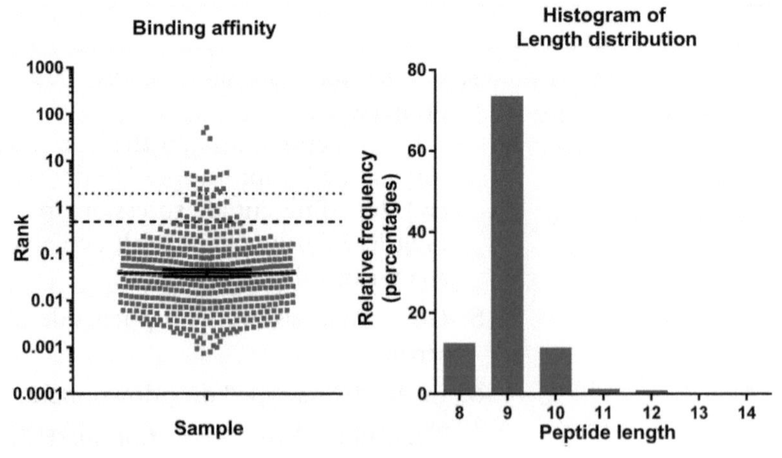

Fig. 4 Representative quality control graphs for length distribution (**a**) and binding affinity scores (**b**). Data from [18]

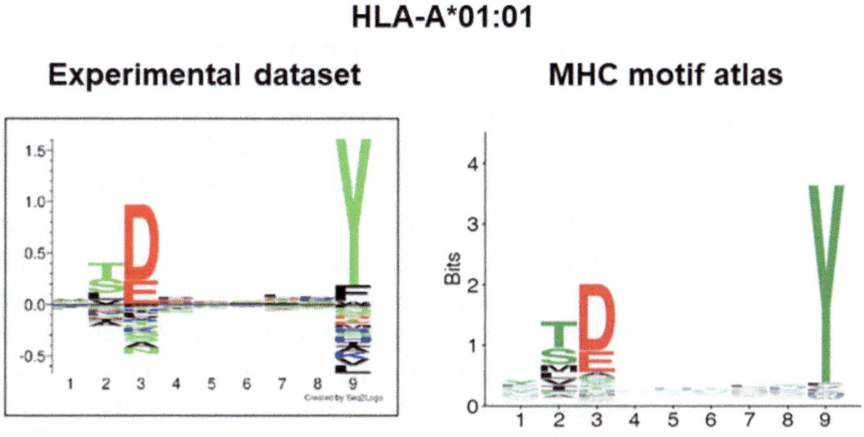

Fig. 5 Representative quality control for motif analysis. Data from [17]

for the isolated peptides, which can be compared to known HLA- preferred motifs found in the MHC motif atlas [28] (Fig. 5).

4 Notes

1. The size of the pellet is dependent on the available mass spectrometry instrumentation. For example, approximately $5*10^8$ cells are needed for a Q Exactive HF-X Orbitrap mass spectrometer [16] but amounts as low as 10^6 cells might be needed for last-generation instruments [29].
2. We used a Q Exactive HF-X Mass Spectrometer supplemented with a nanoESI source using a Pepsep emitter with liquid junction I.D. 30 um, operating in both data-dependent (DDA) and data-independent acquisition mode (DIA). The mass spectrometer was coupled on-line to an Ultimate 3000 RSLCnano HPLC system, configured in a peptide direct injection mode. The system was operating with the XCalibur Software (all Thermo Scientific).
3. HB-95 is a suspension cell line, which we routinely culture in cell culture flasks (T25 to T175). Prior to subculturing, check the confluency under the microscope. If cells are not confluent but the medium has slightly turned orange or if they are mostly floating instead of loosely attached, remove the medium partially without disturbing the attached cells and add the same amount of fresh medium. If cells are confluent, detach cells by taping and pipetting up and down, count them in a cell counting chamber using trypan blue to assess viability and sub-culture at a density close to $2*10^5$ cells/ml.

4. Before passing to serum-free, it is advisable to have at least 10 T175 flasks each time, or more if possible, depending on the available space in the incubator. In case you need to keep producing antibody, make sure to keep a T75 flask in complete medium as well, from which you will start obtaining more flasks after you finish with the first collection phase.
5. After the dialysis, 0.02% NaN_3 can be added as a preservative.
6. For more than 500 ml supernatant, use 2 ml Protein G beads. After purification, used beads can be washed with 20 bed volumes ultrapure water and stored at 20% ethanol at 4 °C.
7. All procedures with glass chromatography columns are based on gravity. Place the columns in stands that allow collection tubes to be placed below them. Stop the flow at the end of the PBS wash by placing the column caps to avoid drying of the beads.
8. For large supernatant volumes the procedure can be performed overnight. Keep 20 ul of the dialysate prior to loading for the SDS-PAGE that will be performed later.
9. If antibody bands are observed in the flow-through, repeat the immunopurification with the flow-through, after washing the column with 20 bed volumes of PBS.
10. For antibody fractions at a concentration below 1 mg/ml, it is recommended to use centrifugal filters to increase concentration prior to storage.
11. For columns and pre-columns, use 1 ml beads' slurry per column. For columns only: aim for a coupling of 2 mg of antibody per ml of slurry. Always thaw 1–2 mg extra to compensate for losses during dialysis.
12. During column/pre-column preparation, calculate for 1 ml more than needed.
13. If needed, add a bit of coupling buffer to make sure that the beads remain hydrated.
14. For long-term storage, add 0.02% NaN_3 to the lysis buffer.
15. Keep a 20 ul sample before the immunoaffinity purification for a Western blot analysis against beta-2-microglobulin.
16. Keep a 10 ul sample before purification for a Western blot analysis against beta-2-microglobulin.
17. After activating the columns, work should be performed relatively fast to avoid drying of the resin.
18. Depending on the protein content of the samples, a 60 min linear gradient may also be considered. In such cases the transition from 7% to 35% Buffer B is achieved over 40 min and is followed by an increase to 45% in 5 min and a second increase to 99% in 0.5 min. The gradient is then kept constant for equilibration for 14.5 min.

19. Tables summarizing selected DDA parameters.
20. Table summarizing selected DIA parameters.
21. The human proteome database provided by UniProt [30], which we also used in our analysis [18], is a good starting point for searching the immunopeptidome of annotated proteins. Alternatively, proteogenomics derived databases, such as the Ribo-seq derived nuorfdb [31] or any other custom database, can be used. If you decide to use such a database, a group specific FDR is recommended to control false identifications and properly account for the lower possibility of identifying non-canonical sequences [32, 33].

Conflict of Interest Statements

The authors have no conflicts of interest to declare that are relevant to the content of this chapter.

References

1. Neefjes J, Jongsma MLM, Paul P, Bakke O (2011) Towards a systems understanding of MHC class I and MHC class II antigen presentation. Nat Rev Immunol 11:823–836. https://doi.org/10.1038/nri3084
2. Yewdell JW, Bennink JR (2001) Cut and trim: generating MHC class I peptide ligands. Curr Opin Immunol 13:13–18. https://doi.org/10.1016/S0952-7915(00)00175-8
3. Admon A (2023) The biogenesis of the immunopeptidome. Semin Immunol 67:101766. https://doi.org/10.1016/j.smim.2023.101766
4. Klein L, Kyewski B, Allen PM, Hogquist KA (2014) Positive and negative selection of the T cell repertoire: what thymocytes see (and don't see). Nat Rev Immunol 14:377–391. https://doi.org/10.1038/nri3667
5. Admon A (2021) Are there indeed spliced peptides in the Immunopeptidome? Mol Cell Proteomics 20:100099. https://doi.org/10.1016/j.mcpro.2021.100099
6. Admon A, Bassani-Sternberg M (2011) The human Immunopeptidome project, a suggestion for yet another Postgenome next big thing. Mol Cell Proteomics 10. https://doi.org/10.1074/mcp.O111.011833
7. Pongcharoen S, Kaewsringam N, Somaparn P, Roytrakul S, Maneerat Y, Pintha K, Topanurak S (2024) Immunopeptidomics in the cancer immunotherapy era. Exploration of Targeted Anti-tumor Therapy 5:801–817. https://doi.org/10.37349/etat.2024.00249
8. Arshad S, Cameron B, Joglekar AV (2025) Immunopeptidomics for autoimmunity: unlocking the chamber of immune secrets. NPJ Syst Biol Appl 11:10. https://doi.org/10.1038/s41540-024-00482-x
9. Illing PT, Ramarathinam SH, Purcell AW (2022) New insights and approaches for analyses of immunopeptidomes. Curr Opin Immunol 77:102216. https://doi.org/10.1016/j.coi.2022.102216
10. Shapiro IE, Bassani-Sternberg M (2023) The impact of immunopeptidomics: from basic research to clinical implementation. Semin Immunol 66:101727. https://doi.org/10.1016/j.smim.2023.101727
11. Bassani-Sternberg M, Coukos G (2016) Mass spectrometry-based antigen discovery for cancer immunotherapy. Curr Opin Immunol 41:9–17. https://doi.org/10.1016/j.coi.2016.04.005
12. Guasp P, Barnea E, González-Escribano MF, Jiménez-Reinoso A, Regueiro JR, Admon A, López De Castro JA (2017) The Behçet's disease-associated variant of the aminopeptidase ERAP1 shapes a low-affinity HLA-B*51 peptidome by differential subpeptidome processing. J Biol Chem 292:9680–9689. https://doi.org/10.1074/jbc.M117.789180

13. Bassani-Sternberg M (2018) Mass spectrometry based Immunopeptidomics for the discovery of cancer Neoantigens. In: Methods in molecular biology. Springer New York, New York, NY, pp 209–221
14. Koumantou D, Barnea E, Martin-Esteban A, Maben Z, Papakyriakou A, Mpakali A, Kokkala P, Pratsinis H, Georgiadis D, Stern LJ, Admon A, Stratikos E (2019) Editing the immunopeptidome of melanoma cells using a potent inhibitor of endoplasmic reticulum aminopeptidase 1 (ERAP1). Cancer Immunol Immunother 68:1245–1261. https://doi.org/10.1007/s00262-019-02358-0
15. Temponeras I, Stamatakis G, Samiotaki M, Georgiadis D, Pratsinis H, Panayotou G, Stratikos E (2022) ERAP2 inhibition induces cell-surface presentation by MOLT-4 leukemia cancer cells of many novel and potentially antigenic peptides. Int J Mol Sci 23:1913. https://doi.org/10.3390/ijms23031913
16. Temponeras I, Samiotaki M, Koumantou D, Nikopaschou M, Kuiper JJW, Panayotou G, Stratikos E (2023) Distinct modulation of cellular immunopeptidome by the allosteric regulatory site of ER aminopeptidase 1. Eur J Immunol 53:2350449. https://doi.org/10.1002/eji.202350449
17. Nikopaschou M, Samiotaki M, Stylianaki E-A, Król K, Gragera P, Raja A, Aidinis V, Chroni A, Fruci D, Panayotou G, Stratikos E (2025) ERAP1 Activity Modulates the Immunopeptidome but Also Affects the Proteome, Metabolism, and Stress Responses in Cancer Cells. Mol Cell Proteomics. 24(5):100964. https://doi.org/10.1016/j.mcpro.2025.100964
18. Nikopaschou M, Samiotaki M, Kannavou A, Angelis NV, Tsitsilonis O, Panayotou G, Stratikos E (2025) ERAP1 Allotypes 2 and 10 Differentially Regulate the Immunopeptidome of Melanocytes. Immunology. https://doi.org/10.1111/imm.70056
19. Hughes CS, Foehr S, Garfield DA, Furlong EE, Steinmetz LM, Krijgsveld J (2014) Ultrasensitive proteome analysis using paramagnetic bead technology. Mol Syst Biol 10:757. https://doi.org/10.15252/msb.20145625
20. Yu F, Teo GC, Kong AT, Fröhlich K, Li GX, Demichev V, Nesvizhskii AI (2023) Analysis of DIA proteomics data using MSFragger-DIA and FragPipe computational platform. Nat Commun 14:4154. https://doi.org/10.1038/s41467-023-39869-5
21. Demichev V, Messner CB, Vernardis SI, Lilley KS, Ralser M (2020) DIA-NN: neural networks and interference correction enable deep proteome coverage in high throughput. Nat Methods 17:41–44. https://doi.org/10.1038/s41592-019-0638-x
22. Hsiao Y, Zhang H, Li GX, Deng Y, Yu F, Valipour Kahrood H, Steele JR, Schittenhelm RB, Nesvizhskii AI (2024) Analysis and visualization of quantitative proteomics data using FragPipe-analyst. J Proteome Res 23:4303–4315. https://doi.org/10.1021/acs.jproteome.4c00294
23. Sarkizova S, Klaeger S, Le PM, Li LW, Oliveira G, Keshishian H, Hartigan CR, Zhang W, Braun DA, Ligon KL, Bachireddy P, Zervantonakis IK, Rosenbluth JM, Ouspenskaia T, Law T, Justesen S, Stevens J, Lane WJ, Eisenhaure T, Lan Zhang G, Clauser KR, Hacohen N, Carr SA, Wu CJ, Keskin DB (2020) A large peptidome dataset improves HLA class I epitope prediction across most of the human population. Nat Biotechnol 38:199–209. https://doi.org/10.1038/s41587-019-0322-9
24. Reynisson B, Alvarez B, Paul S, Peters B, Nielsen M (2020) NetMHCpan-4.1 and NetMHCIIpan-4.0: improved predictions of MHC antigen presentation by concurrent motif deconvolution and integration of MS MHC eluted ligand data. Nucleic Acids Res 48:W449–W454. https://doi.org/10.1093/nar/gkaa379
25. Kaabinejadian S, Barra C, Alvarez B, Yari H, Hildebrand WH, Nielsen M (2022) Accurate MHC motif deconvolution of Immunopeptidomics data reveals a significant contribution of DRB3, 4 and 5 to the Total DR Immunopeptidome. Front Immunol 13:835454. https://doi.org/10.3389/fimmu.2022.835454
26. Andreatta M, Alvarez B, Nielsen M (2017) GibbsCluster: unsupervised clustering and alignment of peptide sequences. Nucleic Acids Res 45:W458–W463. https://doi.org/10.1093/nar/gkx248
27. Crooks GE, Hon G, Chandonia J-M, Brenner SE (2004) WebLogo: A Sequence Logo Generator: Figure 1. Genome Res 14:1188–1190. https://doi.org/10.1101/gr.849004
28. Tadros DM, Eggenschwiler S, Racle J, Gfeller D (2023) The MHC motif atlas: a database of MHC binding specificities and ligands. Nucleic Acids Res 51:D428–D437. https://doi.org/10.1093/nar/gkac965
29. Gomez-Zepeda D, Arnold-Schild D, Beyrle J, Declercq A, Gabriels R, Kumm E, Preikschat A, Łącki MK, Hirschler A, Rijal JB, Carapito C, Martens L, Distler U, Schild H, Tenzer S (2024) Thunder-DDA-PASEF enables high-coverage immunopeptidomics and is boosted by MS2Rescore with MS2PIP timsTOF

fragmentation prediction model. Nat Commun 15. https://doi.org/10.1038/s41467-024-46380-y

30. The UniProt Consortium, Bateman A, Martin M-J, Orchard S, Magrane M, Adesina A, Ahmad S, Bowler-Barnett EH, Bye-A-Jee H, Carpentier D, Denny P, Fan J, Garmiri P, Gonzales LJDC, Hussein A, Ignatchenko A, Insana G, Ishtiaq R, Joshi V, Jyothi D, Kandasaamy S, Lock A, Luciani A, Luo J, Lussi Y, Marin JSM, Raposo P, Rice DL, Santos R, Speretta E, Stephenson J, Totoo P, Tyagi N, Urakova N, Vasudev P, Warner K, Wijerathne S, Yu CW-H, Zaru R, Bridge AJ, Aimo L, Argoud-Puy G, Auchincloss AH, Axelsen KB, Bansal P, Baratin D, Batista Neto TM, Blatter M-C, Bolleman JT, Boutet E, Breuza L, Gil BC, Casals-Casas C, Echioukh KC, Coudert E, Cuche B, De Castro E, Estreicher A, Famiglietti ML, Feuermann M, Gasteiger E, Gaudet P, Gehant S, Gerritsen V, Gos A, Gruaz N, Hulo C, Hyka-Nouspikel N, Jungo F, Kerhornou A, Mercier PL, Lieberherr D, Masson P, Morgat A, Paesano S, Pedruzzi I, Pilbout S, Pourcel L, Poux S, Pozzato M, Pruess M, Redaschi N, Rivoire C, Sigrist CJA, Sonesson K, Sundaram S, Sveshnikova A, Wu CH, Arighi CN, Chen C, Chen Y, Huang H, Laiho K, Lehvaslaiho M, McGarvey P, Natale DA, Ross K, Vinayaka CR, Wang Y, Zhang J (2025) UniProt: the universal protein knowledgebase in 2025. Nucleic Acids Res 53: D609–D617. https://doi.org/10.1093/nar/gkae1010

31. Ouspenskaia T, Law T, Clauser KR, Klaeger S, Sarkizova S, Aguet F, Li B, Christian E, Knisbacher BA, Le PM, Hartigan CR, Keshishian H, Apffel A, Oliveira G, Zhang W, Chen S, Chow YT, Ji Z, Jungreis I, Shukla SA, Justesen S, Bachireddy P, Kellis M, Getz G, Hacohen N, Keskin DB, Carr SA, Wu CJ, Regev A (2022) Unannotated proteins expand the MHC-I-restricted immunopeptidome in cancer. Nat Biotechnol 40:209–217. https://doi.org/10.1038/s41587-021-01021-3

32. Ferreira HJ, Stevenson BJ, Pak H, Yu F, Almeida Oliveira J, Huber F, Taillandier-Coindard M, Michaux J, Ricart-Altimiras E, Kraemer AI, Kandalaft LE, Speiser DE, Nesvizhskii AI, Müller M, Bassani-Sternberg M (2024) Immunopeptidomics-based identification of naturally presented non-canonical circRNA-derived peptides. Nat Commun 15: 2357. https://doi.org/10.1038/s41467-024-46408-3

33. Nesvizhskii AI (2014) Proteogenomics: concepts, applications and computational strategies. Nat Methods 11:1114–1125. https://doi.org/10.1038/nmeth.3144

Chapter 7

Expression and Functional Characterization of Recombinant HLA-DM

Miguel Álvaro-Benito, Jakob Holzapfel, and Christian Freund

Abstract

HLA-DM is a nonclassical MHC class II molecule that plays an essential role in antigen presentation by catalyzing peptide exchange on MHC class II molecules, ensuring the presentation of stable peptide-MHC complexes to CD4+ T cells. This chapter describes detailed protocols for the production and functional analysis of recombinant HLA-DM and HLA-DR proteins. Using a baculovirus expression system in *Spodoptera frugiperda* (Sf9) cells, we outline step-by-step methods for bacmid preparation, baculovirus amplification, and large-scale protein expression. Comprehensive purification strategies, including affinity chromatography and size exclusion chromatography, are provided to ensure high-purity protein yields. Additionally, we describe functional assays, utilizing fluorescence polarization-based peptide exchange to evaluate HLA-DM activity. These protocols enable the reproducible production of recombinant HLA-DM and HLA-DR, providing a reliable foundation for researchers conducting functional and mechanistic studies on catalyzed peptide exchange.

Key words HLA-DM, Peptide editing, Peptide exchange, Antigen presentation, MHC II, Fluorescence polarization

1 Introduction

The Human Leukocyte Antigen (HLA) DM acts as a molecular chaperone stabilizing the peptide-free state of MHCII molecules, hence catalyzing peptide exchange and thereby editing the peptide repertoires displayed to CD4+ T cells [1, 2]. DM activity ensures the selection of kinetically stable, long-lived peptide-MHCII complexes (pMHCII). These complexes are also referred to as DM-resistant, and their features make them more amenable to immune surveillance [3, 4].

DM's function can be studied at very different levels, from cellular assays to molecular and mechanistic analyses. Early research primarily relied on indirect observations of DM-mediated peptide exchange, such as measuring T cell responses to specific antigenic determinants [5, 6]. The use of flow cytometry to assess Class II

Efstratios Stratikos (ed.), *Antigen Processing and Presentation: Methods and Protocols*, Methods in Molecular Biology, vol. 3007, https://doi.org/10.1007/978-1-0716-5092-9_7,

Invariant chain-derived Peptide (CLIP) retention, detected by the Cer.CLIP1 monoclonal antibody, has also been widely adopted in this regard [7]. More recently, advancements in peptidomics on mono-allelic cell lines have provided extraordinary insights into the peptide repertoires generated by DM's activity [8, 9]. While these approaches have offered valuable information about DM's role in shaping the immunopeptidome, they fall short of providing the molecular insights needed to understand its mechanistic details. Such insights are critical for addressing fundamental questions about DM function and its impact on immune responses.Two contemporary questions drive the need for molecular-level investigation of DM function, particularly after the basic mechanism of DM-catalyzed peptide exchange was outlined in 2012 by Pos et al. [10]. These questions refer to the determinants of DM susceptibility on MHCII molecules, and whether we could define the impact of DM variations.

A remarkable aspect of DM function is the so-called DM susceptibility. This parameter measures the relative increase in peptide dissociation from an MHCII-peptide complex in the presence of DM compared to its absence [11–13]. Studies have sought to identify the molecular determinants of DM susceptibility for decades. Pos et al. highlighted the significance of occupancy in the P1 pocket [10], while Stern et al. later demonstrated the influence of P1-distant regions [13]. Since polymorphisms in MHCII molecules affect the conformation and dynamics of DM-sensitive regions, it is not surprising that DM-susceptibility varies between MHC allotypes. Recent studies suggest that the conformational plasticity of specific MHCII regions correlate with catalyzed exchange, allowing us to connect molecular dynamics with disease-relevant DM susceptibilities [14, 15]. However, these findings only represent the tip of the iceberg, given the thousands of different alleles present in the human population. Furthermore, since DM susceptibilities largely depend on the available antigens, understanding peptide association and dissociation, and its modulation by HLA-DM is of equal importance. For instance, a study by Reyes-Vargas investigated the binding of 25 peptides to the HLA-DRB1*01:01 allotype and found that the dissociation rates of these peptides vary by a factor of 10^5 for uncatalyzed, and by a factor of 10^3 for catalyzed peptide exchange [16]. Conceptualization of this data by Ghünter et al. [17] then led to the conclusion that DM-resistant peptides have large basal association rates, a parameter that is notoriously difficult to determine experimentally, given the instability of most MHCII molecules in the absence of peptide. To complicate matters, emerging evidence suggests that natural variations in DM can significantly influence its peptide-editing activity [9, 18]. These variations may be linked to specific health and disease conditions, adding clinical relevance to the study of DM function [2, 19]. In the following sections, we will thus describe the materials required for recombinant HLA-DM production, the methodologies for its expression and purification, and the assays used to evaluate its peptide-editing activity. By providing this resource, we aim to empower researchers with the tools necessary to explore the mechanistic details of HLA-DM function, answer outstanding questions in the field, and ultimately advance our understanding of its role in the immune system.

2 Materials

All chemicals described in this protocol are commercially available, and with a handful of exceptions referred to in the text, altering the brand has no influence on our results, unless otherwise stated. Critical materials such as recombinant molecules need to be produced prior to setting up the experiments. Most of these critical materials are stored at −80 °C or − 20 °C (see specific cases), thawed and kept on ice prior to their use. We strongly recommend avoiding repetitive freezing and thawing of recombinant molecules.

2.1 Equipment

1. Temperature-Controlled Equipment for living organisms: A water bath capable of precise temperature regulation and orbital shakers for bacterial and insect cell culture at 37 °C and 27 °C, respectively.
2. Chromatography Equipment: A size-exclusion chromatography column, such as Superdex S200, coupled with a chromatography system for protein purification.
3. Plate Reader: A fluorescence polarization-compatible microplate reader for monitoring peptide exchange.
4. Software: Data analysis can be performed under many different software packages; altogether one has fit any of the equations provided below to the experimental data. The open-access software R-Rstudio and/or Origin are two recommended alternatives.

2.2 Protein Expression and Purification

1. cDNA Constructs: Plasmids encoding HLA-DR and HLA-DM heterodimers. These constructs have been previously described and should be obtained or generated according to standard cloning techniques (*see* **Note 1**).
2. Plasmid DNA Isolation Kits: Mini- and Midi-preparation kits for isolating plasmid DNA from bacterial cultures.
3. Bacmid Isolation Kit: For bacmid isolation we recommend using a mix of commercial kit preparations and isopropanol precipitation (*see* **Note 2**).

4. Competent Cells and Insect Cells: Chemo-competent *E. coli* DH10B for Bacmid preparation and *Spodoptera frugiperda* (Sf9) insect cells for recombinant protein expression (*see* **Note 3**).
5. Culture Media and Supplements: Appropriate media for bacterial and insect cell cultures, including supplements such as antibiotics for selection in bacterial cultures and antibiotics +antimycotics for protein-free media for insect cells.
6. Transfection Reagents: Cellfectin for transfecting Sf9 cells with Bacmid DNA. Use the reagent as described in the manufacturer's instructions with modifications suited to this protocol (*see* **Note 4**).
7. Affinity Resins: Immuno-affinity resins, e.g., M2 Sepharose (FLAG-tagged HLA-DM) or L243 Sepharose (HLA-DR) for MHCII protein purification (*see* **Note 5**).
8. Enzymes: Thrombin for cleavage of the covalent link between MHCII chains and associated peptides, which in our design is CLIP (*see* **Note 6**).

2.3 Fluorescence Polarization

1. Reaction Buffer: 50 mM Citric-Phosphate buffer pH 5.3 including 150 mM NaCl. Alternatively, PBS with adjusted pH values can be used.
2. Reporter Peptides: Fluorescently labeled peptides, such as fluorescein isothiocyanate (FITC)-labeled peptides, for functional assays (*see* **Note** 7).
3. 384 well plates: the plates should be black, with a flat bottom and from a non-adsorbent surface.
4. Mineral Oil: To prevent evaporation during long-term incubations in functional assays.

3 Methods

An overview of the entire protocol in which the five stages that we considered essential is depicted in Fig. 1. Average times required for each of them are provided as well as the corresponding numbering to each of the methods section that needs to be considered. When cDNA constructs are available, the entire process should take around 5 weeks.

3.1 Recombinant Bacmid Preparation (Fig. 2a)

3.1.1 Preparation of Donor Plasmid

1. Construct donor plasmids (e.g., pFastBac Dual) encoding HLA-DR and HLA-DM heterodimers using standard molecular cloning techniques (*see* **Note 1**).
2. Verify the integrity of the plasmid by restriction digestion and sequencing.

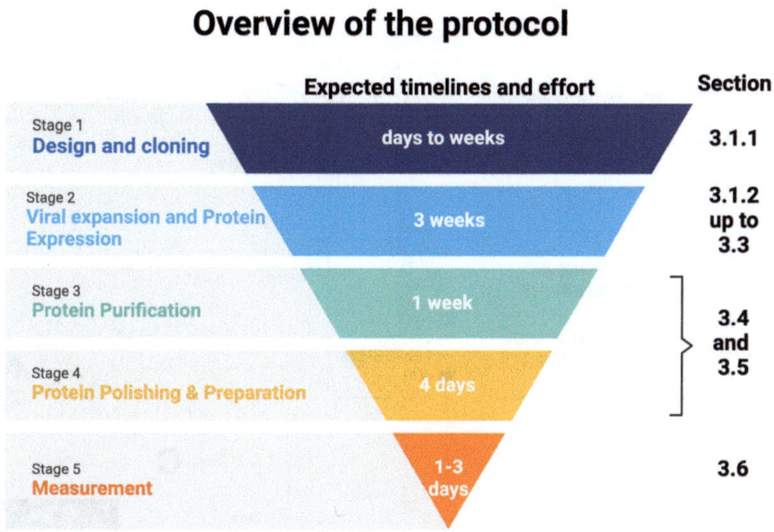

Fig. 1 Overview of the entire protocol, including main steps and estimated times. The departure situation is that of already having at hand the corresponding cDNA constructs in the pFastBacDual plasmid. However, if you need to generate these constructs, it may take a few weeks. Different stages described through the protocol are considered. Note that the expected timelines are only indicative

3. Purify high-quality plasmid DNA using a Mini- or Midi-prep kit as per the manufacturer's instructions.

3.1.2 Transformation of Competent E. coli Host Cells

1. Thaw competent *E. coli* cells suitable for bacmid generation (e.g., DH10EMBacY, DH10Bac) on ice (50 μL per transformation) (*see* **Note 3**).
2. Add 100 ng of donor plasmid to the thawed cells and gently mix.
3. Incubate on ice for 30 min.
4. Perform a heat shock at 42 °C for 45 seconds.
5. Immediately place the tubes on ice for 5 min.
6. Add 500 μL SOC medium, then incubate at 37 °C with shaking (225 rpm) for 5 h to allow expression of antibiotic resistance markers.
7. Plate 100–200 μL of the transformed culture onto LB agar plates containing the appropriate selection antibiotics and screening reagents (Kanamycin (50 μg/mL), Gentamicin (7 μg/mL), Tetracycline (10 μg/mL), X-Gal (100 μg/mL) and IPTG (40 μg/mL)).
8. Incubate at 37 °C for 48 h in the dark to allow colony development.

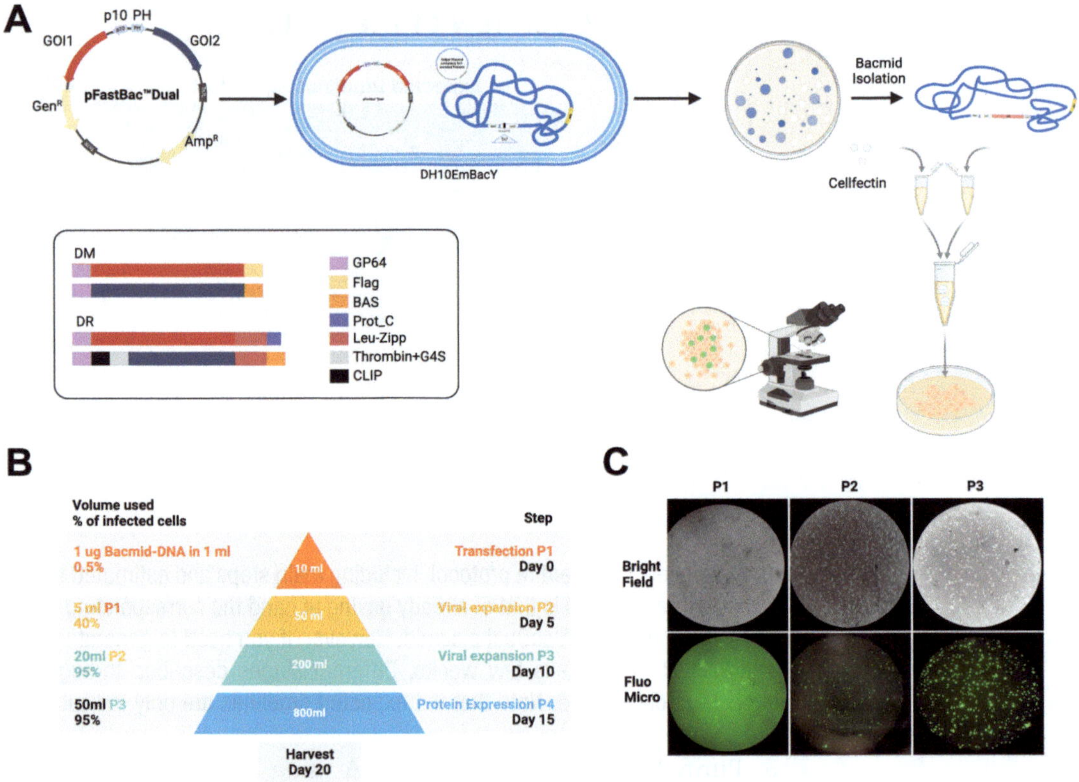

Fig. 2 Overview of the entire Baculovirus expression system. (**a**) Different constructs based on the pFastBac-Dual vector bear the alpha and beta chains of DM and DR, following the scheme depicted and including the corresponding elements. Transformation of DH10EmBacY cells favors the integration of the expression cassette into a Bacmid containing a YFP reporter protein. Upon integration and selection via blue-white screening, positive clones are used to extract their Bacmids. Bacmids are complexed with Cellfectin and used to transfect Sf9 cells plated on the same morning. Infected cells become fluorescent and thus, infection can be monitored by tracking fluorescence via microscope observation or flow cytometry. (**b**) Overview of protocol indicating the steps and required times (right side) and reagents, culture volumes and expected infection rates for each step (left side). (**c**) SF9 Cells after Transfection and Virus Amplification—Brightfield vs. Fluorescence Imaging: The upper row shows brightfield images of SF9 cells observed under a normal light microscope, displaying all cells present in the field of view. The lower row presents corresponding fluorescence images taken using a fluorescent filter, highlighting successfully transfected or infected cells expressing fluorescent proteins. P1-P3 represent the successive passages after transfection (P1) and virus amplification (P2 & P3)

3.1.3 Colony Screening and Bacmid Isolation

1. Identify white colonies, indicating successful recombination by disruption of lacZ gene, and re-streak them onto fresh selection plates for confirmation.
2. Pick 3–5 white colonies and inoculate 10 mL of LB broth supplemented with the appropriate antibiotics and incubate cultures overnight at 37 °C with orbital shaking (225 rpm).
3. Extract high molecular weight recombinant bacmid DNA using a modified plasmid mini-prep protocol:

- Follow the standard plasmid isolation procedure up to the neutralization step.
- Centrifuge at 12,000× *g* for 10 min and carefully pour the supernatant into a fresh tube (avoid pipetting to prevent shearing the DNA).
- Precipitate bacmid DNA by adding ice-cold 800 μL isopropanol, then centrifuge at maximum speed (min. 12,000× *g*) for 10 min.
- Remove the supernatant, wash the pellet with 70% ethanol (200 μL) and repeat centrifugation.
- Dry the pellet and dissolve it in 100 μL nuclease-free water under sterile conditions.

4. Confirm bacmid recombination using PCR screening with appropriate primers (e.g., M13 forward and reverse primers).
5. Analyze PCR products via agarose gel electrophoresis (0.8% gel with ethidium bromide or SYBR Safe staining).
6. Validate the integrity of the bacmid before proceeding with transfection into insect cells.

3.2 Recombinant Baculovirus Production

3.2.1 Preparation of Sf9 Cells

1. Maintain Sf9 cells (or alternative insect cell lines) in serum- and protein-free medium at 27 °C with orbital shaking at 120 rpm.
2. Supplement the medium with antibiotic-antimycotic solution to a final 1x concentration to prevent bacterial or fungal contamination (*see* **Note 8**).
3. Ensure that cell viability exceeds 95% and cell density is between 0.8 and 1.5 × 10^6 cell/mL before proceeding with transfection.

3.2.2 Transfection of Sf9 Cells with Recombinant Bacmid DNA

1. Seed 0.8–1.0 × 10^6 Sf9 cells per well (1 mL) using serum-free medium, protein or antibiotics in an adhesive 6-well plate. Allow cells to attach for 10 min at room temperature.
2. Prepare the transfection mix (~110 μL per well) in a sterile tube:
   - 1.5 μg of recombinant Bacmid DNA (prepared in Sect. 3.1) per well.
   - 8 μL of Cellfectin (or alternative transfection reagent) diluted in 100 μL of serum-free medium (*see* **Note 4**).

3.2.3 Amplification of Recombinant Baculovirus Stocks (Fig. 2b)

1. Collect the P1 (primary) viral stock by transferring the supernatant (10 mL) to a sterile tube.
2. Harvest by centrifugation at 500 × *g* for 5 min to remove cell debris, then sterile filter through a 0.2 μm filter to ensure clarity and sterility.

3. Proceed directly to infection of another culture to attain the P2, or store P1 virus at 4 °C in the dark until further amplification (*see* **Note 10**).
4. First Amplification (P2 Stock, 50 mL):
   - Infect 50 mL of fresh Sf9 culture (1.5–2.0 × 10^6 cells/mL) with 5 mL of P1 virus in a sterile 250 mL Erlenmeyer flask with a vented cap.
   - Incubate cells at 27 °C for 3–5 days while shaking and monitor for signs of viral infection (e.g., stop of growth, increased cell size, reduced viability or fluorescence if DH10EmBacY cells were used).
   - Harvest when cell viability drops below 85%.
5. Second Amplification (P3 Stock, 200 mL):
   - Infect 200 mL of fresh Sf9 culture (1.5–2.0 × 10^6 cells/mL) with 10–20 mL of P2 virus in a sterile 500 mL Erlenmeyer flask with a vented cap.
   - Incubate cells at 27 °C for 3–5 days while shaking and monitor for signs of viral infection (e.g., stop of growth, increased cell size, reduced viability or fluorescence if DH10EmBacY cells were used) (*see* **Note 11**).
   - Harvest the supernatant when cell viability drops below 85%.

3.2.4 Quality Control of Baculovirus Stocks

1. Monitor virus amplification success using (*see* **Notes 9, 10,** and **11**):
   - Plaque assays [20] or real-time PCR [21] to determine infectious viral titers (pfu/mL).
   - Fluorescence markers if present in the Bacmid construct via flow cytometry or microscopy assessment.
2. Once viral amplification is confirmed, store viral stocks at 4 °C protected from light for short-term use. Manuals from different vendors recommend storing viruses at −80 °C for long-term storage (*see* **Note 10**).

3.3 Expression of Recombinant MHCII Proteins in Sf9 Cells

Large-Scale Infection of Sf9 Cells (Fig. 2b)

1. Prepare 800 mL of Sf9 cell culture at a density of 1.8–2.0 × 10^6 cells/mL and cell viability >95% in a sterile 2 L shake flask.
2. Infect the culture with 40–80 mL of P2 baculoviral stock (adjust volume as needed based on infection efficiency from previous passages).
3. Incubate cells at 27 °C for 3–5 days while shaking (120–140 rpm) and monitor for signs of viral infection (e.g., stop of growth, increased cell size, reduced viability or

fluorescence if DH10EmBacY cells were used). Harvest when cell viability drops below 80% (Fig. 2c).

4. Harvesting and Filtration. Harvest by centrifugation at 500 × g for 5 minutes to remove cell debris, then filter through a 0.2 μm filter to ensure that you remove any cellular debris present in the sample. Proceed immediately with protein purification.

3.4 Protein Purification (Fig. 3a)

3.4.1 HLA-DM (Fig. 3a Left)

Buffer Exchange and Concentration

1. Concentrate the supernatant to 100–200 mL using an ultrafiltration system (e.g., Vivaflow 200 or equivalent, MWCO10KDa).
2. Perform buffer exchange by diluting the concentrate up to 1 L with 1× PBS, pH 7.4, with 0.1% sodium azide, then reconcentrating to 100 mL.
3. Repeat the dilution and concentration twice more to ensure thorough buffer exchange.

Affinity Purification (*see* Note 5)

1. Load the 100 mL concentrated protein solution onto an Anti-FLAG M2 immunoaffinity column (Sigma-Aldrich) at 4 °C overnight.
2. Wash the column thoroughly with 1× PBS, pH 7.4 to remove unbound proteins.

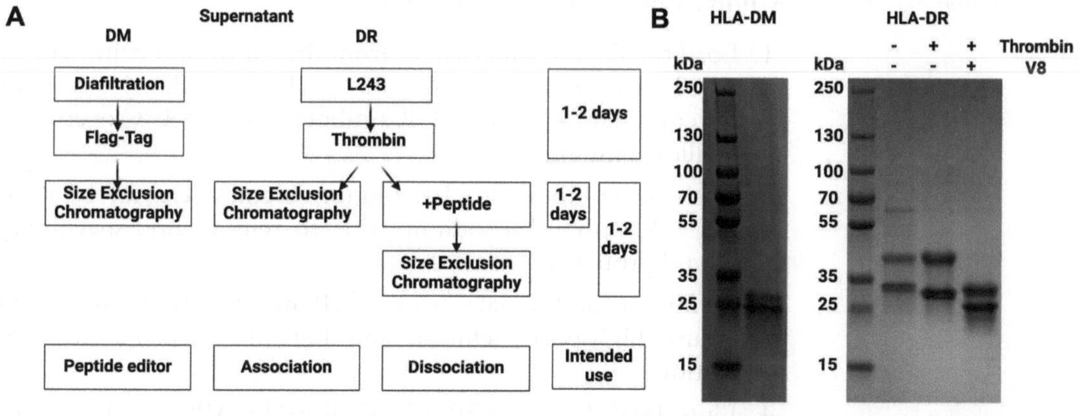

Fig. 3 Protein purification schemes and expected outcome depending on the intended uses. (**a**) In all cases Sf9 supernatants are the starting material used to purify the corresponding proteins. Each protein, whose function or intended use is indicated in the lower row, follows a slightly different protocol in all cases based on immunoaffinity purification (gray shaded boxes), and Size Exclusion Chromatography (SEC). Expected time-frames required for each step are indicated on the right side. (**b**) SDS-PAGE outcome of HLA-DM (left) and HLA-DR (right) purification and treatment with proteases as indicated. Molecular weight references from the used marker are shown on the left of each picture

3. Elute the bound protein with 14 mL of 0.1 M glycine, pH 3.5, into 2 mL collection tubes containing 100 μL of 1 M Tris-HCl, pH 8.0, to neutralize pH (*see* **Note 13**).
4. Analyze elution fractions by SDS-PAGE, and pool those containing the desired protein.

Size-Exclusion Chromatography (SEC)

1. Concentrate the pooled protein fractions to <1 mL using 30-kDa MWCO spin filters.
2. Load the concentrated protein onto a size-exclusion chromatography (SEC) column (e.g., Superdex S200 10/300 GL or equivalent) and run SEC using 1× PBS, pH 7.4, as the mobile phase.
3. Collect fractions corresponding to the expected molecular weight of the protein and check purity and integrity by SDS-PAGE.
4. Pool pure fractions and re-concentrate using 30-kDa MWCO spin filters.

Storage

1. Aliquot the final purified HLA-DR-peptide complexes into small volumes.
2. Snap freeze aliquots in liquid nitrogen and store at −80°C in the dark for long-term use.

3.4.2 HLA-DR (Fig. 3a Right)

Affinity Purification of HLA-DR

1. Load the filtered supernatant from the expression culture onto an L243 antibody (HB-55, ATCC) affinity column (or equivalent anti-HLA-DR antibody resin) at 4 °C overnight to allow binding.
2. Wash the column with 10 column volumes (CV) of 1× PBS, pH 7.4, with 0.1% sodium azide to remove non-specifically bound proteins.
3. Perform an additional wash with 10 mM Na_2HPO_4, pH 7.4 to ensure high-purity elution and fast pH exchange for the elution.
4. Elute HLA-DR with 50 mL of 50 mM CAPS, pH 11.5, collecting into a tube containing 10 mL of 200 mM Na_2HPO_4, pH 6.0, for immediate neutralization.

Peptide Cleavage and Final Purification (Fig. 3B, Note 14)

1. Concentrate the eluted protein to <1 mL using 30-kDa MWCO spin filters and perform buffer exchange to 1× PBS, pH 7.4 by diluting the concentrate back to 15 mL and repeat the concentration.
2. Take a small aliquot for SDS-PAGE.
3. Treat the purified HLA-DR protein with thrombin (20 U/mg) in 1× PBS, pH 7.4, at 37 °C for 2 hours to cleave the linker of covalently bound CLIP peptide to allow peptide dissociation. Validate cleavage by SDS-PAGE mobility shift on the DRB1* chain before going to the next step (compare with an aliquot taken before cleavage).
4. Load the protein mix onto a size-exclusion chromatography (SEC) column (e.g., Superdex S200 10/300 GL or equivalent) and run SEC using 1× PBS, pH 7.4, as the mobile phase.
5. Collect fractions corresponding to the expected molecular weight of the protein. Pool those fractions containing mainly, or exclusively, the protein of interest and concentrate using 30-kDa MWCO spin filters.

Storage

1. Aliquot the purified protein into small volumes.
2. Snap freeze aliquots in liquid nitrogen and store at −80 °C for long-term use.

3.5 Peptide Exchange on HLA-DR Proteins to Follow Dissociation

Peptide Exchange Reaction

1. Using the thrombin-cleaved HLA-DR from the previous purification step (Sect. 3.4.2), ensure the protein is in 1× PBS, pH 5.8 (**Note 14**).
2. Add a ten-fold molar excess of a fluorescently labeled peptide (e.g., FITC-$CLIP_{102–120}$) to the thrombin-treated HLA-DR.
   - Alternative labeled peptides (e.g., TAMRA, Cy5, Alexa Fluor) may be used depending on detection requirements.
   - The choice of peptide and label depends on experimental design (e.g., fluorescence-based assays, spectroscopic analysis).
3. Incubate the mixture at 37 °C for 48 hours to allow efficient peptide exchange.

Size-Exclusion Chromatography (SEC)

1. Load the protein/peptide mixture onto a size-exclusion chromatography (SEC) column (e.g., Superdex S200 10/300 GL or equivalent) and run SEC using 1× PBS, pH 7.4, as the mobile phase, in the dark (**Note 15**).
2. Collect fractions corresponding to the expected molecular weight of the protein and check purity and integrity by SDS-PAGE.
3. Pool pure fractions and re-concentrate using 30-kDa MWCO spin filters.

Storage

1. Aliquot the final purified HLA-DR-peptide complexes into small volumes.
2. Snap freeze aliquots in liquid nitrogen and store at −80 °C in the dark for long-term use.

3.6 Measuring Peptide Exchange via Fluorescence Polarization (FP)

Fluorescence polarization (FP) is a widely used technique for studying peptide-MHC class II (MHCII) interactions, providing quantitative measurements of peptide dissociation (off-rate) and apparent association (on-rate). Both assays rely on the same general principles and workflow, requiring fluorescently labeled peptides, recombinant MHCII proteins, and an FP-compatible plate reader (Fig. 4A).

Association assays determine apparent on-rates since the events measured include the dissociation of the pre-bound peptide and the association of the labeled reporter peptide. Thus, even if using the same departure situation as we report here, with MHCII-CLIP as the initial complex, association rates depend on basal peptide association and re-binding rates, which cannot be determined experimentally so far. In these assays, MHCII-CLIP, or any other complex generated in vitro, is incubated with a fluorescently labeled peptide, and FP is measured over time to monitor the incorporation of the peptide into the MHCII binding groove. This allows the determination of binding kinetics and affinity.

In dissociation assays, preformed MHCII-fluorescent peptide complexes are monitored for the release of the labeled peptide over time. These experiments can be performed in the presence or absence of HLA-DM, allowing for the characterization of DM susceptibility and peptide affinity.

Both methods require careful protein preparation, buffer selection, and reaction setup to ensure reliable and reproducible measurements. Below, we outline the general protocol for FP-based peptide exchange studies, highlighting common steps and optimizations for both association and dissociation assays.

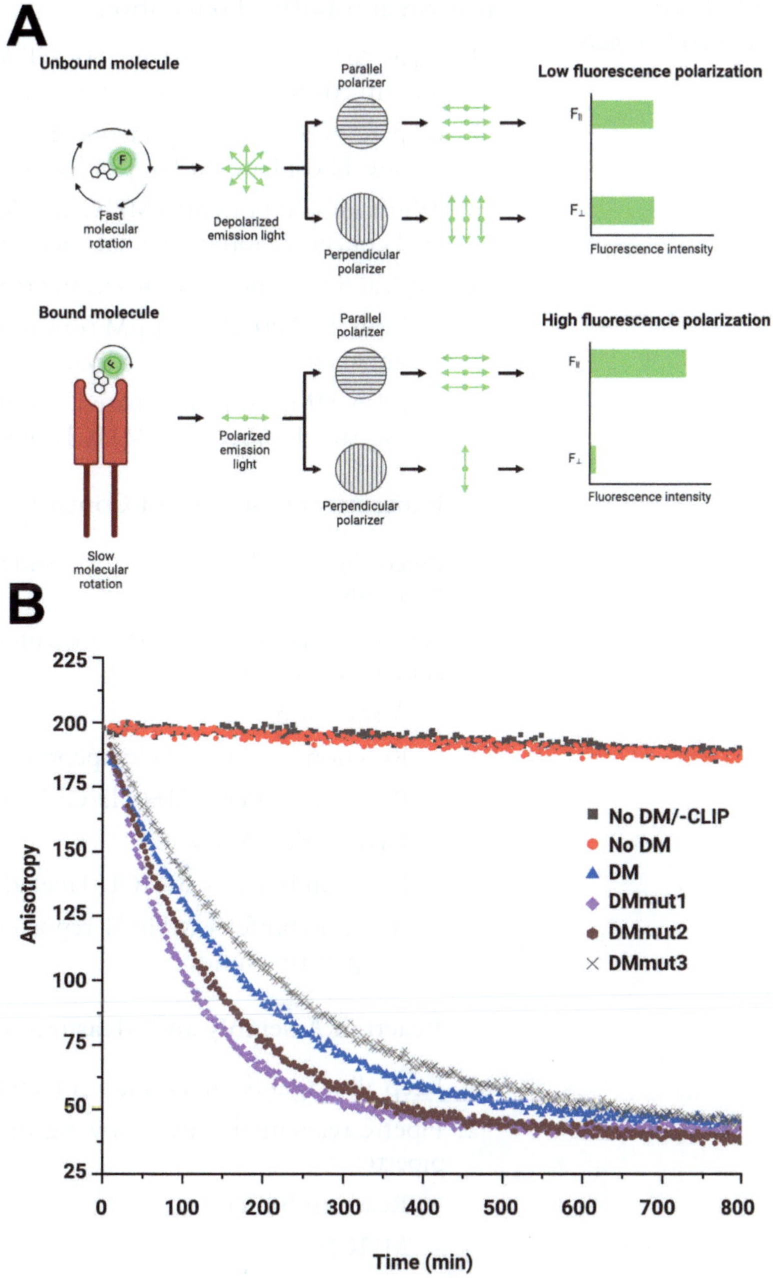

Fig. 4 Fluorescence Polarization. (**a**) Biophysical principle of FP applied to peptide binding to MHCII molecules. (**b**) Exemplary data of a peptide dissociation experiment using a pre-loaded MHCII-reporter peptide complex, DM and three of its mutants

3.6.1 General Preparation for Peptide Exchange FP Assay

Protein and Buffer Preparation

1. Thaw MHCII proteins and HLA-DM (if used) on ice. Check protein concentration and verify integrity by SDS-PAGE.
2. Prepare reaction components in Reaction Buffer (50 mM Citrate-Phosphate buffer, pH 5.3, with 150 mM NaCl).
3. Adjust all components (MHCII, DM and peptide) to 4× the final concentration to facilitate accurate dilution in the assay.
4. Typical final concentrations in the reaction chamber:
   - MHCII: 100 nM to 1 μM (optimized based on background signal and device sensitivity).
   - HLA-DM: Adjusted based on the experiment (typically sub-stoichiometric to MHCII but may be optimized).

Experimental Setup and Controls

1. Preconfigure the FP plate reader for the appropriate fluorophore.
2. Set up the plate map in the instrument software and include appropriate controls:
   - Association assay:

     Reaction buffer + labeled peptide.

     Reaction buffer + MHCII-CLIP + labeled peptide.
   - Dissociation Assay:

     Reaction buffer + MHCII-labeled peptide complex.

     Reaction buffer + MHCII-reporter peptide complex + competitor peptide.

Reaction Assembly and Measurement

1. Keep all reagents on ice and protected from light.
2. Pipette reagents in the following order using a multichannel pipette:
   - Reaction buffer.
   - MHCII.
   - HLA-DM (if applicable).
   - Fluorescently labeled/competitor peptide (*see* **Note 16**).
3. For dissociation assays, overlay wells with 10 μL mineral oil to prevent evaporation (*see* **Note 17**).
4. Place the plate directly into the prewarmed FP reader and begin real-time fluorescence polarization measurement at the desired temperature (usually between 25 °C and 37 °C).
5. Monitor fluorescence polarization over time and analyze the data (Fig. 4b).

3.6.2 Analysis

FP values were transformed into anisotropy (A) units (already done by the plate reader), as these values are strictly linear with regard to labeled free:bound peptide amounts [22]. Provided that.

Using the OriginPro 2023 software, anisotropy values of the first 1000 min were plotted and fit to a one-phase exponential association or decay function (Eq. 1 for association and Eq. 2 for dissociation) with time constant parameter.

$$y = y_{max}\left(1 - e^{-xt_1)}\right) \quad (1)$$

$$y = y_0 + A_1 e^{-x/t_1} \quad (2)$$

y = Anisotropy values at a given time point,

y_{max} = Maximum anisotropy at equilibrium for association experiments,

x = k = observed association/dissociation (rate) constant [min^{-1}],

t_1 = time,

y_0 = Minimum anisotropy values at equilibrium for dissociation experiments

A_1 = amplitude (difference between Anisotropy values at maximum, and at equilibrium,

When fitting the anisotropy values, a fixed value for y_0 was set for all curves of a single DR allotype. If the equilibrium was not reached during the first 1000 min, y_{max} or y_0 (equilibrium) calculated by OriginPro of the fastest exchanging DM haplotype was set for all other haplotypes. The dissociation constants determined were analyzed for significance using one-way ANOVA and Bonferroni correction.

4 Notes

1. cDNA. Plasmids/Constructs. FLAG-tagged HLA-DM and HLA-DR constructs have been described before [16]. DM bears a Biotin Acceptor Site (BAS) on the beta chain and a Flag-tag in the beta chain, whereas DR molecules have dimerization motifs in the C-terminal region of each chain (Leucine Zippers) and an additional BAS in the DRA chain. Note as well that between the DR ectodomains and the zippers, a V8-protease recognition site is included. Likewise, the N-terminal portion of the DRB1 gene encodes the Class II Invariant chain Peptide (CLIP) followed by a Thrombin cleavage sequence and a G4S Linker.
2. For bacmid isolation we recommend using any mini-prep commercial kit/solutions up to the point of neutralizing the lysis solution. Subsequently we perform an isopropanol and ethanol

precipitation to attain the final bacmid. The integrity of the bacmid could be checked in an agarose gel (0.7% agarose), in which the DNA should run at a high molecular weight. Note that the bacmid itself should be larger than 100kpb.

3. Competent bacterial cells for cloning, plasmid amplification and bacmid generation, as well as insect cells, can be purchased from different vendors. We recommend using the following strains: *E. coli* XL1 or DH5alpha for cloning and DH10EmBacY cells for bacmid preparation. In case of recombinant protein expression, we refer in this chapter to Sf9 derived from *Spodoptera frugiperda*. Note that infection can be monitored via YFP fluorescence, which is constitutively expressed through an expression cassette present in the EmBacY backbone.
4. Cellfectin is considered the preferred transfection reagent to transfer the bacmid to Sf9 cells growing in adherence. We recommend using 1 well per transfection as a negative control. This negative control should exclude the bacmid DNA and will only make sure that cells do not die upon transfection and do not get contaminated.
5. Considering our cDNA constructs, we perform protein purification on the basis of a conformation-specific antibody for DRB1 molecules (L243) cast in-house using NHS-activated fast-flow Sepharose, or an anti-flag M2 resin (Anti-flag M2 Affinity Gel, Sigma). We had to adjust the elution volume from 6 mL (suggested by the manufacturer) to 14 mL. Using only 6 mL for elution resulted in significant protein loss.
6. Thrombin from bovine plasma diluted to 1 U/μl is used to cleave the linker between DRB1 and the placeholder peptide CLIP. We usually apply 20 U for every mg of recombinant MHCII molecule. Additionally, Endoproteinase Glu-C (V8 protease) can be used to remove the existing leucine zippers.
7. The selection of reporter peptides is of utmost relevance for measuring peptide exchange. The reporter peptide should be a peptide with a certain affinity towards the measured complex. We typically use CLIP, HA, and MBP. Label (FITC, TAMRA, Cy5, or any other commercial fluorescent label that can be followed by the measuring device) could be incorporated at the N- or C-termini if additional residues from the first or last anchor residue are considered, respectively. Alternatively, if the binding register is known, P5 residues could be substituted by "K" that can then be labeled. It is worth checking and avoiding, if possible, potential proteolytic signals, especially the Thrombin recognition site.
8. Commercial Antibiotic-antimycotic solutions exist as a stock concentration of 100x, which should be directly diluted in the

culture medium at a final 1x concentration. Commercial stock solutions contain 10.000 U/ml penicillin, 10.000 µg/ml Streptomycin and 25 µg/ml Anphotericin B.

9. In case cells do not show clear signs of viral infection as described in the main text, we recommend collecting an aliquot of the supernatant and performing a quick expression test via dot-blot following standard procedures. Proteins can be detected with the antibodies used in the affinity chromatography purification steps and listed below. Using commercial HRP-labeled antibodies and luminol should give a quick answer to whether the protein is expressed or not. Mind using the negative control supernatant to avoid false positives. Alternative methods include plaque assays or PCR from the cell supernatant as described here (Refs plaque assays and/or PCR). However, in our experience a dot blot is usually faster and provides a reliable check. If no expression is observed after day 3, discard the transfection and start over again.
10. We directly infect cells with the collected supernatant on the same day, but if needed, P1 stock can be stored up to 1 week at 4 °C in the dark. Longer storage times result in considerable loss of efficiency at this point. P2 viral stock, on the other hand, can be stored at 4 °C in the dark for up to 6 months with no considerable loss of protein expression efficiency. We have never stored a virus at −80 °C; hence, we cannot report on how reliable the storage under these conditions is.
11. When using DH10EMBacY cells, approximately 30–50% of cells in P2 should exhibit fluorescence, while in P3, fluorescence should be nearly 100%, indicating successful viral amplification.
12. Viral titers can be used to calculate the multiplicity of infection (MOI), with 0.05–0.1 MOI used for viral stock amplification. We replaced time-consuming titer determination with a 1:10 dilution for P2 and 1:10–1:20 for P3 and large-scale expressions, adjusted based on fluorescence observed in the previous passage.
13. If the protein is not going to be subject to SEC, we recommend pooling the fractions containing HLA-DM and dialyzing the resulting solution prior to concentration to less than 1 mL using a 30-kDa MWCO spin filter. This sample should then be stored at −80 °C. Once these samples are thawed, they should be subject to protease treatments (MHCII) prior to SEC, or directly SEC (HLA-DM) prior to their use.
14. If peptide exchange is planned immediately after HLA-DR purification, the final SEC step in Sect. 3.5 can be skipped. Instead, the labeled peptide can be added directly into the HLA-DR/thrombin reaction mix after adjusting the pH to

5.8. Ensure that the fluorescently labeled peptide does not contain a thrombin cleavage site, as this could lead to unintended cleavage.

15. Preserve from light exposure as much as possible the protein loaded with the labelled peptide to prevent damage of the fluorophore.
16. We recommend measuring the time between peptide addition (labeled or competitor) and the first measurement, as some reactions occur rapidly. Including background controls and recording this time allows for accurate extrapolation of the true FP value at the start of the reaction.
17. Certain assays take from a few hours up to several days. Therefore, mineral oil is needed to prevent evaporation.
18. Figures have been created with BioRender.

Acknowledgments

M.Á-B is supported by the DKMS JHRG2022-01, CAM-2022-T1/23572, ISCIII PI24-00605, and is holder of a grant of the Fundación Ramón Areces. C.F. is supported by DFG grants FR-1325/20-1 and TRR186 (Project A21). Some figures were created in https://BioRender.com.

Conflict of Interest The authors have no conflicts of interest to declare that are relevant to the content of this chapter.

References

1. Mellins ED, Stern LJ (2014) HLA-DM and HLA-DO, key regulators of MHC-II processing and presentation. Curr Opin Immunol 26: 115–122
2. Álvaro-Benito M, Freund C (2020) Revisiting nonclassical HLA II functions in antigen presentation: peptide editing and its modulation. HLA 96:415–429
3. Sant AJ, Chaves FA, Jenks SA, et al (2005) The relationship between immunodominance, DM editing, and the kinetic stability of MHC class II:peptide complexes., https://pubmed.ncbi.nlm.nih.gov/16181342/
4. Yin L, Calvo-Calle JM, Dominguez-Amorocho O et al (2012) HLA-DM constrains epitope selection in the human CD4 T cell response to vaccinia virus by favoring the presentation of peptides with longer HLA-DM–mediated half-lives. 189:3983–3994
5. Mellins E, Smith L, Arp B et al (1990) Defective processing and presentation of exogenous antigens in mutants with normal HLA class II genes. Nature 343:71–74
6. Morris P, Shaman J, Attaya M et al (1994) An essential role for HLA–DM in antigen presentation by class II major histocompatibility molecules. Nature 368:551–554
7. Denzin LK, Cresswell P (1995) HLA-DM induces clip dissociation from MHC class II αβ dimers and facilitates peptide loading. Cell 82:155–165
8. Álvaro-Benito M, Morrison E, Abualrous ETET et al (2018) Quantification of HLA-DM-dependent major histocompatibility complex of class II Immunopeptidomes by the peptide landscape antigenic epitope alignment utility. Front Immunol 9

9. Álvaro-Benito M, Morrison E, Ebner F et al (2020) Distinct editing functions of natural HLA-DM allotypes impact antigen presentation and CD4+ T cell activation. Cell Mol Immunol 17:133–142
10. Pos W, Sethi DK, Call MJ et al (2012) Crystal structure of the HLA-DM–HLA-DR1 complex defines mechanisms for rapid peptide selection. Cell 151:1557–1568
11. Belmares MP, Busch R, Wucherpfennig KW et al (2002) Structural factors contributing to DM susceptibility of MHC class II/peptide complexes. J Immunol 169:5109–5117
12. Painter CA, Negroni MP, Kellersberger KA et al (2011) Conformational lability in the class II MHC 3 10 helix and adjacent extended strand dictate HLA-DM susceptibility and peptide exchange. Proc Natl Acad Sci USA 108: 19329–19334
13. Yin L, Trenh P, Guce A et al (2014) Susceptibility to HLA-DM protein is determined by a dynamic conformation of major histocompatibility complex class II molecule bound with peptide. 289:23449–23464
14. Wieczorek M, Sticht J, Stolzenberg S et al (2016) MHC class II complexes sample intermediate states along the peptide exchange pathway. Nat Commun 7:13224
15. Abualrous ET, Stolzenberg S, Sticht J et al (2023) MHC-II dynamics are maintained in HLA-DR allotypes to ensure catalyzed peptide exchange. Nat Chem Biol:1–9
16. Reyes-Vargas E, Barker AP, Zhou Z et al (2020) HLA-DM catalytically enhances peptide dissociation by sensing peptide-MHC class II interactions throughout the peptide-binding cleft. J Biol Chem 295:2959–2973
17. Günther M, Sticht J, Freund C et al (2025) Antigen presentation by MHC-II is shaped by competitive and cooperative allosteric mechanisms of peptide exchange. Structure 33:389
18. Álvaro-Benito M, Wieczorek M, Sticht J, et al (2014) HLA-DMA polymorphisms differentially affect MHC class II peptide loading
19. Alvaro-Benito M, Morrison E, Wieczorek M et al (2016) Human leukocyte Antigen-DM polymorphisms in autoimmune diseases. Open Biol 6
20. Roldão A, Oliveira R, Carrondo MJT et al (2009) Error assessment in recombinant baculovirus titration: evaluation of different methods. J Virol Methods 159:69–80
21. Lo HR, Chao YC (2004) Rapid titer determination of baculovirus by quantitative real-time polymerase chain reaction. Biotechnol Prog 20:354–360
22. Nicholson MJ, Moradi B, Seth NP et al (2006) Small molecules that enhance the catalytic efficiency of HLA-DM. J Immunol 176:4208–4220

Chapter 8

Utilizing TAPBPR for Peptide Loading, Dissociation, and Exchange on Plasma Membrane-Expressed MHC-I

Jack L. Morley and Louise H. Boyle

Abstract

TAPBPR has previously been identified as a homolog of tapasin, though the two proteins serve as mutually exclusive peptide editors. While tapasin functions solely as a constituent of the peptide loading complex, TAPBPR can function alone and independently of other chaperones and cofactors. An additional characteristic of TAPBPR is its lack of an endoplasmic-reticulum retention motif, which enables it to leak to the surface of particular cell types when overexpressed as well as an ability to promote peptide exchange at the cell surface as a recombinant soluble protein. The aforementioned features of TAPBPR provide the protein with unique capabilities for the characterization of its function, as well as the ability to dissect other properties of peptide loading such as peptide affinity for major histocompatibility complex class I and immune response to the presentation of immunoreactive peptide. Here, we describe the key methods used to decorate cells with peptides, permitting the assessment of the function of TAPBPR and its variants: peptide loading, peptide dissociation, and peptide exchange assays. The use of these assays confers the ability to dissect the catalytic function of TAPBPR and its variants, as well as conducting subsequent experiments utilizing the efficient decoration of cells with immunoreactive peptide.

Key words TAPBPR, Peptide, MHC class I, Assays

1 Introduction

The discovery of the TAPBPR gene initially shed some light on the function of the protein; sharing 22% genetic identity with tapasin it was hypothesized that TAPBPR may also perform some role within the antigen processing and presentation pathway [1]. This was later confirmed, with TAPBPR being described as an additional component of the major histocompatibility complex class I (MHC-I) presentation pathway in the absence of the peptide loading complex (PLC), suggesting that it did not require chaperone proteins nor cofactors to bind MHC-I, as well as the fact that interactions between TAPBPR and MHC-I occurred in a mutually exclusive manner alongside tapasin [2, 3]. It was also established that, like tapasin, the interactions of TAPBPR with MHC-I promoted

Efstratios Stratikos (ed.), *Antigen Processing and Presentation: Methods and Protocols*, Methods in Molecular Biology, vol. 3007, https://doi.org/10.1007/978-1-0716-5092-9_8,

peptide editing, with its binding resulting in the ejection of low-affinity peptides from the peptide-binding cleft of MHC-I and the association of high-affinity peptides [4, 5]. Two crystal structures of the TAPBPR: MHC-I complex later revealed the mechanism for this peptide editing function, with TAPBPR causing an outward shift of the α_{2-1} helix of MHC-I, which in turn reduces interactions between class I and the bound peptide [6, 7]. These structures raised dispute; however, with Jiang and colleagues unable to resolve any clear structure in the K22-D35 loop due to low electron density within the region, while Thomas and Tampé suggested that a short α-helix exists, coined the "scoop loop," which inserted itself within the F-pocket of MHC-I and catalyzed the dissociation of peptide sterically [6, 7]. Due to this discrepancy and alongside the lack of a consistent structure of this K22-D35 loop, further experiments have been conducted to assess the function of this structure within the peptide editing function of TAPBPR. It has been posited that the loop acts as a "peptide trap," floating above the F-pocket of MHC-I and lowering the affinity requirements for early incoming peptide intermediates [8]. Research from the Boyle laboratory has focused on functional cell-based assays to divulge the function of the K22-D35 loop of TAPBPR, with the lack of an endoplasmic reticulum (ER)-retention motif and its subsequent ability to leak to the cell surface when overexpressed, permitting the use of on-cell assays to assess its function [1, 9]. These assays have revealed that the K22-D35 loop of TAPBPR is essential for its peptide function, with a mutant variant of the protein (labeled $TAPBPR^{\emptyset}$—a variant of the protein with all residues within the K22-D35 loop mutated to a combination of glycine, serine and alanine) unable to perform peptide loading. Additionally, the residue L30 was found to be essential for peptide dissociation function, characterized by an on-cell peptide dissociation assay [10]. Furthermore, the use of the soluble luminal domain of TAPBPR (sTAPBPR) and the recent generation of a TAPBPR fusion protein (a chimeric protein consisting of the luminal domain of TAPBPR and a nanobody against pDisplay GFP joined by a short GGGGS triplicate motif—sTAPBPR-αGFP) has permitted the use of peptide exchange assays to further investigate TAPBPR function on the surface of target cells, as well as its preference for particular HLA allotypes [9, 11, 12]. This chapter will describe the three on-cell assays that are currently utilized to assess TAPBPR function and harnessed to explore peptide affinity for MHC-I and the immune response to the efficient decoration of peptide onto the surface of cells, namely, the peptide loading, peptide dissociation and peptide exchange assays.

2 Materials

2.1 Peptide Loading Assay

1. HeLaM cells (wildtype and overexpressing TAPBPR variants) (*see* **Notes 1** and **2**).
2. PBS.
3. OptiMEM.
4. Fluorescently labeled peptide, e.g., ETVSK*QSNV peptide for binding to HLA-A*68:02 (a TAMRA-labeled derivative of the known neoantigen ETVSEQSNV, with K* representing the TAMRA-labeled lysine residue used to replace glutamate of the original epitope) (*see* **Note 3**).
5. Trypsin-ethylenediaminetetraacetic acid (EDTA) solution (0.25%).
6. Dulbecco's Modified Eagle's Medium (DMEM).
7. FACs buffer (PBS and 2% fetal bovine serum (FBS)).

2.2 Peptide Dissociation Assay

1. HeLaM cells expressing pDisplay GFP (*see* **Notes 2** and **4**).
2. PBS.
3. sTAPBPR-αGFP fusion protein (*see* **Notes 5** and **6**).
4. Fluorescently labeled peptide, e.g., ETVSK*QSNV peptide for binding to HLA-A*68:02 (*see* **Note 3**).
5. Fluorescently labeled high-affinity competitor peptide with alternate fluorophore, e.g., YVVPFVAK*V peptide (a Cy5-labeled derivative of the known neoantigen YVVPF-VAKV, with K* representing the Cy5-labeled lysine residue) (*see* **Note 3**).
6. Trypsin-EDTA solution (0.25%).
7. DMEM.
8. FACs buffer (PBS and 2% FBS).

2.3 Peptide Exchange Assay

1. HeLaM cells expressing pDisplay GFP (*see* **Notes 2** and **4**).
2. PBS.
3. sTAPBPR-αGFP fusion protein (*see* **Notes 5** and **6**).
4. Fluorescently labeled peptide, e.g., ETVSK*QSNV peptide (*see* **Note 3**).
5. Trypsin-EDTA solution (0.25%).
6. DMEM.
7. FACs buffer (PBS and 2% FBS).

3 Method

3.1 Peptide Loading Assay (Outlined in Fig. 1)

1. Seed 25,000 HeLaM cells (both wildtype and overexpressing TAPBPR) in a 12-well plate 48 hours before experimentation.
2. On the day of experimentation, wash cells with 500 μL PBS to remove media, using a pipette tip to remove all excess PBS from each well.
3. Dilute varying concentrations of fluorescently labeled peptide in 300 μL OptiMEM (typically varying from 1pM to 1 μM) and add 300 μL of peptide solution to each concordant well (*see* **Note 3**).
4. Incubate cells with peptide for 15 minutes at 37 °C in an environment of 5% CO_2 (*see* **Note 7**).
5. Wash cells three times with 500 μL PBS to remove unbound peptide, removing all excess PBS with a pipette tip on the final wash.
6. Add 100 μL of trypsin-EDTA solution to each well and incubate for 5 minutes at 37 °C in an environment of 5% CO_2.
7. Add 100 μL DMEM to each well to inhibit the trypsinization reaction. Remove the 200 μL cell solution from each well of the 12-well plates and transfer to a 96-well plate for analysis by flow cytometry.

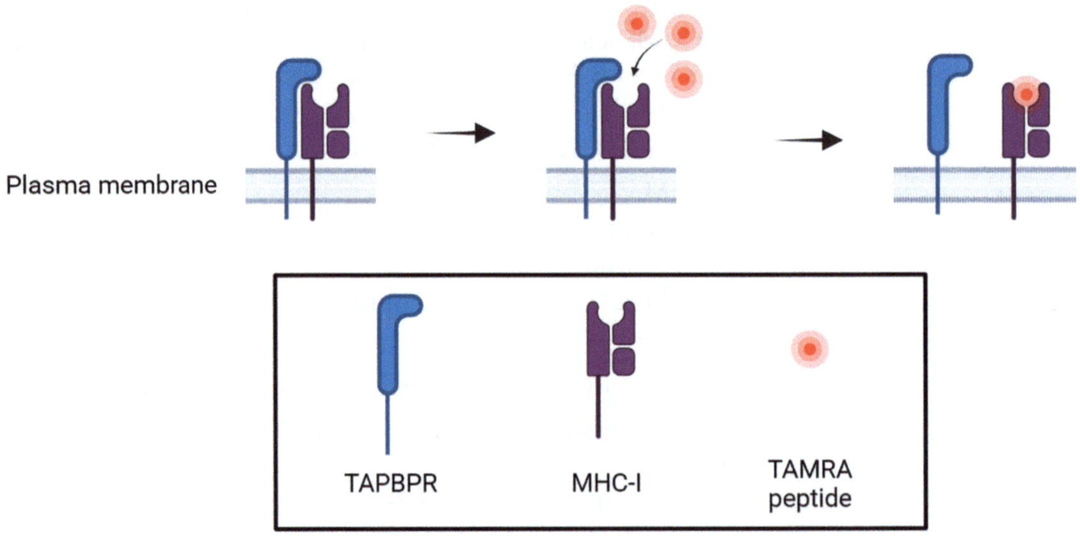

Fig. 1 A graphical overview of the peptide loading assay. When TAPBPR is overexpressed in cells, it can traffic to the plasma membrane with MHC-I, stabilizing the MHC-I in a peptide-receptive state. Following incubation of the cells with fluorescently-labeled peptide (e.g. TAMRA-labeled) with specificity for MHC-I, TAPBPR assists in loading peptide onto the plasma membrane-expressed MHC-I molecules. Upon peptide loading, TAPBPR dissociates from MHC-I, resulting in MHC-I loaded with the antigenic peptide, which can be monitored by flow cytometry

8. Spin down cells in 96-well plate at 1250 rpm and wash twice with 150 μL PBS to remove the DMEM-trypsin solution. Resuspend in 100 μL FACS buffer before analysis by flow cytometry to quantify peptide loading by TAMRA fluorescence.

3.2 Peptide Dissociation Assay (Outlined in Fig. 2)

1. Seed 25,000 HeLaM cells expressing pDisplay GFP and a single HLA allele in 12-well plate 48 hours prior to experimentation.
2. On the day of experimentation, wash cells with 500 μL PBS to remove media, using a pipette tip to remove all excess PBS from each well.
3. Dilute 100 nM sTAPBPR-αGFP in 300 μL OptiMEM and add the fusion protein to each concordant well.

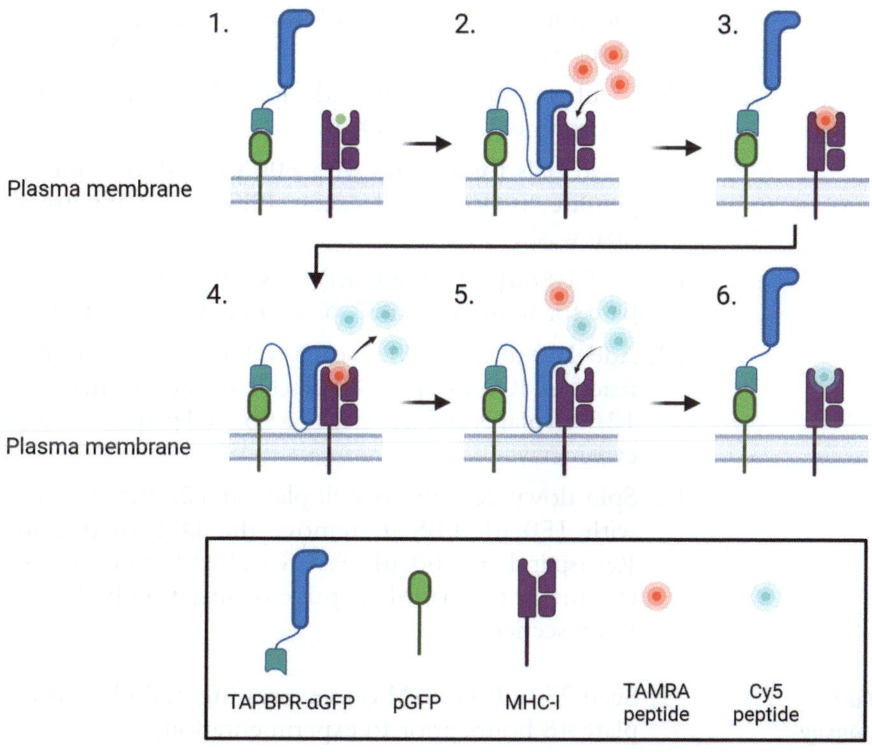

Fig. 2 A graphical overview of the peptide dissociation assay (1) Cells transfected with pDisplay-GFP (pGFP) express GFP on their plasma membrane. Incubation with a TAPBPR fusion protein linked to a GFP-specific nanobody (TAPBPR-αGFP) results in recombinant TAPBPR being tethered to the plasma membrane. (2) TAPBPR binds to plasma membrane-expressed MHC-I, dissociating the endogenously presented peptide and stabilization of MHC-I in a peptide-receptive state. Following incubation with TAMRA-labeled peptide specific for MHC-I (3), TAPBPR promotes its loading onto MHC-I. (4) When cells are subsequently incubated with a Cy5-labeled MHC-I binding competitor peptide (5), the dissociation of the TAMRA-labeled peptide and (6) also the loading of Cy5-labeled peptide to MHC-I can be measured by flow cytometry

4. Incubate cells for 15 minutes at 37 °C in an environment of 5% CO_2.
5. Following incubation, wash cells three times with PBS to remove unbound sTAPBPR-αGFP, using a pipette tip on the final wash to remove all excess PBS from each well.
6. Dilute 1 μM of fluorescently labeled peptide in 300 μL Opti-MEM and add 300 μL of peptide solution to each concordant well (*see* **Note 3**).
7. Incubate cells with peptide for 15 minutes at 37 °C in an environment of 5% CO_2 (*see* **Note 7**).
8. Wash cells three times with 500 μL PBS to remove unbound peptide, removing all excess PBS with a pipette tip on the final wash.
9. Dilute varying concentrations of fluorescently labeled high-affinity competitor peptide (with alternate fluorophore) in 300 μL OptiMEM (typically varying from 1pM to 1 μM) and add 300 μL of peptide solution to each concordant well (*see* **Note 3**).
10. Incubate cells with peptide for 15 minutes at 37 °C in an environment of 5% CO_2.
11. Wash cells three times with 500 μL PBS to remove unbound peptide, removing all excess PBS with a pipette tip on the final wash.
12. Add 100 μL of trypsin-EDTA solution to each well and incubate for 5 minutes at 37 °C in an environment of 5% CO_2.
13. Add 100 μL DMEM to each well to inhibit the trypsinization reaction. Remove the 200 μL cell solution from each well of the 12-well plates and transfer to a 96-well plate for analysis by flow cytometry.
14. Spin down cells in 96-well plate at 1250 rpm and wash twice with 150 μL PBS to remove the DMEM-trypsin solution. Resuspend in 100 μL FACS buffer before analysis by flow cytometry to quantify peptide dissociation by loss of TAMRA fluorescence.

3.3 Peptide Exchange Assay (Outlined in Fig. 3)

1. Seed 25,000 HeLaM cells expressing pDisplay GFP in 12-well plate 48 hours prior to experimentation.
2. On the day of experimentation, wash cells with 500 μL PBS to remove media, using a pipette tip to remove all excess PBS from each well.
3. Dilute 100 nM sTAPBPR-αGFP in 300 μL OptiMEM and add the fusion protein to each concordant well.
4. Incubate cells for 15 minutes at 37 °C in an environment of 5% CO_2.

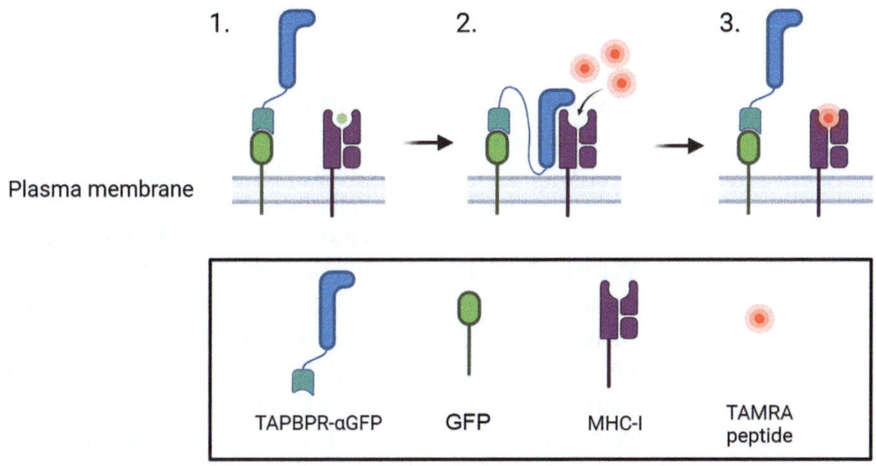

Fig. 3 A graphical overview of the peptide exchange assay (1) Cells expressing GFP on their plasma membrane are incubated with a TAPBPR fusion protein linked to a GFP-specific nanobody (TAPBPR-αGFP), resulting in recombinant TAPBPR being tethered to the plasma membrane. (2) TAPBPR then binds to MHC-I, dissociating the endogenously presented peptide and stabilizing MHC-I in a peptide-receptive state. Following incubation with TAMRA-labeled peptide specific for MHC-I (3), TAPBPR promotes its loading onto MHC-I, which can be measured using flow cytometry

5. Following incubation, wash cells three times with PBS to remove unbound sTAPBPR-αGFP, using a pipette tip on the final wash to remove all excess PBS from each well.
6. Dilute 1 μM of fluorescently labeled peptide in 300 μL Opti-MEM and add 300 μL of peptide solution to each concordant well (*see* **Note 3**).
7. Incubate cells with peptide for 15 minutes at 37 °C in an environment of 5% CO_2 (*see* **Note 7**).
8. Wash cells three times with 500 μL PBS to remove unbound peptide, removing all excess PBS with a pipette tip on the final wash.
9. Add 100 μL of trypsin-EDTA solution to each well and incubate for 5 minutes at 37 °C in an environment of 5% CO_2.
10. Add 100 μL DMEM to each well to inhibit the trypsinization reaction. Remove the 200 μL cell solution from each well of the 12-well plates and transfer to a 96-well plate for analysis by flow cytometry.
11. Spin down cells in 96-well plate at 1250 rpm and wash twice with 150 μL PBS to remove the DMEM-trypsin solution. Resuspend in 100 μL FACS buffer before analysis by flow cytometry to quantify peptide exchange by TAMRA fluorescence.
12. **Note 8** gives more information regarding further applications of this peptide exchange method.

4 Notes

1. HeLaM cells with TAPBPR wildtype and TAPBPR variants overexpressed were generated with the transduction of HeLaM cells using a lentiviral method. Briefly, HEK293T cells were seeded in 6-well plates 24 hours prior to transfection. 0.8 μg of the lentiviral vector pHRSIN-SRRV-MCS(+)-WPRE-PGK-Puro containing the TAPBPR gene, along with packaging vector pCMVΔR8.91 and the envelope vector pMD.G were combined to produce lentivirus, and the mixture was transfected into HEK293T cells with 4.5 μL of the transfection reagent FuGENE HD. Lentiviral supernatant was harvested at both 48 and 72 hours after transfection, with 2 ml of supernatant filtered and applied to the target HeLaM cells with 8 μg/ml of Polybrene.
2. HeLaM cells may express endogenous HLA-I, or if preferred, HLA-I can be knocked out with a single HLA-I allele transduced using the lentiviral method as described above, resulting in the expression of just one HLA-I allotype. In this case, the subsequent peptide used for loading, dissociation or exchange must be of high affinity for the chosen HLA-I allele.
3. As stated above, the peptide selected for use in the loading, dissociation and exchange assays must be of a high affinity for the HLA-I allotype expressed by the HeLaM cells. Here, we have provided ETVSK*QSNV and YVVPFVAK*V as examples due to their high affinity for HLA-A*68:02 (which is expressed endogenously by HeLaM cells). However, if a particular HLA-I allele is selected for testing, peptides must be chosen accordingly.
4. pDisplay GFP was transduced into HeLaM cells using the lentiviral transduction method described in **Note 1**.
5. The sTAPBPR-αGFP fusion protein is a chimeric protein consisting of the luminal domain of TAPBPR bound to an anti-GFP nanobody by a short GGGGS triplicate motif. The anti-GFP nanobody was raised against GFP amplified by PCR from the pGEX6P1-GFP-Nanobody construct (Kazuhisa Nakayama, Addgene plasmid #61836). Primers containing a 6xHis tag, the GGGGS triplicate linker sequence and PmlI and SpeI restriction sites were used to amplify the anti-GFP nanobody, and permitted its digestion and ligation into the PiggyBac PB-T-PAF vector along with sTAPBPR to generate sTAPBPR-αGFP. The PiggyBac PB-T-PAF vector permits the stable expression of sTAPBPR-αGFP using the PiggyBac system [13].
6. sTAPBPR alone (the luminal domain of TAPBPR only) can be used instead of the sTAPBPR-αGFP fusion protein, though we

have found it to be less effective at quantifying peptide dissociation and exchange due to the more transient interactions between sTAPBPR and surface HLA-I, which result in a significant degree of sTAPBPR loss during washing stages.

7. In the case that HLA-I alleles other than HLA-A*68:02 are being investigated, the concordant high-affinity peptide should be incubated with the target HeLaM cells for 1 hour at 37 °C and 5% CO_2 instead of 15 minutes as is the case for HLA-A*68:02 and ETVSK*QSNV and YVVPFVAK*V. This is due to the less optimal affinity of TAPBPR for HLA alleles other than HLA-A*68:02, reducing the efficiency of peptide loading, dissociation and exchange [11].
8. This method enables the efficient loading of immunoreactive peptide onto the surface of target cells, permitting the subsequent exploration of peptide characteristics. For example, the ability of peptide to stimulate an immune response can be investigated with CD8+ T cell Fluorospot or activation and degranulation assays following the loading of peptide onto the target cell surface by TAPBPR. It is worth noting that for immune activation assays such as this, a non-fluorescent peptide (e.g., ETVSEQSNV for binding to HLA-A*68:02) must be loaded onto the cell surface. Peptide affinity for MHC-I can also be examined, with different peptide binding to surface MHC-I following loading by the TAPBPR fusion protein quantified by fluorophore fluorescence.

Conflict of Interest

This study was funded by Wellcome grants [222632/Z/21/Z and 219479/Z/19/Z]. The authors have no conflicts of interest to declare that are relevant to the content of this chapter.

References

1. Teng MS, Stephens R, Du Pasquier L, Freeman T, Lindquist JA, Trowsdale J (2002) A human TAPBP (TAPASIN)-related gene. TAPBP-R Eur J Immunol 32(4):1059–1068. https://doi.org/10.1002/1521-4141(200204)32:4<1059::AID-IMMU1059>3.0.CO;2-G
2. Boyle LH, Hermann C, Boname JM, Porter KM, Patel PA, Burr ML, Duncan LM, Harbour ME, Rhodes DA, Skjodt K, Lehner PJ, Trowsdale J (2013) Tapasin-related protein TAPBPR is an additional component of the MHC class I presentation pathway. Proc Natl Acad Sci USA 110(9):3465–3470. https://doi.org/10.1073/pnas.1222342110
3. Hermann C, Strittmatter LM, Deane JE, Boyle LH (2013) The binding of TAPBPR and Tapasin to MHC class I is mutually exclusive. J Immunol 191(11):5743–5750. https://doi.org/10.4049/jimmunol.1300929
4. Hermann C, van Hateren A, Trautwein N, Neerincx A, Duriez PJ, Stevanovic S, Trowsdale J, Deane JE, Elliott T, Boyle LH (2015) TAPBPR alters MHC class I peptide presentation by functioning as a peptide exchange catalyst. elife 4:e09617. https://doi.org/10.7554/eLife.09617
5. Morozov GI, Zhao H, Mage MG, Boyd LF, Jiang J, Dolan MA, Venna R, Norcross MA, CP

MM, Hildebrand W, Schuck P, Natarajan K, Margulies DH (2016) Interaction of TAPBPR, a tapasin homolog, with MHC-I molecules promotes peptide editing. Proc Natl Acad Sci U S A 113(8):E1006–E1015. https://doi.org/10.1073/pnas.1519894113

6. Jiang J, Natarajan K, Boyd LF, Morozov GI, Mage MG, Margulies DH (2017) Crystal structure of a TAPBPR-MHC I complex reveals the mechanism of peptide editing in antigen presentation. Science 358(6366): 1064–1068. https://doi.org/10.1126/science.aao5154
7. Thomas C, Tampe R (2017) Structure of the TAPBPR-MHC I complex defines the mechanism of peptide loading and editing. Science 358(6366):1060–1064. https://doi.org/10.1126/science.aao6001
8. McShan AC, Devlin CA, Morozov GI, Overall SA, Moschidi D, Akella N, Procko E, Sgourakis NG (2021) TAPBPR promotes antigen loading on MHC-I molecules using a peptide trap. Nat Commun 12(1):3174. https://doi.org/10.1038/s41467-021-23225-6
9. Ilca FT, Neerincx A, Wills MR, de la Roche M, Boyle LH (2018) Utilizing TAPBPR to promote exogenous peptide loading onto cell surface MHC I molecules. Proc Natl Acad Sci USA 115(40):E9353–E9361. https://doi.org/10.1073/pnas.1809465115
10. Ilca FT, Neerincx A, Hermann C, Marcu A, Stevanovic S, Deane JE, Boyle LH (2018) TAPBPR mediates peptide dissociation from MHC class I using a leucine lever. elife 7: 40126. https://doi.org/10.7554/eLife.40126
11. Ilca FT, Drexhage LZ, Brewin G, Peacock S, Boyle LH (2019) Distinct polymorphisms in HLA class I molecules govern their susceptibility to peptide editing by TAPBPR. Cell Rep 29(6):1621–1632. e1623. https://doi.org/10.1016/j.celrep.2019.09.074
12. Neerincx A, Aflalo A, Hafstrand I, Torres AA, McDowell A, Altenburg A, Satti R, Ilca F, Okkenhaug K, Wills M, H. BL (2022) Antibody-mediated delivery of TAPBPR enables the redirection of virus-specific T cell responses to tumour cells. Res Square. https://doi.org/10.21203/rs.3.rs-2357899/v1
13. Li Z, Michael IP, Zhou D, Nagy A, Rini JM (2013) Simple piggyBac transposon-based mammalian cell expression system for inducible protein production. Proc Natl Acad Sci USA 110(13):5004–5009. https://doi.org/10.1073/pnas.1218620110

Chapter 9

Expression, Purification, and Functional Analysis Methods of Antigen-Processing Aminopeptidases ERAP1, ERAP2, and IRAP

Anastasia Mpakali

Abstract

Endoplasmic reticulum aminopeptidases 1 and 2 (ERAP1, ERAP2), along with Insulin-Regulated Aminopeptidase (IRAP), play key roles in antigen processing by trimming precursor peptides for major histocompatibility complex (MHC) class I presentation, thereby contributing to adaptive immune regulation. Dysregulation of these enzymes has been implicated in various diseases, including autoimmune disorders and cancer. This chapter provides a comprehensive, step-by-step methodological guide for the expression, purification, and functional characterization of recombinant ERAP1, ERAP2, and IRAP. Detailed protocols include baculovirus-mediated expression in insect cells, affinity chromatography for protein purification, and a suite of functional assays to study enzymatic activity, substrate specificity, and inhibitor screening. Fluorogenic, colorimetric, and HPLC-based assays are described, offering versatile tools for kinetic analysis and inhibitor evaluation. These protocols are designed to yield high-purity, functionally active enzymes suitable for biochemical, structural, mechanistic, and drug discovery studies.

Key words Endoplasmic reticulum aminopeptidases (ERAP1, ERAP2), Insulin-regulated aminopeptidase (IRAP), Recombinant protein expression, Baculovirus-mediated expression, Affinity chromatography, Aminopeptidase enzymatic assays, Inhibitor screening, Antigen processing, Enzyme kinetics

Abbreviations

ACN	Acetonitrile
AMC	7-Amino-4-methylcoumarin
DMSO	Dimethyl sulfoxide
Em	Emission
ERAP1	Endoplasmic Reticulum Aminopeptidase 1
ERAP2	Endoplasmic Reticulum Aminopeptidase 2
Ex	Excitation

The original version of the chapter has been revised. A correction to this chapter can be found at https://doi.org/10.1007/978-1-0716-5092-9_20

Efstratios Stratikos (ed.), *Antigen Processing and Presentation: Methods and Protocols*, Methods in Molecular Biology, vol. 3007, https://doi.org/10.1007/978-1-0716-5092-9_9,

Corrected Publication 2026

HPLC	High Performance Liquid Chromatography
IRAP	Insulin-Regulated Aminopeptidase
L-AMC	Leucine 7-Amino-4-methylcoumarin
Leu-pNA/L-pNA	L-Leucine-p-nitroanilide
MWCO	Molecular Weight Cut-Off
NaP	Sodium Phosphate
PMSF	Phenylmethylsulfonyl Fluoride
R-AMC	Arginine 7-Amino-4-methylcoumarin
RFU	Relative Fluorescence Units
RT	Room Temperature
SEC	Size-Exclusion Chromatography
TFA	Trifluoroacetic Acid

1 Introduction

Endoplasmic reticulum aminopeptidases (ERAP1 and ERAP2) and Insulin-Regulated aminopeptidase (IRAP) are zinc-dependent metallopeptidases that play pivotal roles in antigen processing by trimming peptides for presentation on MHC class I molecules [1, 2]. These enzymes exhibit substrate specificity, which directly influences the generation of the antigenic peptide repertoire and, consequently, immune responses. Given their central role in immunity, ERAP1, ERAP2, and IRAP have emerged as promising targets for therapeutic intervention in autoimmune disorders, cancer, and infectious diseases. Understanding their enzymatic mechanisms and regulatory properties is therefore essential for the development of targeted therapies. Recombinant expression of these enzymes is a cornerstone for functional and structural studies. Insect cell-based expression systems, utilizing baculovirus vectors, provide an efficient platform for producing active and properly folded enzymes. The subsequent purification process, typically involving Ni-NTA affinity chromatography, yields high-purity recombinant proteins suitable for downstream applications. Functional analysis of ERAPs and IRAP encompasses a variety of assays, including fluorogenic, colorimetric, and HPLC-based methods. Among these, fluorogenic assays are the most widely used due to their rapid and sensitive quantification of enzymatic activity, making them ideal for high-throughput inhibitor screening and enzyme kinetics studies. Colorimetric methods provide a cost-effective alternative for studying substrate specificity and inhibitor interactions, while HPLC-based assays enable precise monitoring of peptide trimming and are well-suited for mechanistic and kinetic studies.

This chapter outlines detailed, step-by-step protocols for the recombinant expression of ERAP1, ERAP2, and IRAP in insect cells using baculovirus-mediated transduction, followed by purification via affinity chromatography. Additionally, a series of functional assays is described, including fluorogenic, colorimetric, and

HPLC-based methods, to characterize the enzymatic activity, substrate specificity, and inhibitor interactions of these enzymes. These protocols are optimized for high yield, purity, and reproducibility, making them suitable for both basic research and drug discovery applications.

2 Materials

2.1 Bacmid Generation and Isolation

2.1.1 Buffers & Solutions for Recombinant Bacmid Generation

1. Plasmid expression vectors (e.g., pDEST8, pFastBac1).
2. DH10Bac™ *E. coli* competent cells.
3. Antibiotics stocks: kanamycin (50 mg/mL), gentamicin (7 mg/mL), tetracycline (10 mg/mL)—(*see* Table 1 and **Note 1**).
4. LB broth & LB agar plates supplemented with antibiotics suitable for blue/white selection (kanamycin 50 μg/mL, gentamicin 7 μg/mL, tetracycline 10 μg/mL, IPTG 40 μg/mL and X-gal or Bluo-gal 0.02%).
5. SOC medium (commercially available or home-made).

2.1.2 Buffers & Solutions for Recombinant Bacmid Isolation

1. Buffers for bacmid isolation:

   Resuspension buffer: 100 mM Tris-HCl, pH 8.0; 10 mM EDTA, pH 8.0; 100 μg/mL RNase A (filter sterilize and store at 4 °C; RNase should be added fresh before use).

   Alkaline lysis buffer: 0.2 N NaOH, 1% SDS (always prepare fresh, as its effectiveness decreases over time).

   Neutralization buffer: 3 M Potassium Acetate, pH 5.5 (autoclave and store at room temperature, RT).
2. TE buffer: 10 mM Tris-HCl, pH 8.0; 1 mM EDTA.

2.1.3 Materials for PCR Validation of Bacmid

1. M13/pUC forward and reverse primers (10 μM stocks).
2. PCR master mix (e.g., Taq polymerase, dNTPs, Taq buffer).
3. Nuclease-free water.
4. Agarose.

Table 1
Stock solutions of the essential antibiotics for recombinant bacmid generation

Solution	Stock concentration
Kanamycin	50 mg/mL in water
Gentamicin	7 mg/mL in water
Tetracycline	10 mg/mL in 70% ethanol
X-gal/ Bluo-gal	200 mg/mL in DMSO
IPTG	40 mg/mL in water

5. GelRed™, SYBR Safe™, or ethidium bromide (EtBr).
6. DNA ladder (e.g., 1 kb).
7. Gel loading dye.

2.1.4 Equipment

1. Pipettes, sterile tips & Eppendorf tubes.
2. Incubator with shaking capability set to 37 °C for bacterial plates and cultures.
3. Microcentrifuge.
4. Water bath or heat block (42 °C).
5. Agarose gel electrophoresis apparatus.
6. Thermocycler.

2.2 Baculovirus Production and Amplification

2.2.1 Media, Reagents, and Buffers for Baculovirus Production & Amplification

1. Sf900II™ SFM medium (or equivalent serum-free medium).
2. IPL-41 medium (for transfection).
3. Transfection reagent (e.g., Cellfectin® II).
4. Trypan Blue solution (0.4%) for cell viability assessment.
5. Gentamicin stock solution 10 mg/mL (filter-sterilized).

2.2.2 Cell Lines

1. *Spodoptera frugiperda* (Sf9) cells: used for virus production and initial steps of virus amplification (adherent cells).
2. *Trichoplusia ni* (High Five/Hi5) cells: used for protein expression (suspension culture).

   Passaging guidelines:

   - Subculture cells every 3–4 days.
   - For Sf9: Split to 0.3–0.5 × 10^6 cells/mL when density reaches 3–4 × 10^6 cells/mL.
   - For Hi5: Split to 0.3–0.5 × 10^6 cells/mL when density reaches 2.5–3 × 10^6 cells/mL.
   - A critical factor for high-yield recombinant protein production is the healthy status of the cell culture. Ensure that the cells are in the log phase at the time of infection.
   - Viability check: use Trypan Blue exclusion to ensure cell viability remains > 95% at the time of infection.

2.2.3 Equipment

1. Erlenmeyer tissue culture flasks with vented caps (working volume < 30% of flask capacity).
2. Shaking incubator with humidity control (27 °C, 120 rpm, no CO_2).
3. Laminar flow hood.
4. Sterile culture flasks and 6-well plates.
5. Benchtop centrifuge.

6. Inverted Light Microscope.
7. Hemocytometer or automated cell counter.
8. 0.22 μm sterile syringe filters.

2.3 Small-Scale Protein Expression in Hi5 Cells

2.3.1 Reagents and Media for Small-Scale Expression

1. Sf900II™ SFM medium (or equivalent serum-free medium).
2. P2 baculovirus stock (from Subheading 3.2.4).
3. Trypan Blue (0.4%) for cell viability assessment.

2.3.2 Equipment for Small-Scale Expression

1. Erlenmeyer tissue culture flasks with vented caps (working volume < 30% of flask capacity).
2. Shaking incubator with humidity control (27 °C, 120 rpm, no CO_2 required).
3. Laminar flow hood.
4. Centrifuge.
5. Inverted Light Microscope.
6. Hemocytometer or automated cell counter.

2.4 Purification of Aminopeptidases ERAP1, ERAP2 and IRAP from Insect Cell Supernatant

2.4.1 Reagents for Purification of Aminopeptidases

1. Sodium phosphate buffer (NaP) stock solution, 500 mM, pH 8.0.
2. NaCl stock solution, 2 M.
3. Imidazole stock solution, 2 M, pH 8.0 (filtered).
4. Dialysis Buffer **A**: 10 mM sodium phosphate pH 8.0, 100 mM NaCl.
5. Dialysis Buffer **B**: 10 mM Hepes pH 7.5, 150 mM NaCl.
6. Phenylmethylsulfonyl fluoride (PMSF) in alcohol (prepare fresh).
7. Complete protease inhibitor cocktail.
8. Glycerol.
9. Ni-NTA agarose resin.
10. SDS-PAGE reagents.
11. **Purification buffers**:
    - Equilibration/binding buffer: 50 mM NaP, pH 8.0; 300 mM NaCl; 20 mM imidazole.
    - Wash buffer **1**: 50 mM NaP, pH 8.0; 300 mM NaCl; 30 mM imidazole.
    - Wash buffer **2**: 50 mM NaP, pH 8.0; 300 mM NaCl; 40 mM imidazole.
    - Wash buffer **3**: 50 mM NaP, pH 8.0; 300 mM NaCl; 300 mM imidazole.

- Elution buffer: 50 mM NaP, pH 8.0; 300 mM NaCl; 150 mM imidazole.

2.4.2 Equipment for Purification of Aminopeptidases

1. Dialysis tubing (MWCO 10 kDa).
2. Gravity flow columns.
3. Amicon Ultra-centrifugal filters (10 or 30 kDa MWCO; used for protein concentration if needed).
4. Centrifuge.
5. 50 mL Falcon conical tubes.
6. SDS-PAGE electrophoresis apparatus.

2.4.3 Solutions & Equipment for Size-Exclusion Chromatography (SEC)

1. SEC Buffer: 50 mM Hepes pH 7.5, 150 mM NaCl (filtered and degassed).
2. Superdex 200 or equivalent SEC column.
3. FPLC system (e.g., ÄKTA Pure).
4. Amicon Ultra-centrifugal filters (10 or 30 kDa MWCO).

2.5 Fluorogenic Activity Assays

2.5.1 Reagents for Fluorogenic Assays

1. Purified recombinant ERAP1, ERAP2, or IRAP.
2. Fluorogenic AMC-conjugated peptide substrates (e.g., Leu-AMC for ERAP1 & IRAP, Arg-AMC for ERAP2 & IRAP).
3. 7-Amino-4-methyl-coumarin (AMC) reference standard.
4. Inhibitor(s) of interest.
5. Assay buffer: 50 mM Hepes pH 7.0, 100 mM NaCl, 0.01% Tween-20.
6. Dimethyl sulfoxide (DMSO) for substrate and inhibitor solubilization.

2.5.2 Equipment for Fluorogenic Assays

1. Pipettes and low-binding tips.
2. 1.5 mL Eppendorf tubes.
3. Analytical balance.
4. Vortex mixer.
5. Black, flat-bottom 96-well fluorescence-compatible microplates (low-binding).
6. Fluorescence microplate reader (Ex: 380 nm, Em: 460 nm, adjustable gain).
7. Data analysis software (e.g., Microsoft Excel, GraphPad Prism).

2.5.3 Gain Settings

1. **Importance of gain:**
   - The gain setting on the microplate reader is critical for accurate fluorescence measurements. It determines the

sensitivity of the detector and must be optimized for each enzyme to ensure the signal is within the linear range.

- Too low gain: Signals may be weak and difficult to distinguish from background noise.
- Too high gain: Signals may saturate, leading to inaccurate measurements.

2. **Optimizing gain:**
   - For each enzyme, adjust the gain manually to ensure the fluorescence signal is strong but not saturated. Record the gain setting used for each experiment to maintain consistency.

2.5.4 Substrate Preparation

1. Dissolve the appropriate amount of fluorogenic substrate in DMSO to prepare a 50 mM stock solution.
2. Vortex thoroughly to ensure complete dissolution of the substrate.
3. Aliquot and store the stock solution at −20 °C or − 80 °C, protected from light, to prevent degradation. Avoid repeated freeze-thaw cycles.

2.6 Colorimetric Assays

2.6.1 Reagents for Colorimetric Assays

1. Assay Buffer: 50 mM Hepes pH 7.0, 100 mM NaCl.
2. Chromogenic substrate: L-leucine-p-nitroanilide (Leu-pNA).
3. Purified enzyme (ERAP1, ERAP2, or IRAP).
4. pNA reference standard (optional, for calibration).

2.6.2 Equipment for Colorimetric Assays

1. Pipettes, tips, Eppendorf tubes.
2. Analytical balance.
3. Vortex mixer.
4. 96-well microplates (polystyrene, flat-bottom, transparent, low binding, spectrophotometry compatible).
5. Spectrophotometer or microplate reader (equipped with a 405 nm filter).

2.7 HPLC-Based Assays

2.7.1 Reagents for HPLC-Based Assays

1. Purified recombinant human aminopeptidase (ERAP1, ERAP2, IRAP).
2. Peptide substrate (>95% purity): e.g., MHCI antigenic precursors, synthetic analogs, or peptide hormones.
3. Assay buffer: 50 mM Hepes, pH 7.0, 100 mM NaCl.
4. Acid Quenching Solution: 0.1% (v/v) Trifluoroacetic acid (TFA).
5. HPLC-gradient solvents (e.g., Acetonitrile- ACN, water).

2.7.2 Equipment for HPLC-Based Assays

1. Pipettes, tips & Eppendorf tubes.
2. HPLC system equipped with a UV detector.
3. Analytical reversed-phase column (e.g. Chromolith C18, 3–5 μm particle size, 100–250 mm length).
4. Water bath (set at 37 °C).
5. Microcentrifuge.

2.7.3 HPLC Conditions

1. Flow rate: 1–2 mL/min (depending on the specifications of the column used for the experiment)
2. Mobile Phase: *Solvent A*: 0.05% TFA, 5–10% ACN in water; *Solvent B*: 0.05% TFA, 50% ACN in water (*see* **Notes 2** and **3**).
3. Gradient: Linear gradient 5–50% ACN for 20–30 min (*see* **Note 2** and Table 3).
4. UV detection: at 220 nm for peptide bond detection, at 257 nm for peptides containing aromatic residues, at 280 nm for Tryptophan (W)-containing peptides.

2.7.4 Software

1. Peak integration & quantification of substrate/product conversion (e.g., Empower, Chromeleon etc.).
2. Conversion of peak areas to substrate concentrations, calculation of % substrate depletion (e.g., Excel).
3. Kinetic analysis and IC_{50} determination (e.g., GraphPad Prism).

3 Methods

A. **Expression and Purification of Aminopeptidases ERAP1, ERAP2 & IRAP**

The protocol outlines the recombinant expression of ERAP1, ERAP2, and IRAP (the soluble form of the enzyme) in insect cells using a baculovirus-mediated transduction system. The coding sequences of the aminopeptidases can be cloned into suitable expression vectors such as pDEST8 (via LR recombination using the Gateway® System) or pFastBac1 (using the Bac-to-Bac® System), for bacmid generation and subsequent protein expression [3–6]. Alternatively, other commercially available kits for bacmid generation may be used. For comprehensive details on the baculovirus expression system, refer to the relevant Invitrogen manual [4].

All plasmid constructs should include the following features:

(i) A 21-amino acid lobster sequence at the N-terminus to enhance protein yield [7].

(ii) A signal sequence (intrinsic or engineered) for protein secretion into the cell culture supernatant.

(iii) A C-terminal deca-histidine (His_{10}) tag for purification by affinity chromatography.

3.1 Generation and Isolation of Recombinant Bacmid DNA

3.1.1 Transformation of Recombinant Plasmid into DH10Bac *E. coli* Cells

1. Prepare in advance DH10Bac *E. coli* competent cells and LB agar plates supplemented with antibiotics/IPTG/X-gal as indicated in the Materials section.
2. Thaw DH10Bac competent cells on ice. Use approximately 50 μL of cells per transformation.
3. Mix 10–100 ng of plasmid DNA with 50 μL of competent cells and gently tap the bottom of the Eppendorf tube to mix.
4. Incubate the mixture on ice for 30 min.
5. Heat-shock the cells at 42 °C for 45 s, then immediately place on ice for 2 min (*see* **Note 4**).
6. Add 180 μL of SOC medium to the transformation mix and incubate at 37 °C with shaking (200 rpm) for 4–5 h (*see* **Note 5**).
7. Prepare a 1:10 dilution of the cells with SOC medium.
8. Spread both diluted and undiluted cells evenly onto LB agar plates. Incubate the plates at 37 °C for 48 h or until the color of the colonies is discernible.
9. Select white colonies (indicative of successful transposition) and streak them onto a fresh LB plate. Incubate the plates at 37 °C for 48 h (*see* **Note 6**).

3.1.2 Bacmid Isolation

1. Inoculate a single colony confirmed as having a clear white phenotype (using a sterile pipette tip) into 3 mL LB broth medium supplemented with kanamycin 50 μg/mL, gentamicin 7 μg/mL, and tetracycline 10 μg/mL. Grow overnight at 37 ° C with shaking (220–250 rpm).
2. Transfer 2 mL of the bacterial culture to a microcentrifuge tube and pellet the cells by centrifugation at 10,000 rpm for 20 min. (*Optional: prepare a glycerol stock by mixing 0.5 mL of the overnight culture with 0.2 mL of sterile 50% glycerol in a cryovial. Store at −80 °C*).
3. Discard the supernatant and resuspend the cell pellet in 250 μL of Resuspension Buffer.
4. Add 250 μL of Alkaline Lysis Buffer. Mix gently by inverting the tube 4–5 times and incubate at room temperature for 5 min. The suspension should become translucent, indicating cell lysis.
5. Add 250 μL of Neutralization Buffer. Mix gently and incubate on ice for 15 min. A thick white precipitate containing protein and *E. coli* genomic DNA will form.

6. Centrifuge at 10,000 rpm for 20 min at 4 °C. During centrifugation, label a new Eppendorf tube and add 0.8 mL of absolute isopropanol.
7. Carefully transfer the supernatant to the tube containing isopropanol, avoiding the white precipitate. Mix gently by inverting the Eppendorf tube a few times and incubate on ice for 20 min.
8. Centrifuge the sample at 12,000 rpm for 20 min at 4 °C. Discard the supernatant and wash the pellet with 0.7 mL of 70% ethanol. Centrifuge for 5 min at 10,000 rpm at RT.
9. Carefully remove the supernatant without disturbing the pellet.
10. Air dry the pellet briefly (avoid over-drying) and resuspend the DNA in 40 μL of TE buffer (*see* **Note 7**).
11. Store the bacmid DNA at 4 °C for up to 2 weeks or aliquot and store at −20 °C (*see* **Note 8**). Confirm the presence of the bacmid by agarose gel electrophoresis, which should reveal a high molecular weight band corresponding to the recombinant bacmid DNA. For optimal resolution of bacmid DNA, use a 0.8% agarose gel and run at low voltage (50–70 V).

3.1.3 Validation of Successful Transposition

To confirm successful transposition of the target gene into the bacmid, perform PCR using M13/pUC primers. These primers flank the Tn7 transposition site and amplify a region that includes the target inserted gene. A successful PCR will yield a product of the expected size (expected product: gene of interest + 400 bp of flanking vector sequence), confirming the presence of recombinant bacmid. In contrast, colonies without an insert result in a 400 bp band, or no PCR product.

1. Prepare the PCR reaction mix as follows:

   1 μL of bacmid DNA (from Subheading 3.1.2, **step 10**, typically at a concentration of 50–100 ng/μL).

   10 μM each of M13 forward and reverse primers.

   PCR master mix (including dNTPs, Taq polymerase, and buffer).

2. Run the PCR with the following cycling conditions:
   - Initial denaturation: 95 °C for 5 min.
   - 30 cycles of: 95 °C for 30 s, 55 °C for 30 s, 72 °C for 1 min per kb of expected product.
   - Final extension: 72 °C for 10 min.
3. Analyze the PCR products by agarose gel electrophoresis. A single band of the expected size confirms successful transposition.

3.2 Baculovirus Production and Amplification

Successful recombinant protein expression in insect cells depends on the generation of high-quality baculovirus stocks. This section describes the steps for producing and amplifying recombinant baculovirus, from the transfection of Sf9 cells to the generation of high-titer P2 virus stocks. Proper handling of cell cultures, optimization of infection conditions, and careful monitoring of viral amplification are critical for achieving high yields of recombinant protein.

3.2.1 Transfection of Sf9 Cells with Recombinant Bacmid DNA for P0 (Passage Zero) Virus Production

Seed adherent Sf9 cells in a 6-well plate at a density of 1×10^6 cells/mL in 2 mL Sf900II™ SFM medium per well. Ensure cells are healthy (>95% viability) and in the log phase of growth at the time of transfection.

1. Prepare transfection mix:
   - Dilute 1 μg of recombinant bacmid DNA (ERAP1, ERAP2, or IRAP) in 100 μL of IPL-41 medium.
   - In a separate tube, dilute 6 μL of Cellfectin® II in 100 μL of IPL-41 medium.
   - Combine the diluted DNA and Cellfectin® II solutions. Mix gently and incubate at RT for 30 min to allow complex formation.
2. Aspirate medium from cells and replace with 1 mL of fresh IPL-41 medium.
3. Add the transfection mixture dropwise to the Sf9 cells. Gently swirl the plate to ensure even distribution. Incubate the plate at 27 °C for 5 h.
4. After 5 h, replace the transfection mix with 2 mL of fresh filtered Sf900II™ SFM medium containing 10 μg/mL gentamicin. Incubate the cells at 27 °C for 5–7 days, monitoring daily for cytopathic effects (cell rounding, detachment and enlargement) (*see* **Note 9**).

3.2.2 Harvest P0 Virus Stock

1. After 5–7 days, collect the supernatant from the transfected Sf9 cells. This supernatant contains the P0 virus stock.
2. Centrifuge the supernatant at 1000 rpm for 5 min to pellet cell debris.
3. Filter the supernatant through a 0.22 μm sterile filter to remove any remaining debris.
4. Store the P0 virus stock at 4 °C for short-term use (up to 1 month) or aliquot and store at −80 °C for long-term use.

3.2.3 Production of P1 (Passage 1) Virus Stock

1. Seed Sf9 cells in a flask at a density of 1×10^6 cells/mL in Sf900II™ SFM medium.

2. Infect cells with the filtered P0 virus stock at a 1:50 dilution (e.g., 1 mL of P0 virus per 49 mL of cell culture). It is recommended to use more than one P0 virus stock, as some constructs may require further amplification.
3. Incubate the infected cells at 27 °C for 72 h, until the cytopathic effects are observed.
4. Collect the supernatant and centrifuge at 1000 rpm for 10 min to remove cell debris.
5. Filter the supernatant through a 0.22 μm sterile filter. Store the P1 virus stock at 4 °C or aliquot and store at −80 °C.

3.2.4 Production of High-Titer P2 (Passage 2) Virus Stock (See Note 10)

1. Seed Hi5 cells in a 250 mL Erlenmeyer flask at a density of 1×10^6 cells/mL in 50 mL of Sf900II™ SFM medium. Ensure cells are healthy and in the log phase of growth.
2. Infect cells with P1 virus: Add 1 mL of filtered P1 virus stock (1:50 dilution) to the Hi5 cells. Alternatively, use a MOI (multiplicity of infection) of 2–5 for optimal infection. It is recommended to test different MOIs to determine the best conditions for your protein of interest. Gently swirl the flask to ensure even distribution of the virus.
3. Incubate the infected Hi5 cells at 27 °C with gentle shaking (120 rpm) for 48–72 h (*see* **Note 11**). Monitor cell viability and morphology daily.
4. Harvest P2 virus: After 48–72 h, collect the culture supernatant and centrifuge at 1000 rpm for 10 min to pellet cells.
5. Filter the supernatant through a 0.22 μm sterile filter to remove any remaining debris.
6. Store the P2 virus stock at 4 °C for short-term use (up to 1 month) or aliquot and store at −80 °C for long-term storage.

3.3 Small-Scale Protein Expression in Hi5 Cells

Small-scale protein expression in Trichoplusia ni (Hi5) cells is a critical step for evaluating recombinant protein production before scaling up. This section describes the process of infecting Hi5 cells with recombinant baculovirus, optimizing expression conditions, and harvesting the expressed protein. Proper handling of cells, optimization of infection parameters, and monitoring of protein expression are essential for achieving high yields.

1. Maintain Hi5 cells: Grow Hi5 cells in suspension culture in Sf900II™ SFM medium at 27 °C with gentle shaking (120 rpm).
2. Subculture cells every 3–4 days, maintaining a density between 0.5×10^6 and 3×10^6 cells/mL to ensure cells remain in exponential growth.
3. Seed cells for infection: On the day before infection, seed Hi5 cells in a 250 mL Erlenmeyer flask at a density of 0.5×10^6

cells/mL in 50 mL of Sf900II™ SFM medium. Ensure cells are healthy and in the log phase of growth (viability >95%).

4. Thaw the P2 virus stock on ice (if frozen) immediately before use.
5. Infect Hi5 cells: Add 1–2 mL of P2 virus stock (1:50 to 1:25 dilution) to the Hi5 cells. Alternatively, use a MOI (multiplicity of infection) of 2–5 for optimal infection.
6. Gently swirl the flask to ensure even distribution of the virus.
7. Incubate the infected Hi5 cells at 27 °C with gentle shaking (120 rpm) for 72 h.
8. Use a microscope and Trypan Blue exclusion to assess cell viability. A sharp decline in viability (e.g., <50%) after 72 h typically indicates successful infection and recombinant protein expression (*see* **Note 12**).

3.4 Purification of Aminopeptidases ERAP1, ERAP2 & IRAP from Insect Cell Supernatant

Following expression in Hi5 insect cells, ERAP1, ERAP2 and IRAP are secreted into the culture medium due to the presence of the N-terminal signal sequence. The presence of a C-terminal His_{10} tag enables efficient purification using Ni-NTA affinity chromatography. When higher purity is needed, an additional step of Size-Exclusion Chromatography (SEC) can be performed. The following protocol provides a detailed workflow for purifying ERAP1, ERAP2, and IRAP from insect cell culture supernatant, yielding high-purity and active recombinant enzymes suitable for downstream biochemical and structural studies (*see* Fig. 1).

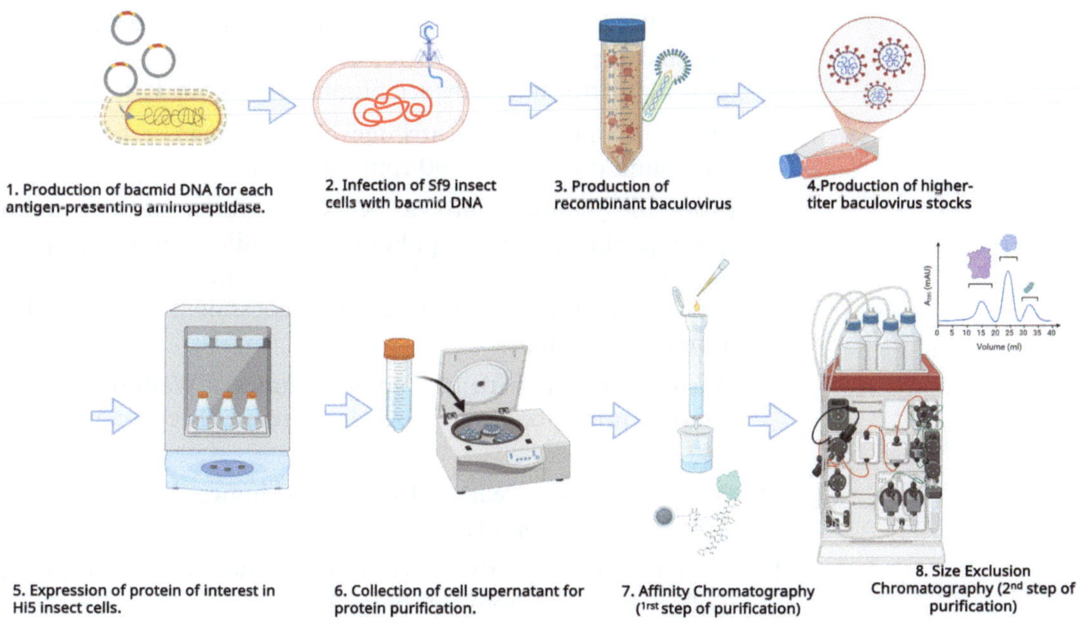

Fig. 1 Workflow of antigen presenting aminopeptidases expression and purification. (Created in BioRender. https:/BioRender.com/c01l066)"

3.4.1 Purification of Aminopeptidases ERAP1, ERAP2 & IRAP

1. Transfer the cell culture supernatant to a 50 mL Falcon tube. Centrifuge at 4500 rpm for 15 min at 4 ° C to pellet cells and clarify the culture medium.
2. Collect the supernatant containing the secreted protein and transfer it to a fresh 50 mL Falcon tube. Centrifuge again at 4500 rpm for 45 min at 4 °C to remove any remaining particles. Optionally, filter the supernatant through a 0.45-micron (μm) filter.
3. Supplement the clarified supernatant with PMSF to a final concentration of 0.1 mM and ¼ tablet of complete protease inhibitor cocktail per 50 mL to protect the protein from degradation.
4. Dialyze supernatant against 2 L of Dialysis Buffer **A** at 4 °C (3 changes, 2 h each, with the final dialysis step performed overnight). Use 10 kDa MWCO dialysis tubing. Retain pre- and post-dialysis samples for SDS-PAGE and enzymatic activity assessment.
5. Transfer the dialysate from the tubing to a new Falcon tube. Adjust the composition to the following final concentrations:
   - 50 mM NaP pH 8.0
   - 300 mM NaCl
   - 20 mM imidazole
6. Ni-NTA resin preparation: Equilibrate Ni-NTA resin by washing with 10 resin volumes of Equilibration buffer. Allow the buffer to drain by gravity.
7. Batch Binding: Incubate the Ni-NTA agarose resin with the adjusted dialysate at 4 °C for 2 h with gentle mixing on a rotating platform to enhance binding (*see* **Note 13**).
8. After binding, gently centrifuge Ni-NTA agarose beads with the bound protein at 1000 rpm for 2 min.
9. Load the sample onto a Ni-NTA gravity flow column (pre-equilibrated with equilibration/binding buffer) to pack the resin. Collect the flow through and analyze it by SDS-PAGE and enzymatic activity assessment to check for unbound protein.
10. Wash the column with 10 column volumes of Wash Buffer **1** to remove non-specifically bound proteins (keep sample for analysis).
11. Wash the column with 10 column volumes of Wash Buffer **2** (keep sample for analysis).
12. Elute the protein with 3–5 column volumes of Elution Buffer. Collect 1 mL fractions (*see* **Note 14**).

13. Wash the column with 10 column volumes of Wash Buffer **3** (keep sample for analysis).
14. Perform quality evaluation of all the samples collected and elution fractions.

    Quality evaluation:

    - Assess protein purity with SDS-PAGE and Coomassie staining.
    - Assess enzymatic activity by following the hydrolysis of small synthetic fluorogenic substrates (*see* Subheading 3.5).
15. Collect and pool the eluted fractions with the highest enzymatic activity and purity and dialyze against Dialysis Buffer **B** overnight at 4 °C to remove imidazole. Alternatively, you can proceed to buffer exchange using a desalting column.
16. Stabilize protein by adding glycerol to a final concentration of 10% (v/v). Aliquot and store at −80 °C or flash freeze in liquid nitrogen.

3.4.2 Size-Exclusion Chromatography (SEC)

To enhance purity and remove aggregates, perform size-exclusion chromatography after Ni-NTA purification. SEC ensures removal of aggregates and contaminants, improving protein homogeneity for biochemical assays or crystallization.

1. Equilibrate the SEC column with 2–3 column volumes of SEC buffer at a flow rate of 0.5 mL/min.
2. Concentrate the eluted protein from Ni-NTA purification to a final volume of 500 μL using an Amicon Ultrafilter (10–30 kDa MWCO).
3. Centrifuge the sample at 10,000×*g* for 10 min to remove precipitates.
4. Inject 500 μL of concentrated protein onto the column.
5. Monitor UV absorbance at 280 nm to track protein elution.
6. Collect 1 mL fractions containing monomeric protein peaks and analyze by SDS-PAGE. Additionally, assess enzymatic activity by monitoring the hydrolysis of small synthetic fluorogenic substrates.
7. Aliquot purified protein and store at −80 °C with 10% (v/v) glycerol for stability or flash freeze in liquid nitrogen.

B. **Functional Analysis of Aminopeptidases**

Endoplasmic Reticulum Aminopeptidases (ERAPs) and IRAP are critical enzymes in antigen processing and immune regulation, making their functional characterization essential for elucidating their biological roles and evaluating their potential as therapeutic targets. To study their enzymatic activity, substrate specificity, kinetic properties and response to inhibitors, a range of functional

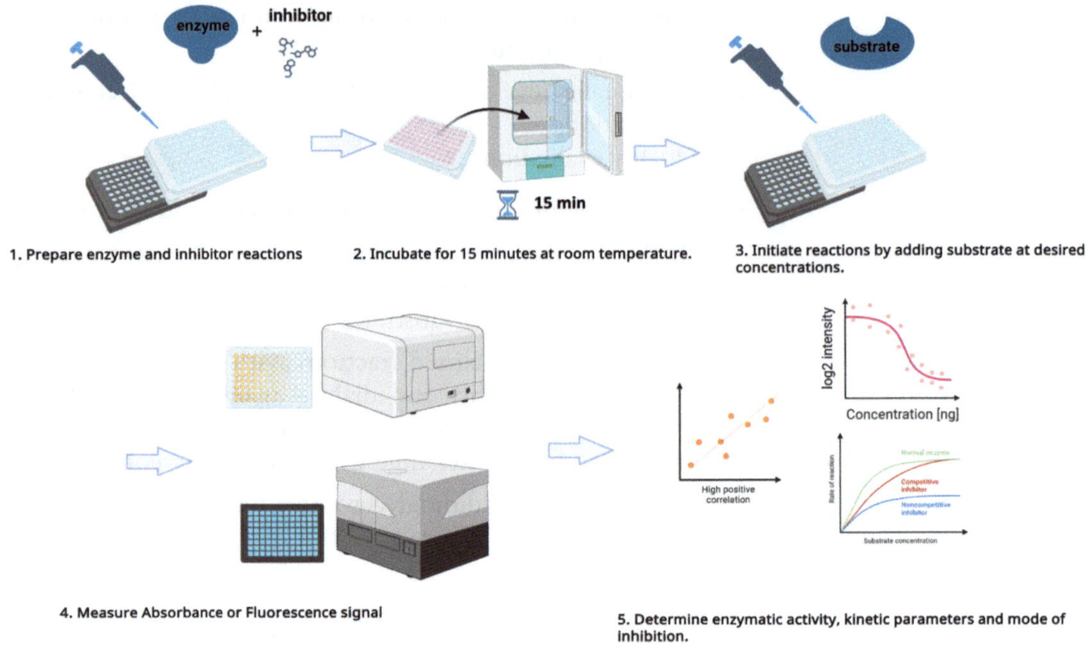

Fig. 2 Brief pipeline of fluorogenic and colorimetric assays. (Created in BioRender. https:/BioRender.com/c01l066)

assays have been developed. This section outlines three widely used methods: fluorogenic assays, colorimetric assays, and HPLC-based assays (*see* Fig. 2). Each approach offers distinct advantages—ranging from high-throughput screening to precise kinetic measurements—and can be used individually or in combination to comprehensively characterize the enzymatic properties of ERAP1, ERAP2, and IRAP.

3.5 Fluorogenic Assays

Principle Fluorogenic assays are a powerful and widely used tool for measuring enzymatic activity and studying inhibitors of aminopeptidases. These assays rely on the hydrolysis of a non-fluorescent substrate to release a fluorescent product (e.g., 7-Amino-4-methylcoumarin, AMC). Fluorescence intensity correlates with product formation, enabling enzyme activity quantification. This protocol covers:

1. *Enzymatic activity determination: Measuring the rate of substrate hydrolysis.*
2. *Specific activity calculation: Normalizing enzyme activity to protein concentration.*
3. *AMC standard curve.*

4. *Inhibitor studies:*
   - *IC_{50} determination: Calculating the concentration of inhibitor required to reduce enzyme activity by 50%.*
   - *Mode of inhibition: Determining the mechanism of inhibitor action.*

3.5.1 Enzymatic Activity Determination

1. In a black 96-well microplate, assemble reactions by adding the following to each well:
   - Working substrate solution: Add to a final concentration near the Km value (e.g., 50 μM for L−/R-AMC).
   - Enzyme solution: Add enzyme solution (1–2 μL, empirically determined) to initiate the reaction. The amount should yield a linear fluorescence increase within the measurement time frame, typically corresponding to <80% substrate hydrolysis.
   - Assay buffer: Adjust volume to achieve the desired final reaction volume (e.g., 150 μL).
   - Include substrate-only controls (no enzyme) to assess background fluorescence.
2. Gently mix the reaction components by pipetting up and down or briefly shaking the plate.
3. Measure the fluorescence intensity using a microplate reader with the following settings:
   - Excitation wavelength (Ex): 380 nm.
   - Emission wavelength (Em): 460 nm.
   - Gain setting: Manually set to ensure signals are within the linear range of detection.
4. Record the relative fluorescence intensity (RFU) for each well.
5. After running the assay, export the fluorescence data (RFU) and corresponding time points for each well from the microplate reader (e.g., Tecan) to Excel.
6. In Excel, highlight the two columns (time and RFU) for a single well.
7. Insert Scatter Plot: Go to the Insert tab in Excel. Select Scatter Plot (choose the option with straight lines and markers).
8. Perform Linear Regression: Right-click on the data points in the scatter plot and select Add Trendline. Choose Linear as the trendline type.
9. Check the box for Display Equation on Chart.
10. Extract the slope. The equation of the trendline will appear on the chart in the form:

$$y = a \cdot x + b,$$

where

y = fluorescence intensity (RFU)

a = slope (RFU/min, representing the initial reaction rate)

x = time (min)

b = y-intercept

11. Record the value of a (the slope) for each well. The resulting value represents the relative enzyme activity in RFU/min.

3.5.2 Specific Activity Calculation

Enzymatic activity is a measure of the catalytic efficiency of an enzyme, defined as the rate at which it converts a substrate into a product. This activity is typically quantified in Relative Fluorescence Units per minute (RFU/min). These values reflect the raw activity of the enzyme under specific experimental conditions and are directly proportional to the amount of product generated over time. However, to compare the purity and efficiency of enzyme preparations across different experimental conditions, a standardized metric known as specific activity is used. Specific activity normalizes enzymatic activity to the total protein concentration and is commonly expressed in units such as nmol/min/μg or μmol/min/μg.

1. Determine the total protein concentration [P] in your sample using a suitable quantification method (e.g., BSA standard curve, BCA protein assay, Nanodrop, interpolation from a standard curve using a known enzyme concentration, etc.).
2. Perform the enzymatic assay to measure the rate of product formation as described in Subheading 3.5.1. Record the activity at RFU/min. Ensure the measurements are within the linear range of both the protein quantification assay and the enzymatic assay to avoid saturation effects.
3. Convert the raw fluorescent signal to AMC concentration by using an AMC standard curve (*see* Subheading 3.5.3). Use the equation from the AMC standard curve ($y = mx + b$) to convert raw fluorescence values into AMC concentration (μM) by solving for x. Proceed with calculating the amount of AMC released per unit time to determine reaction velocity V (μM/min).
4. Normalize the enzymatic activity by dividing the reaction velocity (V) by the total protein concentration ([P]) determined in **step 1**. The specific activity (SA) can be calculated using the following formula:

$$\mathrm{SA} = \frac{V}{[P]},$$

where

SA = specific activity (nmol/min/μg or μmol/min/μg)

V = reaction velocity (μM/min)

$[P]$ = total protein concentration (μg/mL or mg/mL)

3.5.3 AMC Standard Curve

The AMC standard curve is essential for accurately determining aminopeptidase activity in fluorometric assays. Free AMC, a highly fluorescent compound, serves as a reference to correlate Relative Fluorescence Units (RFU) with absolute concentrations of the hydrolysis product. This correlation is useful for:

- Quantifying enzyme activity and specific activity.
- Conducting kinetic studies, such as determining reaction rates, enzyme efficiency, and inhibitor effects.
- Normalizing measurements across different instruments, experimental setups, and reading settings, as fluorescence output can vary significantly.

Preparation of AMC Standard Solutions

1. Stock solution: Prepare a 10 mM AMC stock solution by dissolving AMC in DMSO. Vortex for 30 s to ensure complete AMC dissolution. Aliquot the stock solution into small volumes and store at −20 °C in a desiccated environment, protected from light, to prevent degradation. Avoid repeated freeze-thaw cycles.
2. Prepare three different concentrations of AMC (1, 10 & 100 μM) by serial diluting the stock solution in assay buffer (*see* **Note 15**). Ensure thorough mixing after each dilution step.

Fluorescence Measurement

1. Prepare an "in well" dilution series of AMC standard in the concentration range of 0.01–10 μM (by using the intermediate concentrations) in a 100 μL final volume of assay buffer. Include a blank well containing only assay buffer (0 μM AMC) to account for background fluorescence. Perform reactions at least in duplicates to ensure reproducibility (*see* Table 2 and **Note 16**).
2. Measure fluorescence intensity of the AMC standard solutions containing different amounts of AMC using a microplate reader set at Ex 380 nm/Em 460 nm. Set gain manually (*see* **Note 17**).
3. Average the duplicate readings for each AMC concentration.

Table 2
AMC standard curve calculations

Final AMC concentration (μM)	Volume of intermediate AMC concentration (μL)	Assay buffer (μL)
0 (blank)	0	100
0.01	1 μL from 1 μM AMC stock	99
0.025	2.5 μL from 1 μM AMC stock	97.5
0.05	5 μL from 1 μM AMC stock	95
0.1	10 μL from 1 μM AMC stock	90
0.25	2.5 μL from 10 μM AMC stock	97.5
0.5	5 μL from 10 μM AMC stock	95
1	10 μL from 10 μM AMC stock	90
2.5	2.5 μL from 100 μM AMC stock	97.5
5	5 μL from 100 μM AMC stock	95
10	10 μL from 100 μM AMC stock	90

4. Subtract the average fluorescence intensity of the blank (0 μM AMC) from all readings to account for background fluorescence.
5. Plot the fluorescence intensity (y-axis) against the corresponding AMC concentration (x-axis) using graphing software such as Excel or GraphPad Prism (*see* **Note 18**).
6. Determine slope of the plot by applying linear regression analysis to obtain the equation

$$y = mx + b,$$

   where

   y = fluorescence intensity

   m = slope of the standard curve (fluorescence per μM AMC)

   x = AMC concentration

   b = y—intercept (should be close to zero)

   Ensure the R^2 value is close to 1.0, indicating a strong linear relationship.

- Convert fluorescence intensity from enzymatic assays to AMC concentration using the equation from the regression line: $y = mx + b$.
- Calculate enzyme activity in nmol/min/μg using the converted AMC concentrations and the reaction time.

3.5.4 IC_{50} Determination

The half-maximal inhibitory concentration (IC_{50}) is a key metric used to quantify the potency of enzyme inhibitors. It represents the concentration of inhibitor required to reduce enzyme activity by 50%. This section describes a step-by-step protocol for determining IC_{50}, including inhibitor dilution, activity measurement, and data analysis.

1. Prepare a series of inhibitor dilutions spanning a range above and below the expected IC_{50} value. Use $\log_{10}$ dilutions (e.g., 0.1 nM to 100 μM) to ensure a broad concentration range. Maintain a consistent final solvent concentration (e.g., ≤1% (v/v) DMSO) across all wells to avoid solvent-induced artifacts.
2. In a 96-well plate, mix the inhibitor with the enzyme solution (*1–10 nM, depending on specific activity. As a rule, enzyme concentration should achieve 20–80% substrate conversion in the absence of inhibitor*) and incubate for a fixed period (e.g., 10–30 min) at the assay temperature. This allows sufficient time for inhibitor binding to the enzyme.
3. Add the substrate to the enzyme-inhibitor mixture in each well and initiate the reaction. Ensure the substrate concentration is at or below the Km value to maintain sensitivity to inhibition. Gently mix the plate to ensure uniform distribution of reactants and to avoid edge effects.
4. Include appropriate control wells: substrate-only, enzyme-only, inhibitor-only.
5. Measure the fluorescence intensity (or product formation rate) at regular intervals using a fluorescence microplate reader or spectrophotometer (Ex: 380 nm, Em: 460 nm). Continue measurements until the reaction reaches the linear phase (typically 5–30 min, depending on the enzyme kinetics).
6. Normalize the measured activity to the control (no inhibitor) to calculate percent inhibition, using the formula:

$$\text{Percent (\%) Inhibition} = (1 - \text{Activity of inhibitor}/\text{Activity of control})^{*}100.$$

Data Analysis

7. Plot percent inhibition versus inhibitor concentration on a logarithmic scale (x-axis: inhibitor concentration; y-axis: percent inhibition).
8. Fit the data to a sigmoidal dose-response curve using nonlinear regression analysis in software like GraphPad Prism or equivalent. The standard equation for the dose response curve is:

$$Y = \text{bottom} + \left(\text{Top} - \text{Bottom}/1 + 10^{((\log 10(\text{IC50}) - X) * \text{Hill Slope}}\right).$$

where Y is the enzymatic activity/percent inhibition, X is $\log_{10}$ of the inhibitor concentration and Hill Slope is the slope factor describing the steepness of the curve. Top and Bottom are the maximum and minimum Y values in the graph, constrained as 1 and 0, respectively, and IC_{50} is the inhibitor concentration needed to acquire half-maximal enzyme activity [8].

9. Determine the IC_{50} value (the concentration of inhibitor required to reduce enzyme activity by 50%). This is derived from the midpoint of the fitted curve.

3.5.5 Determination of Mode of Inhibition

1. Prepare a series of substrate concentrations in assay buffer, spanning a range from 0.2× to 5× Km (5–200 μM).
2. Prepare at least four inhibitor concentrations, such as 0 (control), ½ IC_{50}, IC_{50}, and 2× IC_{50}.
3. In a 96-well plate, add the enzyme (concentration depending on specific activity) to each well containing the substrate and inhibitor combinations. Ensure each substrate concentration is tested with each inhibitor concentration and enzyme's concentration maintains a linear reaction rate.
4. Measure the rate of reaction (fluorescence intensity) over time for each substrate and inhibitor combination at regular intervals using a fluorescence plate reader.
5. Record the data in an Excel datasheet for analysis.
6. Plot the data (initial velocities V versus substrate concentration [S]) in GraphPad Prism, using the Michaelis-Menten kinetics equation:

$$V = V\max[S]/Km + [S]$$

to determine Km, Vmax and Kcat values.

- For visual analysis, prepare Lineweaver-Burk (double reciprocal) plots: $1/V$ versus $1/[S]$ for each inhibitor concentration.
- Analyze the pattern of the lines to determine the mode of inhibition.

3.6 Colorimetric Assays

Principle Colorimetric assays are widely used to measure aminopeptidase activity by utilizing chromogenic substrates like L-leucine p-nitroanilide (Leu-pNA). Upon enzymatic cleavage, these substrates release a measurable colored product (yellow pNA), which can be quantified spectrophotometrically [9, 10]. The intensity of the color is directly proportional to the enzymatic activity, allowing for straightforward quantification. While colorimetric assays are less sensitive than fluorogenic assays, they are cost-effective, and are easy to perform and provide reliable results. These assays are particularly useful for determining the mode of action of aminopeptidase inhibitors [11].

3.6.1 Determination of Enzymatic Activity

1. Thaw and dilute the purified enzyme in assay buffer to achieve a working concentration of 1–10 nM (optimize for each enzyme to yield linear product formation). Keep the enzyme on ice until use.
2. In a 96-well plate, set up the reactions as follows:
   (a) Add enzyme and substrate (final concentration: 100–500 μM) in a final reaction volume of 150 μL assay buffer.
   (b) Include a control sample with assay buffer instead of the enzyme to account for any background absorbance.
   (c) Perform each reaction in at least duplicates to ensure reproducibility.
3. Measure the absorbance at 405 nm using a microplate reader.
4. Record the absorbance at regular intervals (e.g., every 1–2 min) for up to 30 min to monitor the reaction progress.

Data Analysis

Initial Velocity (V_0) Calculation

5. Plot the absorbance values versus time.
6. Identify the linear portion of the curve (typically the first 5–10 min of the reaction).
7. Calculate the slope of the linear region to determine the initial velocity (V_0) in units of absorbance change per minute (ΔA/min).

Product Concentration Calculation

8. Convert the absorbance values to product concentration using the molar extinction coefficient of pNA ($\varepsilon = 9900\ M^{-1}\ cm^{-1}$) [12].
9. Use the formula: *Product concentration (M) = Absorbance (405 nm)/[Path length (cm) × Extinction coefficient (ε)].*

   For a 96-well plate, assume a path length of 0.6 cm unless otherwise specified. Alternatively, you can construct a pNA standard curve, as in Subheading 3.5.3.
10. Express the enzymatic activity as the amount of product formed per unit time (e.g., μmol/min).

3.6.2 Kinetic Inhibition Assay

1. Prepare a series of substrate concentrations ranging from 50 μM to 8 mM in assay buffer. These concentrations correspond to 2× concentration of the desired final working concentration (*see* **Note 19**).
2. Prepare inhibitor dilutions to achieve final working concentrations of ½ IC_{50}, IC_{50} and 2× IC_{50} in the reaction mixture.
3. In a 96-well plate add 75 μL assay buffer solution containing enzyme (to achieve a final concentration of 1–10 nM in a 150 μL reaction volume) and inhibitor (or buffer for control) and pre-incubate the mixture at 37 °C for 10–15 min to allow the inhibitor to bind to the enzyme.
4. Initiate reaction by adding 75 μL of substrate solution. Mix gently by pipetting up and down, avoiding the formation of bubbles.
5. Monitor absorbance at 405 nm for 30 min, recording data at regular intervals (e.g., every 1–2 min).

Data Analysis

6. Export absorbance values versus time to Excel or similar software. For each inhibitor and substrate concentration, calculate the slope of the linear portion of the curve to determine the initial velocity (V_0).

$$V^0 = \frac{\Delta Abs}{\Delta t} \times 1/\varepsilon.$$

7. Open GraphPad Prism or a similar software. Select XY analysis and then nonlinear regression.
8. Enter the substrate concentrations on the X-axis and the corresponding V_0 values on the Y-axis.
9. Fit the data to the Michaelis-Menten equation and determine the kinetic parameters.
10. Confirm the mode of inhibition by Lineweaver-Burk analysis. Compare the plots and kinetic parameters (*Km*, *V*max) for reactions with and without the inhibitor.

3.7 HPLC-Based Assays

Principle HPLC-based enzymatic assays provide high specificity and are suitable for detailed analysis of peptide substrates and products. These methods are particularly useful as they allow the direct assessment of the enzymatic activity of ERAPs against physiologically relevant substrates, a tool crucial for understanding their role in antigen processing and autoimmunity. The following protocol describes three HPLC-based applications: (a) monitoring aminopeptidase ability to process peptide substrates through analysis of substrate cleavage patterns, (b) determination of kinetic parameters: measurement of reaction rates and substrate affinity, and (c) assessment of inhibitor potency: evaluation of inhibitor efficacy and mode of action [8, 13–15].

3.7.1 Determination of Trimming Activity

1. Prepare enzyme reactions in a total volume of 100 μL, containing enzyme (5–400 nM) and peptide substrate (10–100 μM) (*see* **Note 20**). Perform all reactions in duplicates or triplicates.
2. Incubate at 37 °C for 30–60 min (optimize time depending on enzyme kinetics).
3. Quench the reaction by adding an equal volume of 0.1% (v/v) TFA (acidification) to lower the pH and denature the enzyme (final pH ~2.5). Flash-freeze the reactions at −20 °C or −80 °C until analysis.
4. Thaw each sample and centrifuge at 10,000 rpm for 5 min to remove precipitates.
5. Inject the reaction mixture onto the HPLC system (*See* **Note 21**).
6. Perform elution using a linear 5–50% ACN gradient (*see* Table 3).
7. Monitor peptide cleavage and product formation by following UV absorbance (*see* Fig. 3).

Data Analysis

8. Quantify substate cleavage or hydrolysis product formation by integrating chromatographic peak areas.
9. Calculate specific activity with GraphPad Prism or similar software by fitting the results in the one-phase decay equation $Y = Y0e - kx$, where x is the reaction time, Y_0 the substrate

Table 3
Representative HPLC method

Time (min)	% Solvent A	% Solvent B
0.00	100	0
8.00	100	0
26.00	0	100
30.00	0	100
30.50	100	0
35.00	100	0
35.50	100	0
36.50	100	0

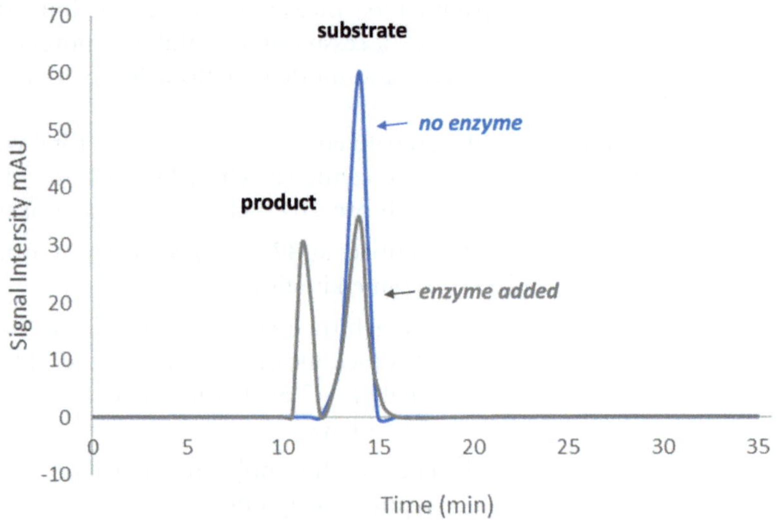

Fig. 3: A typical HPLC chromatogram showing the substrate peak before aminopeptidase addition (blue line) and the substrate and product peaks after aminopeptidase addition (grey line)

fraction left intact after time zero (constrained as $\Upsilon = 1$), Υ the substrate fraction left intact after the end of the reaction, and k the reaction rate constant.

10. For a higher resolution analysis of enzymatic cleavage products or for the identification of novel substrates, HPLC can be combined with LC-MS/MS.

3.7.2 Kinetic Studies

1. Prepare a range of substrate concentrations (e.g., 10–100 μM).
2. Incubate the enzyme with each substrate concentration at 37 ° C for 30–60 min (ensure <30% substrate turnover).
3. Stop the reactions by adding 0.1% (v/v) TFA.
4. Proceed as described in Subheading 3.7.1.

Data Analysis

5. Integrate the chromatographic peaks corresponding to product formation.
6. Calculate initial velocity (V_0) based on product peak areas.
7. Plot initial velocity vs substrate concentration.
8. Fit data to the Michaelis-Menten equation using nonlinear regression.
9. Determine kinetic parameters (Km, Vmax).

3.7.3 Evaluation of Inhibitor Potency

1. Prepare a range of inhibitor concentrations (e.g., serial dilutions in $\log_{10}$ scale, 1 nM–100 μM).
2. Pre-incubate the enzyme with the increasing inhibitor concentrations at 37 °C for 15 min.
3. Add substrate at a fixed concentration (e.g., ~ Km) to initiate the reaction and incubate at 37 °C for 30–60 min.
4. Stop the reactions by adding 0.1% (v/v) TFA.
5. Proceed as described in Subheading 3.7.1.

Data Analysis

6. Integrate the chromatographic peaks corresponding to product formation.
7. Normalize all peak areas to the control reaction without inhibitor (100% activity). Express activity as a percentage of the control reaction.
8. Relative activity (%) = (Peak area sample/Peak area control) × 100.
   - Plot enzyme activity versus inhibitor concentrations (or log [inhibitor]).
   - Fit data to a 4-parameter equation (sigmoidal dose-response curve) with GraphPad Prism or similar software to determine IC_{50} value.
   - Confirm the mode of inhibition by Lineweaver-Burk analysis, comparing kinetic parameters in the presence and absence of the inhibitor.

4 Notes

1. Ensure that antibiotic stock solutions are freshly prepared and properly stored to maintain efficacy. Degraded antibiotics can lead to contamination or failed selection. Antibiotic stock solutions should be filter-sterilized using a 0.22-μm filter. Tetracycline and X-gal/Bluo-gal are light sensitive; store aliquots at – 20 °C and protect from light.
2. Optimize gradient elution and mobile phase buffer composition for optimal separation of substrate and product peaks.
3. Filter and de-gas HPLC buffers before use to protect the integrity of the column and avoid high background signal.
4. For optimal transformation efficiency, use freshly thawed competent cells and avoid prolonged heat-shock times. Overheating can reduce cell viability.

5. SOC medium is recommended for optimal transformation efficiency as it enhances cell recovery after heat shock. If unavailable, LB medium can be used, but transformation efficiency may be reduced by two to threefold (Invitrogen) [4].
6. Select large, well-isolated white colonies to avoid false positives. Avoid gray or dark-centered colonies, as they may contain mixed populations of recombinant and non-recombinant bacmids. If white colonies are scarce, consider increasing the amount of plasmid DNA or optimizing the transformation conditions.
7. Bacmid DNA is large (>135 kb) and prone to shearing. Avoid vigorous pipetting when resuspending the DNA pellet. Gently tap the tube to dissolve the DNA in TE buffer.
8. Store bacmid DNA at 4 °C for short-term use (up to 2 weeks). For long-term storage, store at −20 °C. Repeated freeze-thaw cycles can damage bacmid DNA. Aliquot the DNA before storage to avoid degradation.
9. Virus-infected cells exhibit distinct morphological changes, including enlargement, grainy appearance, and the presence of occlusion bodies (refractive crystals in the nucleus). Over time, infected cells detach from the flask surface and lyse, releasing viral particles into the supernatant. If a cytopathic effect is not observed after 5–7 days, the transfection may have failed. Verify the quality of the bacmid DNA and transfection reagents.
10. The quality of P0, P1 and P2 virus stocks can significantly impact protein expression. If protein yields are low, consider re-isolating the virus or optimizing the infection conditions (e.g., MOI, incubation time). Quality control of baculovirus can be performed by plaque assay or quantitative PCR (qPCR) with primers targeting the baculovirus *gp64* gene to determine the absolute viral titer (genome copies per mL).
11. Harvest times can vary depending on the protein. Monitor cell viability and protein expression levels at 24, 48, and 72 h post-infection to determine the optimal harvest time.
12. A sharp decline in cell viability (e.g., <50%) after 72 h typically indicates successful infection and protein expression. However, it is crucial to avoid complete or excessive cell lysis, as this can lead to the release of proteases and other cellular components that may degrade the recombinant protein or reduce its yield. If cell lysis is excessive, consider reducing the multiplicity of infection (MOI) or shortening the incubation time.
13. If protein binding to the Ni-NTA resin is inefficient, verify the pH and imidazole concentration of the binding buffer. Low pH or high imidazole can reduce binding efficiency.

14. If the protein does not elute efficiently, consider increasing the imidazole concentration in the Elution Buffer (up to 250 mM) or extending the incubation time with the resin.
15. Prepare fresh working solutions, as AMC fluorescence can degrade over time in assay buffer. Additionally, the AMC standard curve must be prepared in the same buffer and conditions as the enzyme reaction to ensure accurate fluorescence conversion.
16. The standard curve should be linear up to ~10 μM AMC; at higher concentrations fluorescence signal saturation may be observed.
17. The change of gain is closely related to changes of the measured values. Therefore, it is important that the standard plate is read at the same gain as the samples, or else the sample values are meaningless.
18. The readings of your experimental values are dependent on the validity of the standard curve. Therefore, it is critical that the R^2 value be close to 1.0.
19. Avoid substrate concentrations >8 mM to prevent solubility issues.
20. Optimization is needed. The substrate should produce a well-formed, detectable peak in the chromatogram. Then adapt enzyme's concentration to ensure that substrate's turnover remains below 20–30% to maintain initial velocity conditions.
21. Ensure proper column equilibration before injection.

Conflicts of Interest

The authors have no conflicts of interest to declare that are relevant to the content of this chapter.

References

1. Hattori A, Tsujimoto M (2013) Endoplasmic reticulum aminopeptidases: biochemistry, physiology and pathology. J Biochem 154: 219–228. https://doi.org/10.1093/jb/mvt066
2. Saveanu L, Van Endert P (2012) The role of insulin-regulated aminopeptidase in MHC class I antigen presentation. Front Immunol 3:1–13. https://doi.org/10.3389/fimmu.2012.00057
3. Goto Y, Hattori A, Ishii Y, Tsujimoto M (2006) Reduced activity of the hypertension-associated Lys528Arg mutant of human adipocyte-derived leucine aminopeptidase (A-LAP)/ER-aminopeptidase-1: 1833–1838 https://doi.org/10.1016/j.febslet.2006.02.041
4. InVitrogen Life Technologies (2006) Bac-to-Bac™ Baculovirus expression system: USER GUIDE, Pub No. MAN0000414
5. Contreras-Gómez A, Sánchez-Mirón A, García-Camacho F, Molina-Grima E, Chisti Y (2014) Protein production using the baculovirus-insect cell expression system. Biotechnol Prog 30:1–18. https://doi.org/10.1002/btpr.1842
6. Sui L, Guo HC (2021) Enhanced recombinant expression and purification of human IRAP for

biochemical and crystallography studies. Biochem Biophys Rep 27:101042. https://doi.org/10.1016/j.bbrep.2021.101042

7. Sano KI, Maeda K, Oki M, Maéda Y (2002) Enhancement of protein expression in insect cells by a lobster tropomyosin cDNA leader sequence. FEBS Lett 532:143–146. https://doi.org/10.1016/S0014-5793(02)03659-1
8. Stamogiannos A, Maben Z, Papakyriakou A, Mpakali A, Kokkala P, Georgiadis D, Stern LJ, Stratikos E (2017) Critical role of interdomain interactions in the conformational change and catalytic mechanism of endoplasmic reticulum aminopeptidase 1. Biochemistry 56:1546–1558. https://doi.org/10.1021/acs.biochem.6b01170
9. Spungin A, Blumberg S (1989) Streptomyces griseus aminopeptidase is a calcium-activated zinc metalloprotein. Eur J Biochem 183:471–477. https://doi.org/10.1111/j.1432-1033.1989.tb14952.x
10. Cahan R, Axelrad I, Safrin M, Ohman DE, Kessler E (2001) A secreted aminopeptidase of Pseudomonas aeruginosa: identification, primary structure, and relationship to other aminopeptidases. J Biol Chem 276:43645–43652. https://doi.org/10.1074/jbc.M106950200
11. Maben Z, Arya R, Rane D, An WF, Metkar S, Bender S, Ali A, Nguyen TT, Evnouchidou I, Stratikos E, Golden J, Stern LJ (2021) Discovery of selective inhibitors of endoplasmic reticulum aminopeptidase 1. J Med Chem 63:103–121. https://doi.org/10.1021/acs.jmedchem.9b00293
12. Greenfield LM, Puissant J, Jones DL (2021) Synthesis of methods used to assess soil protease activity. Soil Biol Biochem 158. https://doi.org/10.1016/j.soilbio.2021.108277
13. Temponeras I, Chiniadis L, Papakyriakou A, Stratikos E (2021) Discovery of selective inhibitor leads by targeting an allosteric site in insulin-regulated aminopeptidase. Pharmaceuticals 14. https://doi.org/10.3390/ph14060584
14. Mpakali A, Georgaki G, Buson A, Findlay AD, Foot JS, Mauvais FX, van Endert P, Giastas P, Hamprecht DW, Stratikos E (2024) Stabilization of the open conformation of insulin-regulated aminopeptidase by a novel substrate-selective small-molecule inhibitor. Protein Sci 33:1–18. https://doi.org/10.1002/pro.5151
15. Evnouchidou I, Berardi MJ, Stratikos E (2009) A continuous fluorigenic assay for the measurement of the activity of endoplasmic reticulum aminopeptidase 1: competition kinetics as a tool for enzyme specificity investigation. Anal Biochem 395:33–40. https://doi.org/10.1016/j.ab.2009.07.032

Chapter 10

Methods and Guidelines for Analyzing the Conformational Dynamics of Proteins in the Antigen Presentation Pathway

Athanasios Papakyriakou

Abstract

Endoplasmic reticulum aminopeptidase 1 (ERAP1) digests antigenic precursors up to the optimum length for binding to the major histocompatibility complex (MHC) class I. This zinc-dependent aminopeptidase has been found in a catalytically active closed conformation and in open conformations that expose its substrate-binding cavity to the bulk solvent. Therefore, its activity has been linked to catalytic cycles between conformational states that accommodate substrate binding, peptide hydrolysis and product release. During the last decade, X-ray crystallography in conjunction with biophysical and enzymatic studies has deciphered the remarkable peptide-length dependence of ERAP1, the allosteric regulation mechanism, and the role of disease-associated polymorphisms. In this regard, molecular dynamics (MD) simulations have contributed to atomic-level investigations of ERAP1 conformational dynamics. This chapter provides a detailed protocol to generate the systems required to carry out and analyze atomistic MD simulations of ERAP1 with a bound inhibitor in explicit solvent, starting from different conformational states. This protocol is also applicable to the analysis of the conformational dynamics of other aminopeptidases in the antigen presentation pathway.

Key words Molecular Dynamics (MD) simulations, ERAP1, ERAP2, IRAP, M1 aminopeptidase, Zinc-dependent enzyme, Conformational dynamics, Open–closed state, Allotype, Allosteric regulation

1 Introduction

A critical step in the antigen presentation pathway by the major histocompatibility complex I molecules (MHC-I) is the preparation of precursor antigenic peptides for loading onto the MHC-I. This step comprises the proteolytic trimming of N-terminally extended precursors down to the optimum for the MHC-I length of 8–10 amino acids by a group of zinc-dependent enzymes that belong to the oxytocinase subfamily of M1 aminopeptidases [1, 2]. The two ER-resident aminopeptidases ERAP1 (107 kDa) and ERAP2 (110 kDa) share 51% sequence identity and have been shown to play a key role in antigen processing and presentation by regulating

Efstratios Stratikos (ed.), *Antigen Processing and Presentation: Methods and Protocols*, Methods in Molecular Biology, vol. 3007, https://doi.org/10.1007/978-1-0716-5092-9_10,

the peptide epitopes that will be presented to cytotoxic T lymphocytes (CD8+ T cells). Another member of the same family is the endosomal insulin-regulated aminopeptidase IRAP (117 kDa) with an additional cytosolic domain for targeting to intracellular vesicles, which displays sequence identity of 46% for ERAP1 and 44% for ERAP2. IRAP is mainly implicated in the final peptide trimming of exogenous antigens for cross-presentation by MHC-I in dendritic cells [3]. These homologous enzymes contain a thermolysin-like domain that comprises the Zn(II) active site (domain II), and a conserved HEAT-repeat-like motif that forms a bowl-shaped C-terminal structure (domain IV). The X-ray structures of ERAP1 and IRAP have revealed that these enzymes exist in at least two distinct conformations regarding the orientation of the active site and C-terminal domain [4–11]. In most of these structures, the enzymes displayed a closed conformation where domains II and IV form an internal cavity without access to the bulk solvent, whereas others have captured the enzymes in conformations where domain IV is reoriented, exposing the internal cavity to the bulk solvent. The degree of the internal cavity exposure is variable; still, the open conformations are required for substrate entrance and product release during the catalytic cycles of the enzyme. The closed states exhibited all active site residues in the optimal orientation for catalysis, especially regarding a conserved tyrosine residue (Tyr438 of ERAP1 and Tyr549 of IRAP), which reorients away from the active site in the open conformations. Therefore, the conformational state of these enzymes has been associated with their catalytic activity, where the closed state is the active state [4, 5]. It should be noted that ERAP2 has been crystallized only in the closed conformation [12–17].

ERAP1 displays a very interesting peptide-length dependence that is well-suited to its role in antigenic precursor processing [18]. In particular, ERAP1 trims efficiently peptides of 10–15 amino acids but spares eight-residue or shorter peptides. The "molecular ruler mechanism" is the main hypothesis to explain this particular length dependence, by which optimal peptides have the appropriate length to bind simultaneously at the active site Zn(II) and at a distal site that activates enzyme closure and catalysis [5, 19]. The wealth of X-ray crystal structures during the last decade, in conjunction with biophysical and enzymatic studies, has revealed the molecular basis of this mechanism to a large extent [10, 20]. In these studies, the examination of the conformational dynamics using molecular dynamics simulations (MDs) has played an evident role in providing atomic-level information [21]. Specifically, MDs in combination with enzymatic analysis of disease-associated polymorphisms of ERAP1, namely K528R and Q730E, have been used to study variations that are distal from the active site, but translate to changes in substrate-dependent function of the enzyme [22]. Snapshots from MD simulations of ERAP1 in conjunction with small-angle X-ray scattering (SAXS), site-directed mutagenesis and enzymatic analysis of ERAP1 were used to understand the role of interdomain salt bridges in the catalytic efficiency and allosteric kinetic behavior of ERAP1 [23]. MD simulations in combination with dynamical network analysis of high-resolution ERAP1 structures in complex with transition-state pseudopeptide analogs have been used to probe the dynamical coupling between the distal C-terminal recognition site and the N-terminal binding site of the substrate [20]. MD-generated snapshots of ERAP1 have also been employed in a more recent study of ERAP1 free in solution, and in the presence of substrates, allosteric activators, and inhibitors [10]. Considering the enhanced peptide trimming activity observed with a recombinant ERAP1/ERAP2 dimer that was stabilized by a leucine zipper region [24], we have also investigated the dimerization potential of the two homologous enzymes using MD simulations and estimated free energy of binding calculations [25]. Taken together, these studies have demonstrated the prominent role of conformational dynamics in deciphering the peptide-length dependence mechanism of ERAP1, which, however, exhibits a broad sequence-specificity that is required for its biological function, as well as the observed ligand-dependent allosteric regulation, and the association of polymorphism with disease.

This chapter contains a step-by-step guide to perform MDs of ERAP1, starting from the crystallographic structure of the enzyme in a closed and three open conformations, with the aim of sampling the conformational space as efficiently as possible in the accessible timescales of hundreds of nanoseconds to tenths of microseconds. Special attention has been given to the appropriate preparation of the starting structures, comprising the modeling of missing residues, preparation of an active-site bound inhibitor and calculation of the protonation state of the ionizable groups of ERAP1 in each state. The next sections provide input files and guidelines to prepare the systems and perform the simulations using the Amber suite of biomolecular simulation programs [26]. The last section provides an overview of a typical analysis of the results obtained, including the extraction of time-dependent features of the system, a principal component analysis and clustering of the trajectories. The guide is intended for beginners that have access to a Linux-based workstation, preferably equipped with an NVIDIA graphics processing unit (GPU), and have little experience in bash commands or Python programming. It should be noted that the guide provided here does not contain any description of the principles underlying classical molecular dynamics simulations and molecular mechanics force fields, for which refs [27, 28]. provide very comprehensive overviews.

2 Materials

To reproduce the MDs described here access to a Linux workstation equipped with a NVIDIA GPU is required. At the time of writing, Ubuntu Linux 24.04 LTS (Noble Numbat) is a good choice considering the ease of installing the required NVIDIA CUDA drivers and the auxiliary software through its repositories. The current MD engine of Amber24 [29], namely *pmemd_cuda* [30], as well as the AmberTools25 [31] for non-commercial use can be obtained free of charge from: https://ambermd.org/GetAmber.php. A guide for building efficient GPU workstations is provided here: https://ambermd.org/GPUHardware.php, whereas operating system-specific installation instructions can be found here: https://ambermd.org/Installation.php. It should be noted that the MD protocol provided here is also applicable to older versions of the code, Amber18 and later, without any modification.

The auxiliary open-source software required for preparation of the systems comprises PyMOL and MODELLER [32], which can be installed via the Ubuntu repository using the bash command:

```
apt-get -y install pymol modeller
```

VMD [33], which will be used for visual analysis of the MD trajectories and normal modes from principal component analysis (PCA), can be readily downloaded after registration here:

https://www.ks.uiuc.edu/Development/Download/download.cgi?PackageName=VMD

The H++ server [34] that will be used to calculate the protonation states of the ionizable groups in our systems can be assessed after creating a user account here: http://newbiophysics.cs.vt.edu/H++/index.php

3 Methods

This section provides a detailed, step-by-step guide for the preparation of the systems and the input files required to perform MD simulations of ERAP1. The first four sections describe the steps required to prepare the X-ray structures in appropriate forms for running the simulations, the fifth section provides the input files required to carry out relaxation of the systems and the production runs, whereas the sixth section provides typical examples of processing and analysis of the MD trajectories. With appropriate modifications, the protocol provided here can be used to sample the conformational dynamics of ERAP2, as well as of other M1 aminopeptidases that have been crystallized in distinct conformations, for example the human insulin-regulated aminopeptidase (IRAP) [6, 8, 35], and aminopeptidase N (APN, CD13) [36], or the archaeon enzyme tricorn interacting factor F3 (TIFF3) [37].

3.1 Preparation of the X-Ray Structures

To sample the conformational space of ERAP1 we will employ a starting structure of the enzyme in the closed state and three starting structures in the open state. For the closed state we will use the 1.60-Å resolution X-ray structure with PDB ID: 6q4r [9] and for the open states, the three chains resolved in the asymmetric unit of the most recent X-ray structure with PDB ID: 6mgq [10]. To obtain comparable systems we will use the same inhibitor for all starting structures, and in particular, a potent phosphinic pseudopeptide inhibitor [38] that is bound at the active site of ERAP1 in the closed state. A major issue in the crystallographic structures of ERAP1 is the number of missing residues within the sequence crystallized, especially those encoded by exon 10 and part of exon 11 of the ERAP1 gene, which comprise a 28-amino acid sequence (residues 486–513). These residues have been substituted by a short loop consisting of a –GSG– sequence to assist crystallization of ERAP1 in either state. In a particular ERAP2 structure, however, the exon 10 loop formed an appendix extending from helix 8 of domain II and was further stabilized through a disulfide bond between Cys503 and Cys514 (PDB ID: 5ab0, [14]). We will employ this X-ray structure of ERAP2 as a template to model the missing residues of ERAP1. To do so, first launch PyMOL and type the following commands within the (PYMOL>) console:

```
fetch 6q4r 5ab0, async=0
remove not chain A
align 5ab0, 6q4r and i. 60-450
select 6q4r and i. 45-937+1008 and not altloc B
save erap1_6q4r.pdb, sele
save erap2_5ab0.pdb, 5ab0
select r. HJ5
copy_to lig, sele
remove lig and hydrogens
unbond lig/G/A/1009/P11, lig/G/A/1009/O12
bond lig/G/A/1009/P11, lig/G/A/1009/O12, 1
unbond lig/G/A/1009/P11, lig/G/A/1009/O13
bond lig/G/A/1009/P11, lig/G/A/1009/O13, 2
alter lig/G/A/1009/N10, formal_charge=1
h_add lig
save HJ5.mol2, lig
fetch 6mgq, async=0
split_chains 6mgq
align 6mgq_A, 6q4r and i. 60-450
align 6mgq_B, 6q4r and i. 60-450
```

```
align 6mgq_C, 6q4r and i. 60-450
remove i. 736+743 and altloc A and 6mgq_*
save erap1_6mgq-A.pdb, 6mgq_A
save erap1_6mgq-B.pdb, 6mgq_B
save erap1_6mgq-C.pdb, 6mgq_C
```

After that you should have five PDB files within the working directory, one for ERAP1 in the closed state and three in the open state, one for ERAP2 in the closed state, as well as the MOL2 file of the inhibitor (HJ5). In lines 7–16 we removed all hydrogen atoms of the ligand and fixed the two P–O bonds, so as to protonate the ligand correctly at physiological conditions before exporting it in MOL2 format (Fig. 1, *see* **Note 1**). In line 22 we removed the alternative conformation of the cysteine pair that does not form a

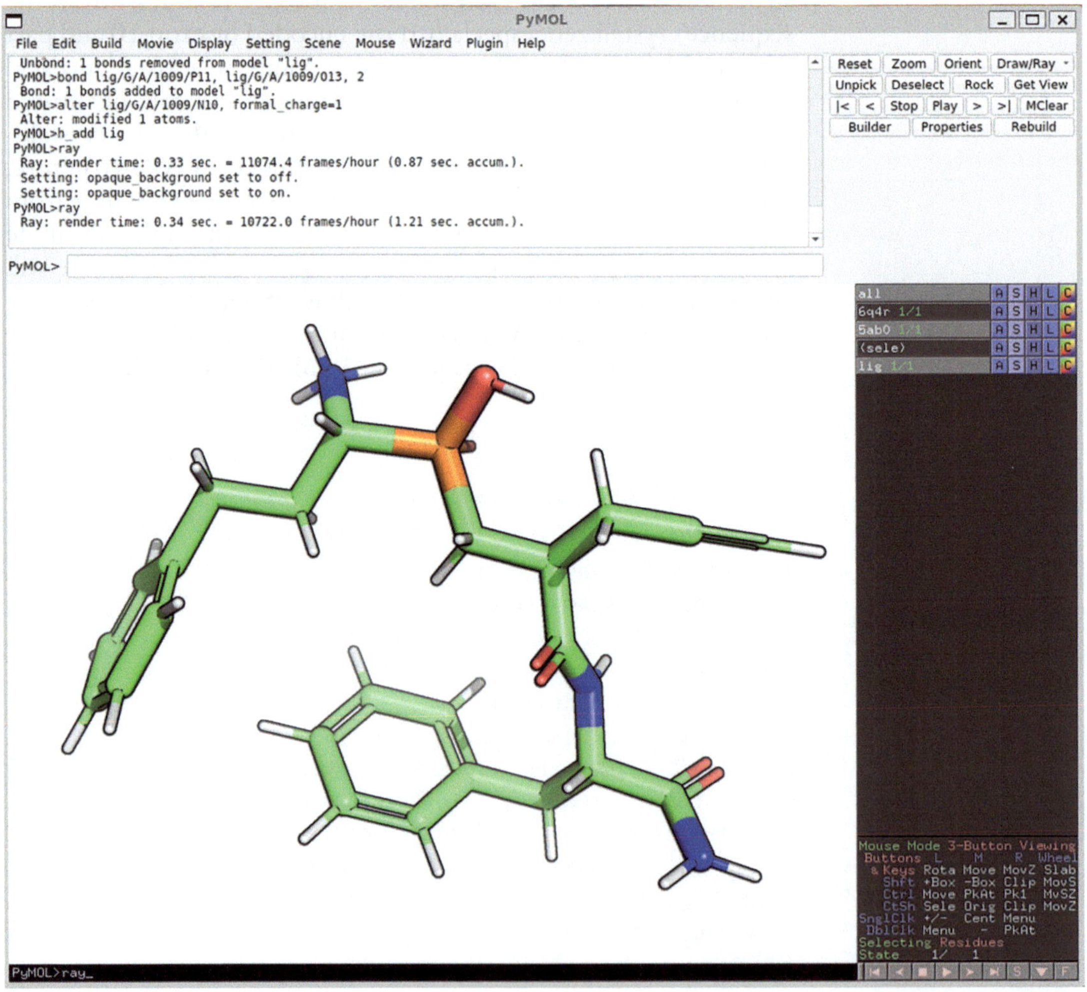

Fig. 1 Illustration of the inhibitor (HJ5) after fixing its protonation state and addition of hydrogen atoms in PyMOL

disulfide bridge, considering that this pair is found to be in the oxidized state in most X-ray structures, including the one employed here for the closed state of ERAP1.

3.2 Modeling of the Missing Residues

The second key step is to add the missing residues, including the exon 10 loop region that is not resolved in the X-ray structures of ERAP1 (residues 486–513). Therefore, MODELLER will be used to add these residues based on the sequence alignment with ERAP2. For this, first prepare the following two input files using a simple text editor, here shown using the GNU nano editor:

```
nano alignment_cl-erap1.pir
```

```
>P1;erap1_6q4r
structureX: erap1_6q4r: 46: A: 1008: A: 6Q4R: HUMAN: 1.60:-1.00
-------PFPWNKIRLPEYVIPVHYDLLIHANLTTLTFWGTTKVEITASQPTSTIILHSH
HLQISRATLRK----RLS--EEPLQVLEHPRQEQIALLAPEPLLVGLPYTVVIHYAGNLS
ETFHGFYKSTYRTKEGELRILASTQFEPTAARMAFPCFDEPAFKASFSIKIRREPRHLAI
SNMPLVKSVTVAEGLIEDHFDVTVKMSTYLVAFIISDFESVSKITKSGVKVSVYAVPDKI
NQADYALDAAVTLLEFYEDYFSIPYPLPKQDLAAIPDFQSGAMENWGLTTYRESALLFDA
EKSSASSKLGITMTVAHELAHQWFGNLVTMEWWNDLWLNEGFAKFMEFVSVSVTHPELKV
GDYFFGKCFDAMEVDALNSSHPVSTPVENPAQIREMFDDVSYDKGACILNMLREYLSADA
FKSGIVQYLQKHSYKNTKNEDLWDSMASI------------------------------V
DVKTMMNTWTLQKGFPLITITVRGRNVHMKQEHYMKG-----------DTGYLWHVPLTF
ITSKSDMVHRFLLKTKTDVLILPEEVEWIKFNVGMNGYYIVHYEDDGWDSLTGLLKGTHT
AVSSNDRASLINNAFQLVSIGKLSIEKALDLSLYLKHETEIMPVFQGLNELIPMYKLMEK
RDMNEVETQFKAFLIRLLRDLIDKQTWTDEGSVSERMLRSQLLLLACVHNYQPCVQRAEG
YFRKWKESNGNLSLPVDVTLAVFAVGAQSTEGWDFLYSKYQFSLSSTEKSQIEFALCRTQ
NKEKLQWLLDESFKGDKIKTQEFPQILTLIGRNPVGYPLAWQFLRKNWNKLVQKFELGSS
SIAHMVMGTTNQFSTRTRLEEVKGFFSSLKENGSQLRCVQQTIETIEENIGWMDKNFDKI
RVWLQSEK----.*

>P1;erap2_5ab0
 structureX: erap2_5ab0: 54: A: 961: A: 5AB0: HUMAN: 2.50:-1.00
PVATNGERFPWQELRLPSVVIPLHYDLFVHPNLTSLDFVASEKIEVLVSNATQFIILHSK
DLEITNATLQSEEDSRYMKPGKELKVLSYPAHEQIALLVPEKLTPHLKYYVAMDFQAKLG
DGFEGFYKSTYRTLGGETRILAVTDFEPTQARMAFPCFDEPLFKANFSIKIRRESRHIAL
SNMPKVKTIELEGGLLEDHFETTVKMSTYLVAYIVCDFHSLSGFTSSGVKVSIYASPDKR
NQTHYALQASLKLLDFYEKYFDIYYPLSKLDLIAIPDFAPGAMENWGLITYRETSLLFDP
KTSSASDKLWVTRVIAHELAHQWFGNLVTMEWWNDIWLNEGFAKYMELIAVNATYPELQF
DDYFLNVCFEVITKDSLNSSRPISKPAETPTQIQEMFDEVSYNKGACILNMLKDFLGEEK
FQKGIIQYLKKFSYRNAKNDDLWSSLSNSCLESDFTS-GGVCHSDPKMTSNMLAFLGENA
EVKEMMTTWTLQKGIPLLVVKQDGCSLRLQQERFLQGVFQEDPEWRALQERYLWHIPLTY
STSSSNVIHRHILKSKTDTLDLPEKTSWVKFNVDSNGYYIVHYEGHGWDQLITQLNQNHT
LLRPKDRVGLIHDVFQLVGAGRLTLDKALDMTYYLQHETSSPALLEGLSYLESFYHMMDR
RNISDISENLKRYLLQYFKPVIDRQSWSDKGSVWDRMLRSALLKLACDLNHAPCIQKAAE
```

```
LFSQWMESSGKLNIPTDVLKIVYSVGAQTTAGWNYLLEQYELSMSSAEQNKILYALSTSK
HQEKLLKLIELGMEGKVIKTQNLAALLHAIARRPKGQQLAWDFVRENWTHLLKKFDLGSY
DIRMIISGTTAHFSSKDKLQEVKLFFESLEAQGSHLDIFQTVLETITKNIKWLEKNLPTL
RTWLMVNTR----*

>P1;cl-erap1
sequence: cl-erap1: 46: A: 940: A: ERAP1: HUMAN: 2.00:-1.00
-------PFPWNKIRLPEYVIPVHYDLLIHANLTTLTFWGTTKVEITASQPTSTIILHSH
HLQISRATLRKGAGERLS--EEPLQVLEHPRQEQIALLAPEPLLVGLPYTVVIHYAGNLS
ETFHGFYKSTYRTKEGELRILASTQFEPTAARMAFPCFDEPAFKASFSIKIRREPRHLAI
SNMPLVKSVTVAEGLIEDHFDVTVKMSTYLVAFIISDFESVSKITKSGVKVSVYAVPDKI
NQADYALDAAVTLLEFYEDYFSIPYPLPKQDLAAIPDFQSGAMENWGLTTYRESALLFDA
EKSSASSKLGITMTVAHELAHQWFGNLVTMEWWNDLWLNEGFAKFMEFVSVSVTHPELKV
GDYFFGKCFDAMEVDALNSSHPVSTPVENPAQIREMFDDVSYDKGACILNMLREYLSADA
FKSGIVQYLQKHSYKNTKNEDLWDSMASICPTDGVKGMDGFCSR-SQHSSSSSHWHQEGV
DVKTMMNTWTLQKGFPLITITVRGRNVHMKQEHYMKGSDG------APDTGYLWHVPLTF
ITSKSDMVHRFLLKTKTDVLILPEEVEWIKFNVGMNGYYIVHYEDDGWDSLTGLLKGTHT
AVSSNDRASLINNAFQLVSIGKLSIEKALDLSLYLKHETEIMPVFQGLNELIPMYKLMEK
RDMNEVETQFKAFLIRLLRDLIDKQTWTDEGSVSERMLRSQLLLLACVHNYQPCVQRAEG
YFRKWKESNGNLSLPVDVTLAVFAVGAQSTEGWDFLYSKYQFSLSSTEKSQIEFALCRTQ
NKEKLQWLLDESFKGDKIKTQEFPQILTLIGRNPVGYPLAWQFLRKNWNKLVQKFELGSS
SIAHMVMGTTNQFSTRTRLEEVKGFFSSLKENGSQLRCVQQTIETIEENIGWMDKNFDKI
RVWLQSEKLERM.*
```

```
nano model_cl-erap1.py
```

```
# Comparative modeling with multiple templates
from modeller import *
from modeller.automodel import *

log.verbose()
env = environ()

# directories for input atom files
env.io.atom_files_directory = ['.', '../atom_files']

# Read in HETATM records from template PDBs
env.io.hetatm = True

a = AutoModel(env,
alnfile = 'alignment_cl-erap1.pir',
knowns = ('erap1_6q4r', 'erap2_5ab0'),
sequence = 'cl-erap1')
```

```
a.starting_model= 1
a.ending_model = 1

a.make()
```

and run MODELLER using the bash command:

```
mod10.6 model_cl-erap1.py
```

To obtain more than one model, the a.ending_model variable can be adjusted to the desired number. To assess the resulting models based on the DOPE and GA341 scores, add the following command within the AutoModel section: assess_-methods = (assess.DOPE, assess.GA341). For the modeling of other common ERAP1 allotypes [39], or using different X-ray structures as templates, *see* **Note 2**. Now, for modeling the three open conformations of ERAP1, we repeat the procedure described for closed ERAP1, first by creating three alignment (pir) files in which only the first block of the X-ray sequence and the input filename are replaced as shown below.

```
alignment_oa-erap1.pir:
```

```
>P1;erap1_6mgq-A
structureX: erap1_6mgq-A: 46: A: 1001: A: 6MGQ: HUMAN: 2.92:-1.00
-------PFPWNKIRLPEYVIPVHYDLLIHANLTTLTFWGTTKVEITASQPTSTIILHSH
HLQISRATLRKGAGERLS--EEPLQVLEHPRQEQIALLAPEPLLVGLPYTVVIHYAGNLS
ETFHGFYKSTYRTKEGELRILASTQFEPTAARMAFPCFDEPAFKASFSIKIRREPRHLAI
SNMPLVKSVTVAEGLIEDHFDVTVKMSTYLVAFIISDFESVSKITKSGVKVSVYAVPDKI
NQADYALDAAVTLLEFYEDYFSIPYPLPKQDLAAIPDFQSGAMENWGLTTYRESALLFDA
EKSSASSKLDITMTVAHELAHQWFGNLVTMEWWNDLWLNEGFAKFMEFVSVSVTHPELKV
GDYFFGKCFDAMEVDALNSSHPVSTPVENPAQIREMFDDVSYDKGACILNMLREYLSADA
FKSGIVQYLQKHSYKNTKNEDLWDSMASICPT-------------------------ERV
DVKTMMNTWTLQRGFPLITITVRGRNVHMKQEHYMKGSDG------APDTGYLWHVPLTF
ITSKSDMVHRFLLKTKTDVLILPEEVEWIKFNVGMNGYYIVHYEDDGWDSLTGLLKGTHT
AVSSNDRASLINNAFQLVSIGKLSIEKALDLSLYLKHETEIMPVFQGLNELIPMYKLMEK
RDMNEVETQFKAFLIRLLRDLIDKQTWTDEGSVSERMLRSELLLLACVHNYQPCVQRAEG
YFRKWKESNGNLSLPVDVTLAVFAVGAQSTEGWDFLYSKYQFSLSSTEKSQIEFALCRTQ
NKEKLQWLLDESFKGDE-----FPQILTLIGRNPVGYPLAWQFLRKNWNKLVQKFELGSS
SIAHMVMGTTNQFSTRTRLEEVKGFFSSL----SQLRCVQQTIETIEENIGWMDKNFDKI
RVWLQSE-----.*
```

```
alignment_ob-erap1.pir:
```

```
>P1;erap1_6mgq-B
structureX: erap1_6mgq-B: 46: B: 1001: B: 6MGQ: HUMAN: 2.92:-1.00
-------PFPWNKIRLPEYVIPVHYDLLIHANLTTLTFWGTTKVEITASQPTSTIILHSH
HLQISRATLRKGAGERLS--EEPLQVLEHPRQEQIALLAPEPLLVGLPYTVVIHYAGNLS
ETFHGFYKSTYRTKEGELRILASTQFEPTAARMAFPCFDEPAFKASFSIKIRREPRHLAI
SNMPLVKSVTVAEGLIEDHFDVTVKMSTYLVAFIISDFESVSKITKSGVKVSVYAVPDKI
NQADYALDAAVTLLEFYEDYFSIPYPLPKQDLAAIPDFQSGAMENWGLTTYRESALLFDA
EKSSASSKLDITMTVAHELAHQWFGNLVTMEWWNDLWLNEGFAKFMEFVSVSVTHPELKV
GDYFFGKCFDAMEVDALNSSHPVSTPVENPAQIREMFDDVSYDKGACILNMLREYLSADA
FKSGIVQYLQKHSYKNTKNEDLWDSMASI----------------------------RV
DVKTMMNTWTLQRGFPLITITVRGRNVHMKQEHYMKGSDG------APDTGYLWHVPLTF
ITSKSDMVHRFLLKTKTDVLILPEEVEWIKFNVGMNGYYIVHYEDDGWDSLTGLLKGTHT
AVSSNDRASLINNAFQLVSIGKLSIEKALDLSLYLKHETEIMPVFQGLNELIPMYKLMEK
RDMNEVETQFKAFLIRLLRDLIDKQTWTDEGSVSERMLRSELLLLACVHNYQPCVQRAEG
YFRKWKESNGNLSLPVDVTLAVFAVGAQSTEGWDFLYSKYQFSLSSTEKSQIEFALCRTQ
NKEKLQWLLDESFKGD---------ILTLIGRNPVGYPLAWQFLRKNWNKLVQKFELGSS
SIAHMVMGTTNQFSTRTRLEEVKGFFSSLKENGSQLRCVQQTIETIEENIGWMDKNFDKI
RVWLQSEK----.*
```

alignment_oc-erap1.pir:

```
>P1;erap1_6mgq-C
structureX: erap1_6mgq-C: 46: C: 1001: C: 6MGQ: HUMAN: 2.92:-1.00
-------PFPWNKIRLPEYVIPVHYDLLIHANLTTLTFWGTTKVEITASQPTSTIILHSH
HLQISRATLRKGAGERLS--EEPLQVLEHPRQEQIALLAPEPLLVGLPYTVVIHYAGNLS
ETFHGFYKSTYRTKEGELRILASTQFEPTAARMAFPCFDEPAFKASFSIKIRREPRHLAI
SNMPLVKSVTVAEGLIEDHFDVTVKMSTYLVAFIISDFESVSKITKSGVKVSVYAVPDKI
NQADYALDAAVTLLEFYEDYFSIPYPLPKQDLAAIPDFQSGAMENWGLTTYRESALLFDA
EKSSASSKLDITMTVAHELAHQWFGNLVTMEWWNDLWLNEGFAKFMEFVSVSVTHPELKV
GDYFFGKCFDAMEVDALNSSHPVSTPVENPAQIREMFDDVSYDKGACILNMLREYLSADA
FKSGIVQYLQKHSYKNTKNEDLWDSMASICPTD-----------------------ERV
DVKTMMNTWTLQRGFPLITITVRGRNVHMKQEHYMKGSDG------APDTGYLWHVPLTF
ITSKSDMVHRFLLKTKTDVLILPEEVEWIKFNVGMNGYYIVHYEDDGWDSLTGLLKGTHT
AVSSNDRASLINNAFQLVSIGKLSIEKALDLSLYLKHETEIMPVFQGLNELIPMYKLMEK
RDMNEVETQFKAFLIRLLRDLIDKQTWTDEGSVSERMLRSELLLLACVHNYQPCVQRAEG
YFRKWKESNGNLSLPVDVTLAVFAVGAQSTEGWDFLYSKYQFSLSSTEKSQIEFALCRTQ
NKEKLQWLLDESFIKTQE----FPQILTLIGRNPVGYPLAWQFLRKNWNKLVQKFEL---
-----VMGTTNQFSTRTRLEEV-----------------QQTIETIEENIGWMDKNFDKI
RVWLQSE-----.*
```

The output model names should be replaced at the header of the last block in the corresponding alignment (pir) files, as shown below:

```
>P1;oa-erap1
sequence: oa-erap1: 46: A: 940: A: ERAP1: HUMAN: 2.00:-1.00


>P1;ob-erap1
sequence: ob-erap1: 46: A: 940: A: ERAP1: HUMAN: 2.00:-1.00


>P1;oc-erap1
sequence: oc-erap1: 46: A: 940: A: ERAP1: HUMAN: 2.00:-1.00
```

Then prepare the three corresponding input files for MODELLER by replacing the required filenames in the *AutoModel* block as shown below:

`model_oa-erap1.py:`

```
a = AutoModel(env,
alnfile = 'alignment_oa-erap1.pir',
knowns = ('erap1_6mgq-A', 'erap2_5ab0'),
sequence = 'oa-erap1')
```

`model_ob-erap1.py:`

```
a = AutoModel(env,
alnfile = 'alignment_ob-erap1.pir',
knowns = ('erap1_6mgq-B', 'erap2_5ab0'),
sequence = 'ob-erap1')
```

`model_oc-erap1.py:`

```
a = AutoModel(env,
alnfile = 'alignment_oc-erap1.pir',
knowns = ('erap1_6mgq-C', 'erap2_5ab0'),
sequence = 'oc-erap1')
```

Then run MODELLER for the three chains of ERAP1 in the open state:

```
mod10.6 model_oa-erap1.py

mod10.6 model_ob-erap1.py

mod10.6 model_oc-erap1.py
```

Upon successful completion, the working directory should contain several MODELLER output files, including the four requested PDB models that will be used at the next stage:

```
cl-erap1.B99990001.pdb oa-erap1.B99990001.pdb
ob-erap1.B99990001.pdb oc-erap1.B99990001.pdb
```

3.3 Calculation of the Protonation States

To calculate the protonation states of the ionizable groups of each model, we will employ the H++ server [34], which can be accessed after creating a user account here: http://newbiophysics.cs.vt.edu/H++/index.php

After logging in, press on **PROCESS A STRUCTURE**, then load the PDB file using **Choose File** and press the **Process File** button. At the next page, change the pH value of 6.5 to 7.4 in order to protonate the proteins at physiological conditions. Leave all other options as is and press the **PROCESS...** button at the bottom. A progress bar and information about each calculation stage is shown. When done, press **VIEW RESULTS** link and you will be redirected to the Results page. From the **Other Useful Files** section, download the AMBER-compatible PDB (PQR) structure in the predicted protonation state, that is, for the four ERAP1 conformations:

```
0.15_80_10_pH7.4_cl-erap1.B99990001.genpdb.cpptraj.pdb
0.15_80_10_pH7.4_oa-erap1.B99990001.genpdb.cpptraj.pdb
0.15_80_10_pH7.4_ob-erap1.B99990001.genpdb.cpptraj.pdb
0.15_80_10_pH7.4_oc-erap1.B99990001.genpdb.cpptraj.pdb
```

To clean up the filenames and add the inhibitor at the active site in the appropriate position, run the following bash shell command within the working directory:

```
for i in 0.15_80_10_pH7.4_*.pdb; do n = ${i##*_}; \
grep -v CONECT $i > ${n%%.*}.prot.pdb; done
```

Then launch PyMOL and execute the following commands within the console:

```
load erap1_6q4r.pdb
load cl-erap1.prot.pdb
load oa-erap1.prot.pdb
load ob-erap1.prot.pdb
load oc-erap1.prot.pdb
alignto erap1_6q4r and i. 60–450
load HJ5.mol2
copy_to cl-erap1.prot, HJ5
copy_to oa-erap1.prot, HJ5
copy_to ob-erap1.prot, HJ5
copy_to oc-erap1.prot, HJ5
alter /oa-erap1.prot///312, resn = "HID"
remove /oa-erap1.prot///312/HE2
save cl-erap1-prot.pdb, cl-erap1.prot
save oa-erap1-prot.pdb, oa-erap1.prot
save ob-erap1-prot.pdb, ob-erap1.prot
save oc-erap1-prot.pdb, oc-erap1.prot
```

This will generate the four input files in the form required by the Amber utility *tLEaP* at the next stage (Fig. 2). Renaming of residue 312 to HID and removal of HE2 in the oa-erap1.prot model is required for binding to the zinc ion.

3.4 Preparation of the Systems for AMBER

First we will generate a parameter file for the ligand using the PyMOL-exported `HJ5.mol2` file (stage 3.1) and the *antechamber* utility of Amber. The net charge of our inhibitor is +1 (−nc 1) and the atomic point charges will be calculated using the AM1-BCC methodology (−c bcc) [40]. Force field parameters will be applied from the General Amber Force Field (−at gaff) [41]:

```
antechamber -i HJ5.mol2 -fi mol2 -o HJ5.prepc -fo prepc /
-nc 1 -c bcc -at gaff
```

Then we estimate the missing force-field parameters according to the Amber GAFF (−s 1) for organic molecules using the *parmchk* utility:

```
parmchk2 -i HJ5.prepc -f prepc -o HJ5.frcmod -s 1
```

For alternative parameterization of the inhibitor, *see* **Note 3**. The next step is to create the following file with force field parameters for the Zn(II)-binding site.

```
zinc.frcmod
```

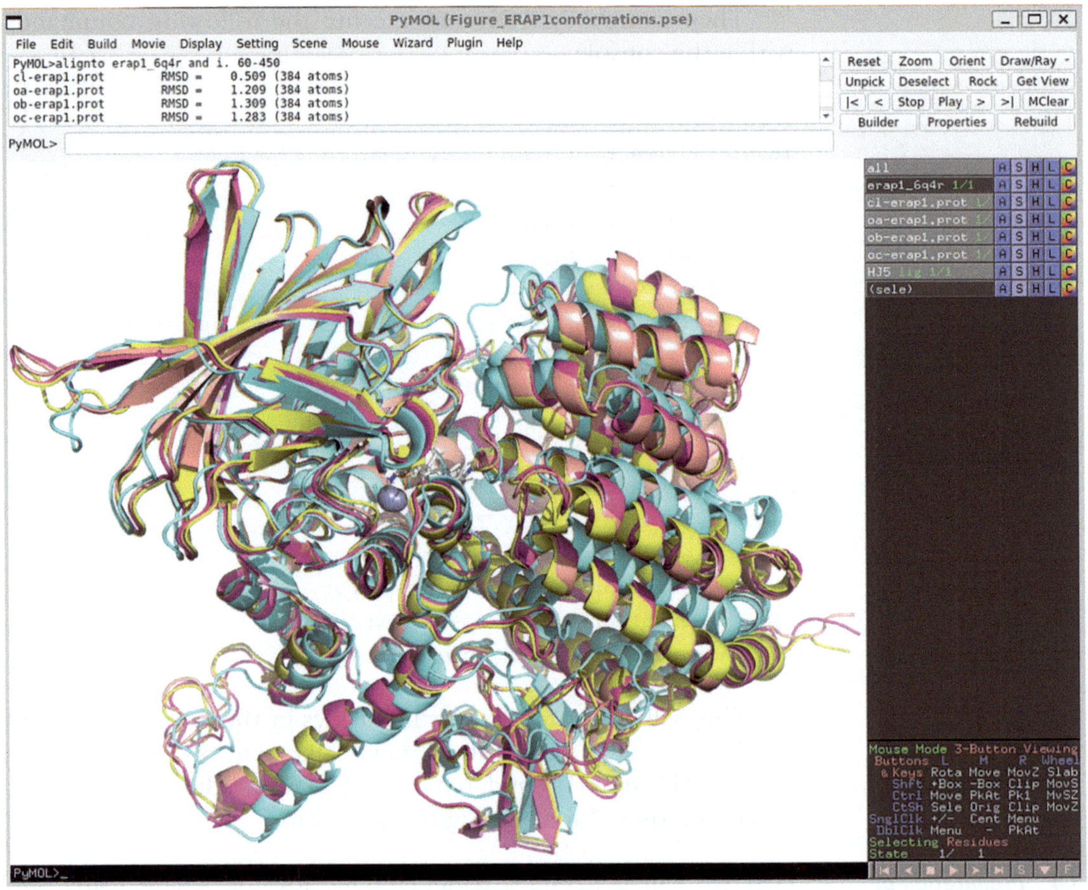

Fig. 2 Illustration of the four initial ERAP1 conformations in PyMOL, after modeling of the missing residues, calculation of the ionizable group protonation states, and addition of the same inhibitor at their active site. The structures have been superimposed with respect to residues 60–450 and ERAP1 in the closed state is the cyan-colored cartoon

```
Extra Parameters for the Zinc Binding Site
MASS
TZ 65.39 0.000

BOND
TZ-NB 150.00 2.150
TZ-O2 150.00 2.200
TZ-o 150.00 2.200

ANGLE
NB-TZ-NB 25.00 109.5
NB-TZ-O2 25.00 109.5
NB-TZ-o 25.00 109.5
O2-TZ-o 25.00 109.5
```

```
TZ-NB-CV 35.00 121.5
TZ-NB-CR 35.00 121.5
TZ-O2-CO 35.00 120.0
TZ-o -p5 35.00 120.0

DIHE
X -TZ-NB-X 2 0.000 0.000 2.000
X -TZ-O2-X 2 0.000 0.000 2.000
X -TZ-o -X 2 0.000 0.000 2.000
TZ-o -p5-c3 2 0.000 0.000 2.000
TZ-o -p5-o 2 0.000 0.000 2.000
O2-CO-O2-TZ 2 0.000 0.000 2.000
2C-CO-O2-TZ 2 0.000 0.000 2.000

IMPROPER

NONBON
TZ 1.10 0.012500
```

These force field parameters treat the tetrahedral coordination sphere of Zn(II) using harmonic potentials and angle terms, that is, explicit bonds that provide the required directionality. The bond and angle force field parameters are based on crystallographic structures of zinc aminopeptidases, as well as Amber force field parameters of similar atoms but with lower force constants to allow for the flexibility of the zinc ligands. All dihedrals involving the metal ion have been set to zero and the non-bonded parameters are taken from ref. [42].

As a next step, make four input command files for *LEaP* that will generate the topology and coordinate files of the solvated systems:

prep-cl-erap1.cmd

```
source leaprc.protein.ff14SB
source leaprc.water.tip3p
source leaprc.gaff

loadamberprep HJ5.prepc
loadamberparams HJ5.frcmod
zinc=loadamberparams zinc.frcmod

mol=loadpdb cl-erap1-prot.pdb

bond mol.897.ZN mol.308.NE2
```

```
bond mol.897.ZN mol.312.NE2
bond mol.897.ZN mol.331.OE1
bond mol.897.ZN mol.898.O13
set mol.897.ZN type TZ
set mol.897 restype protein

set mol.897.ZN charge 0.8000
set mol.308.NE2 charge -0.3727
set mol.312.NE2 charge -0.3727
set mol.331.OE1 charge -0.6188
set mol.331.OE2 charge -0.6188
set mol.898.O12 charge -0.6284
set mol.898.O13 charge -0.6483

bond mol.359.SG mol.398.SG
bond mol.441.SG mol.453.SG
bond mol.691.SG mol.698.SG

charge mol

solvateoct mol TIP3PBOX 10
addions mol Na+ 0

saveamberparm mol cl-erap1.parm7 cl-erap1.rst7
quit
```

The corresponding files for the three open conformations of ERAP1 can be prepared using the bash commands:

```
sed 's/cl-/oa-/g' prep-cl-erap1.cmd > prep-oa-erap1.cmd
sed 's/cl-/ob-/g'prep-cl-erap1.cmd > prep-ob-erap1.cmd
sed 's/cl-/oc-/g' prep-cl-erap1.cmd > prep-oc-erap1.cmd
```

The scripts will import the required force field parameters, then load the initial structures and create the three disulfide bridges and the four required bonds to zinc. The point charge on Zn(II) is set to +0.8 e^- and the remaining +1.2 e^- is split equally to the four coordinating atoms, plus the two non-coordinating oxygen atoms of Glu376 carboxylate and the phosphinic group of the inhibitor (each +0.2 e^-). This is a simple partial charge distribution approach, which is, however, quite similar to DFT-based calculation of partial atomic charges using restrained fitting to the electrostatic potential (RESP, unpublished results). The system is solvated in an octahedral solvent box of TIP3P waters, and then the required number of sodium ions is added to neutralize the system. In this case, we employed the typical combination of the Amber force field *ff14SB* [43] with the TIP3P water model [44], and the corresponding parameters for monovalent ions [45]. For alternative combinations of force field parameters, *see* **Note 4**. To execute tLEaP using the command files prepared, type:

```
tleap -s -f prep-cl-erap1.cmd
```

Wait until it finishes and repeat using the three command files (`prep-*-erap1.cmd`) for the open conformations (`oa- ob- oc-`). After that, the working directory should contain one pair of topology (parm7) and coordinate (rst7) files for each system. At this point it is convenient to create four working directories for each of the four ERAP1 conformations:

```
mkdir cl-erap1/ oa-erap1/ ob-erap1/ oc-erap1/
```

so as to copy or move the parm7/rst7 files in the corresponding folder. You can generate PDB files of the solvated systems with the correct residue numbering for visual investigation (Fig. 3) using the command:

```
ambpdb -offset 45 -p cl-erap1.parm7 < cl-erap1.rst7 /
cl-erap1.amber.pdb
```

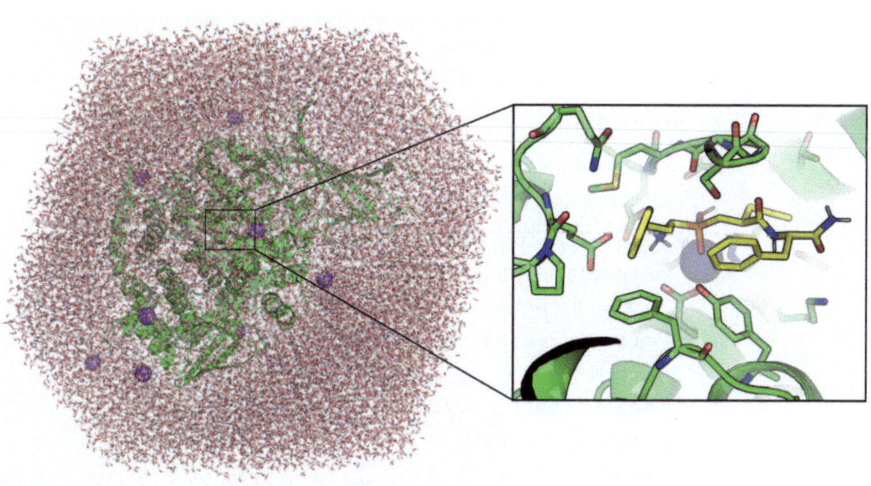

Fig. 3 Illustration of the initial closed ERAP1 system (`cl-erap1.amber.pdb`) in PyMOL (left). Purple spheres represent the Na^+ ions added to neutralize the total charge of the system. Inset on the right is a close-up view of the active site showing the inhibitor with yellow-colored C atoms and the catalytic Zn(II) with a gray sphere

3.5 Running the MD Simulations

This section provides the input files and the command scripts for the relaxation of the systems and for the production runs. The protocol provided for the closed state of ERAP1 will be carried out within the cl-erap1/ folder. The same input and command files can be employed for the three open conformations of ERAP1 in the directories: oa-erap1/ ob-erap1/ oc-erap1/. The protocol provided here is based on an Amber tutorial that can be found here:

https://ambermd.org/tutorials/basic/tutorial13/index.php

and which provides explanation of key parameters, as well as description of every step. Now start by preparing the following input files, which can be used for all four systems without any modification:

minrst1.in

```
Minimization with restraints on all protein atoms
&cntrl
imin=1, ntb=1, ntxo=1,
ntr=1,       restraintmask='!:WAT,Na+       &       !@H=',
restraint_wt=10,
cut=12.0,
maxcyc=5000, ncyc=1500, drms=0.01,
ntpr=50, ntwx=50,
&end
```

minrst2.in

```
Minimization with restraints on backbone atoms
&cntrl
imin=1, ntb=1, ntxo=1,
ntr=1, restraintmask='@CA,C,N', restraint_wt=10,
cut=12.0,
maxcyc=5000, ncyc=500, drms=0.01,
ntpr=50, ntwx=50,
&end
```

minrst3.in

```
Minimization without any restraints
&cntrl
```

```
imin=1, ntb=1, ntxo=1,
cut=12.0,
maxcyc=5000, ncyc=500, drms=0.001,
ntpr=50, ntwx=50,
&end
```

eqnvtrst.in

```
Short relaxation of solvent in the NVT heating from
100 to 200 K
&cntrl
ntb=1, ioutfm=1, iwrap=1, ntxo=1,
nstlim=100000, dt=0.002,
ntpr=2500, ntwx=2500, ntwr=25000,
ntc=2, ntf=2, tol=0.000001,
cut=9,
ntt=3, gamma_ln=1.0,
nmropt=1, ntr=1,
restraintmask='!:WAT,Na+', restraint_wt=100,
/
&wt type='TEMP0', istep1=0, istep2=100000, value1=100,
value2=200, /
&wt type='END' /
/
&ewald
dsum_tol=0.000001,
/
```

eqnptrst.in

```
Short relaxation of solvent in the NPT ensemble at 300 K
&cntrl
ntb=2, ioutfm=1, iwrap=1,
irest=1, ntx=5, ntxo=1,
nstlim=400000, dt=0.002,
ntpr=2500, ntwx=2500, ntwr=25000,
ntc=2, ntf=2, tol=0.000001,
cut=9,
ntt=3, gamma_ln=1.0,
ntp=1, taup=0.5,
nmropt=1, ntr=1,
```

```
restraintmask='!:WAT,Na+', restraint_wt=100,
&end
&wt type='TEMP0', istep1=0, istep2=100000, value1=200,
value2=300, /
&wt   type='TEMP0',    istep1=100001,    istep2=400000,
value1=300, value2=300, /
&wt type='END' /
/
&ewald
dsum_tol=0.000001,
/
```

minrst.in

```
Minimization with restraints only on protein CA atoms
&cntrl
imin=1, ntb=1, ntxo=1,
ntr=1, restraintmask='@CA', restraint_wt=10,
cut=12.0,
maxcyc=10000, ncyc=100, dx0=1.0, drms=0.1,
ntpr=100, ntwx=100,
&end
```

eqnptrst1.in

```
Relaxation in the NPT at 300 K with restraints on CA atoms
&cntrl
ntb=2, ioutfm=1, iwrap=1, ntxo=1,
nstlim=100000, dt=0.002,
ntpr=2500, ntwx=2500, ntwr=25000,
ntc=2, ntf=2, tol=0.000001,
cut=9, tempi=300, temp0=300,
ntt=3, gamma_ln=1.0,
ntp=1, taup=1.0,
ntr=1, restraintmask='@CA', restraint_wt=10,
/
&ewald
dsum_tol=0.000001,
/
```

eqnptrst2.in

```
Relaxation in the NPT at 300 K with lower restraints on CA
atoms
&cntrl
ntb=2, ioutfm=1, iwrap=1,
irest=1, ntx=5, ntxo=1,
nstlim=100000, dt=0.002,
ntpr=2500, ntwx=2500, ntwr=25000,
ntc=2, ntf=2, tol=0.000001,
cut=9, tempi=300, temp0=300,
ntt=3, gamma_ln=1.0,
ntp=1, taup=1.0,
ntr=1, restraintmask='@CA', restraint_wt=1.0,
/
&ewald
dsum_tol=0.000001,
/
```

eqnptrst3.in

```
Relaxation in the NPT at 300 K with very low restraints
&cntrl
ntb=2, ioutfm=1, iwrap=1,
irest=1, ntx=5, ntxo=1,
nstlim=300000, dt=0.002,
ntpr=2500, ntwx=2500, ntwr=25000,
ntc=2, ntf=2, tol=0.000001,
cut=9, tempi=300, temp0=300,
ntt=3, gamma_ln=1.0,
ntp=1, taup=1.0,
ntr=1, restraintmask='@CA', restraint_wt=0.1,
/
&ewald
dsum_tol=0.000001,
/
```

eqnpt.in

```
Relaxation in the NPT at 300 K without restraints
&cntrl
ntb=2, ioutfm=1, iwrap=1,
irest=1, ntx=5, ntxo=2,
nstlim=4500000, dt=0.002,
ntpr=2500, ntwx=2500, ntwr=25000,
ntc=2, ntf=2, tol=0.000001,
cut=9, tempi=300, temp0=300,
ntt=3, gamma_ln=1.0,
ntp=1, taup=1.0,
/
&ewald
dsum_tol=0.000001,
/
```

`mdrun.in`

```
Relaxation in the NPT at 300 K with "leaky pair lists"
&cntrl
ntb=2, ioutfm=1, iwrap=1,
irest=1, ntx=5, ntxo=2,
nstlim=20000000, dt=0.002,
ntpr=2500, ntwx=2500, ntwr=250000,
ntc=2, ntf=2, tol=0.000001,
cut=9, tempi=300, temp0=300,
ntt=3, gamma_ln=1.0,
ntp=1, taup=1.0,
/
&ewald
dsum_tol=0.000001, netfrc = 0, skin_permit = 0.75,
/
```

Then prepare the following script and make it executable using the bash shell command: `chmod +x relaxation.x`

`relaxation.x`

```
#!/bin/bash
#If not in your PATH, uncomment the following line:
#export PATH=$PATH: $AMBERHOME/bin/
pmemd.cuda -O -i minrst1.in -o $1minrst1.out -p $1.parm7
/
```

```
-c $1.rst7 -r $1minrst1.rst7 -ref $1.rst7
pmemd.cuda -O -i minrst2.in -o $1minrst2.out -p $1.parm7
/
-c $1minrst1.rst7 -r $1minrst2.rst7 -ref $1.rst7
pmemd.cuda -O -i minrst3.in -o $1minrst3.out -p $1.parm7
/
-c $1minrst2.rst7 -r $1minrst3.rst7
pmemd.cuda -O -i eqnvtrst.in -o $1eqnvtrst.out -p $1.
parm7 /
-c $1minrst3.rst7 -r $1eqnvtrst.rst7 -x $1eqnvtrst.nc /
-ref $1minrst3.rst7
pmemd.cuda -O -i eqnptrst.in -o $1eqnptrst.out -p $1.
parm7 /
-c $1eqnvtrst.rst7 -r $1eqnptrst.rst7 -x $1eqnptrst.nc /
-ref $1eqnvtrst.rst7
 pmemd.cuda -O -i minrst.in -o $1minrst.out -p $1.parm7 /
-c $1eqnptrst.rst7 -r $1minrst.rst7 -x $1minrst.nc /
-ref $1eqnptrst.rst7
pmemd.cuda -O -i eqnptrst1.in -o $1eqnptrst1.out -p $1.
parm7 /
-c $1minrst.rst7 -r $1eqnptrst1.rst7 -x $1eqnptrst1.nc /
-ref $1minrst.rst7
pmemd.cuda -O -i eqnptrst2.in -o $1eqnptrst2.out -p $1.
parm7 /
-c $1eqnptrst1.rst7 -r $1eqnptrst2.rst7 -x $1eqnptrst2.nc
/
 -ref $1eqnptrst1.rst7
pmemd.cuda -O -i eqnptrst3.in -o $1eqnptrst3.out -p $1.
parm7 /
-c $1eqnptrst2.rst7 -r $1eqnptrst3.rst7 -x $1eqnptrst3.nc
/
-ref $1eqnptrst2.rst7
pmemd.cuda -O -i eqnpt.in -o $1eqnpt.out -p $1.parm7 /
-c $1eqnptrst3.rst7 -r $1eqnpt.rst7 -x $1eqnpt.nc
pmemd.cuda -O -i mdrun.in -o $1mdrun.out -p $1.parm7 /
-c $1eqnpt.rst7 -r $1mdrun.rst7 -x $1mdrun.nc
```

The script can be executed in the background with the command:

```
nohup. /relaxation.x cl-erap1 &
```

Here for the closed ERAP1 state, and in a similar fashion for the three open conformations in the corresponding directory.

For longer production runs, we can use the hydrogen mass repartitioning [46] (HMR) method that redistributes some of the mass from heavy atoms connected to hydrogen into the bonded hydrogen atoms. In this way, instability issues related to high-frequency hydrogen motions are eliminated; therefore, simulations can be carried out using a 4 fs time-step instead of the regular 2 fs time-step. For this reason, first we need to obtain topology files with the new masses using the parmed utility of Amber.

Prepare the following simple input file for each system: `hmr-cl-erap1.in`

```
parm cl-erap1.parm7
HMassRepartition
parmout cl-erap1_hmr.parm7
go
```

and obtain the modified topology files using the command:

```
parmed < hmr-cl-erap1.in
```

This will generate the topology file `cl-erap1_hmr.parm7` that will be used in 4-fs time step production runs. An input file for a 200-ns production run in the NPT ensemble is shown below:

`mdrunp.in`

```
NPT production at 300K with H mass repartitioned and a
4-fs step
&cntrl
ntb=2, ioutfm=1, iwrap=1,
irest=1, ntx=5, ntxo=2,
nstlim=50000000, dt=0.004,
ntpr=2500, ntwx=2500, ntwr=250000,
ntc=2, ntf=2, tol=0.000001,
cut=9, tempi=300, temp0=300,
ntt=3, gamma_ln=1.0,
ntp=1, taup=1.0,
/
&ewald
dsum_tol=0.000001, netfrc = 0, skin_permit = 0.75,
/
```

The following script can be used to carry out long production runs in five chunks of 200-ns simulations, for an aggregate simulation time of 1.0 μs:

```
#!/bin/bash
#If not in your PATH, uncomment the following line:
#export PATH=$PATH: $AMBERHOME/bin/
pmemd.cuda -O -i mdrunp.in -o $1mdrunp1.out -p $1_hmr.
parm7 /
-c $1mdrun.rst7 -r $1mdrunp1.rst7 -x $1mdrunp1.nc
pmemd.cuda -O -i mdrunp.in -o $1mdrunp2.out -p $1_hmr.
parm7 /
-c $1mdrunp1.rst7 -r $1mdrunp2.rst7 -x $1mdrunp2.nc
pmemd.cuda -O -i mdrunp.in -o $1mdrunp3.out -p $1_hmr.
parm7 /
-c $1mdrunp2.rst7 -r $1mdrunp3.rst7 -x $1mdrunp3.nc
pmemd.cuda -O -i mdrunp.in -o $1mdrunp4.out -p $1_hmr.
parm7 /
-c $1mdrunp3.rst7 -r $1mdrunp4.rst7 -x $1mdrunp4.nc
pmemd.cuda -O -i mdrunp.in -o $1mdrunp5.out -p $1_hmr.
parm7 /
-c $1mdrunp4.rst7 -r $1mdrunp5.rst7 -x $1mdrunp5.nc
```

Due to the high conformational space of ERAP1, it is of particular importance to carry out longer simulations (of up to 1–2 μs) starting from the (three) open conformations of the enzyme in order to obtain thermodynamically stable ensembles (*see* **Note 5** for disk size requirements). Simulations starting from the closed state of the enzyme generally converge within shorter (100–200 ns) time scales. Still, more advanced methods of enhanced conformational sampling may be required to obtain convergence, including but not limited to accelerated molecular dynamics simulations (aMD) [22, 47, 48].

3.6 Analysis of the MD Simulations

The first type of analysis can be readily performed at the output files of each simulation by examination of the fluctuations of temperature, density and energies of the system. This can be facilitated using the Amber utility mdout_analyzer.py followed by the requested mdout file(s). In this example we will analyze the simulation of an open ERAP1 state (oc-erap1); however, the procedure can be repeated for every state of the enzyme and any output file (s) by replacing the appropriate file name(s). Within the oc-erap1/ directory type:

```
mdout_analyzer.py oc-erap1mdrun.out
```

and plot the temperature, density and total energy (Etot) of the system, both native and as running averages (Fig. 4).

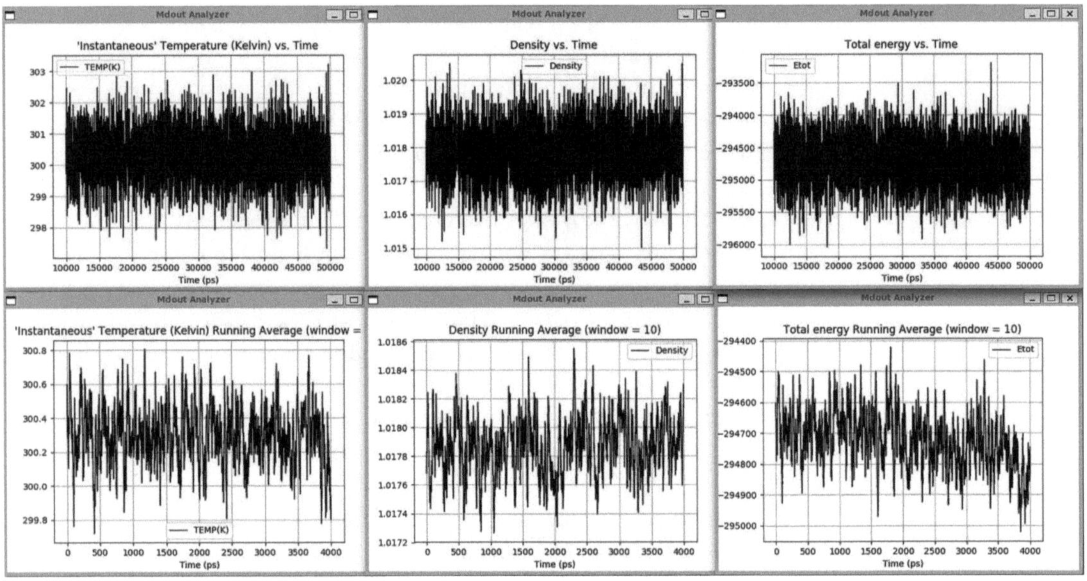

Fig. 4 Graphs of the system temperature, density and total energy during the relaxation period between 10 and 50 ns for the open state of ERAP1 (oc–erap1). The plots below are the running averages for a window of 10 steps (frames)

The next type of analysis will be carried out with the powerful utility of AmberTools *cpptraj* [49], with the aim of obtaining basic information about the system, including geometrical features of interest. In this regard, we will obtain the root-mean-square deviation (rmsd), which measures the deviation of a target set of coordinates to a reference set, the root-mean-square fluctuations (rmsf or atomicfluct), which reports on atomic positional fluctuations with respect to their average position over all input frames, and the radius of gyration (radgyr), which is the root mean square distance of the selected atoms from their center. Then, we will also calculate two metrics of the overall conformational state of ERAP1, the interdomain angle and the interdomain torsion [22], including two distances between the ligand and the enzyme. Lastly, the solvent and counter ions are removed and the trajectory is stored with one frame per 10 ps (see **Note 5**). First, prepare the following input file:

basic-process.in

```
parm oc-erap1.parm7
trajin oc-erap1.rst7
trajin oc-erap1minrst.rst7
trajin oc-erap1eqnptrst1.nc
trajin oc-erap1eqnptrst2.nc
```

```
trajin oc-erap1eqnptrst3.nc
trajin oc-erap1eqnpt.nc
trajin oc-erap1mdrun.nc
autoimage

rms first mass out rmsd_pro_ha.dat ':12-438,473-890 & !
@H='
rms first mass out rmsd_lig_ha.dat ':898 & !@H=' nofit
rms first mass out rmsd_pro_ca.dat :12-438,473-890@CA
atomicfluct out atomfl_ca.dat @CA byres
radgyr :1-898 out radgyr_all.dat mass nomax

angle phi :12-484@CA :485-569@CA :570-893@CA out an-
g_I_II-IV.dat
dihedral psi :12-484@CA :490@CA :524@CA :570-893@CA /
out tors_I_II-IV.dat
distance tyr :393@OH :898@O13 out dist_Y438-PO2.dat
distance ar1 :388@CG,CZ,CD1,CD2,CE1,CE2 :898@C4,C5,C6,C7,
C8,C9 /
out dist_F433-hPhe.dat

strip :WAT,Na+
trajout dry_oc-erap1_100ns.nc netcdf offset 2
```

and execute it with the command:

```
cpptraj < basic-process.in > basic-process.log
```

In this example, we have extended the relaxation period without restraints up to 100 ns in order to monitor convergence of the systems (nstlim in mdrun.in has been set to 45,000,000). A simple 2D plotting tool such as Grace (xmgrace) can be used to plot the extracted features and inspect their fluctuations as a function of the number of frames (Fig. 5). The latter can be converted to simulation time by multiplying by 10 (sampling time is 10 ps per frame, *see* **Note 6**).

Another very powerful analysis that can be carried out is the principal component analysis (PCA), which aims to transform coordinates into a set of orthogonal vectors, the PCs that represent the variance in coordinate space. This analysis is based on a comprehensive tutorial at AMBER-Hub, which can be found here:

https://amberhub.chpc.utah.edu/introduction-to-principal-component-analysis/

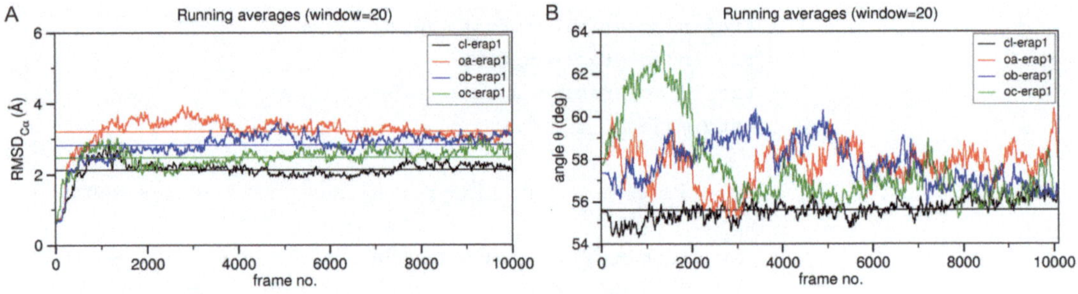

Fig. 5 Plots showing the running averages of protein Cα atoms RMSD (`A, rmsd_pro_ca.dat`) and interdomain angle θ (`B, ang_I_II-IV.dat`) as a function of frame number during relaxation of the closed and the three open states of ERAP1. Frames were extracted every 10 ps (i.e., 100 ns in total), and horizontal lines indicate the average value of selected datasets

To carry out the PCA using *cpptraj*, prepare and run the following script:

`pca_anal.in`

```
parm dry_oc-erap1.parm7
trajin dry_oc-erap1_100ns.nc
rms first :12-438,473-890@CA
average crdset av-oc-erap1
createcrd Traj
run
crdaction Traj rms ref av-oc-erap1 :12-438,473-890@CA
crdaction Traj matrix covar name Covar :12-438,473-890@CA
runanalysis diagmatrix Covar out evecs.dat vecs 3 name
Evecs \
nmwiz nmwizvecs 3 nmwizfile oc-erap1.nmd \
nmwizmask :12-438,473-890@CA
crdaction Traj projection Proj modes Evecs beg 1 end 3 \
:12-438,473-890@CA out pca.dat crdframes 1,last
hist Proj:1 bins 100 out hist.agr norm name projection-1
hist Proj:2 bins 100 out hist.agr norm name projection-2
hist Proj:3 bins 100 out hist.agr norm name projection-3
run
clear all
readdata evecs.dat name Evecs
parm dry_oc-erap1.parm7
parmstrip !(:12-438,473-890@CA)
parmwrite out ca-oc-erap1.parm7
runanalysis modes name Evecs trajout node-1_oc-erap1.nc
pcmin -60\
pcmax 60 tmode 1 trajoutmask :1-845@CA trajoutfmt netcdf
runanalysis modes name Evecs trajout node-2_oc-erap1.nc
```

```
pcmin -60\
pcmax 60 tmode 2 trajoutmask :1-845@CA trajoutfmt netcdf
runanalysis modes name Evecs trajout node-3_oc-erap1.nc
pcmin -60\
pcmax 60 tmode 3 trajoutmask :1-845@CA trajoutfmt netcdf
```

By executing the script with `ccptraj < pca_anal.in > pca_anal.log`,

a 100-ns trajectory of the open ERAP1 conformation (oc-erap1) that was obtained from the `basic-process.in` script is used to calculate the average coordinates (av-oc-erap1) upon which rms-fitting of the trajectory is performed. Then the coordinate covariance matrix (Covar) is generated, eigenvectors and eigenvalues (Evecs) are calculated for the first three PCs, and an output in.nmd format file for viewing with NMWiz module [50] of VMD is generated (`oc-erap1.nmd`, Fig. 6). These calculations have been performed for the Cα atoms of the enzyme excluding the flexible N- and C-termini, as well as without the modelled exon-10 loop (residues 439–472). Consequently, the same atoms and frames are used to project the trajectory along the PCs and generate normalized histograms of the three calculated projections.

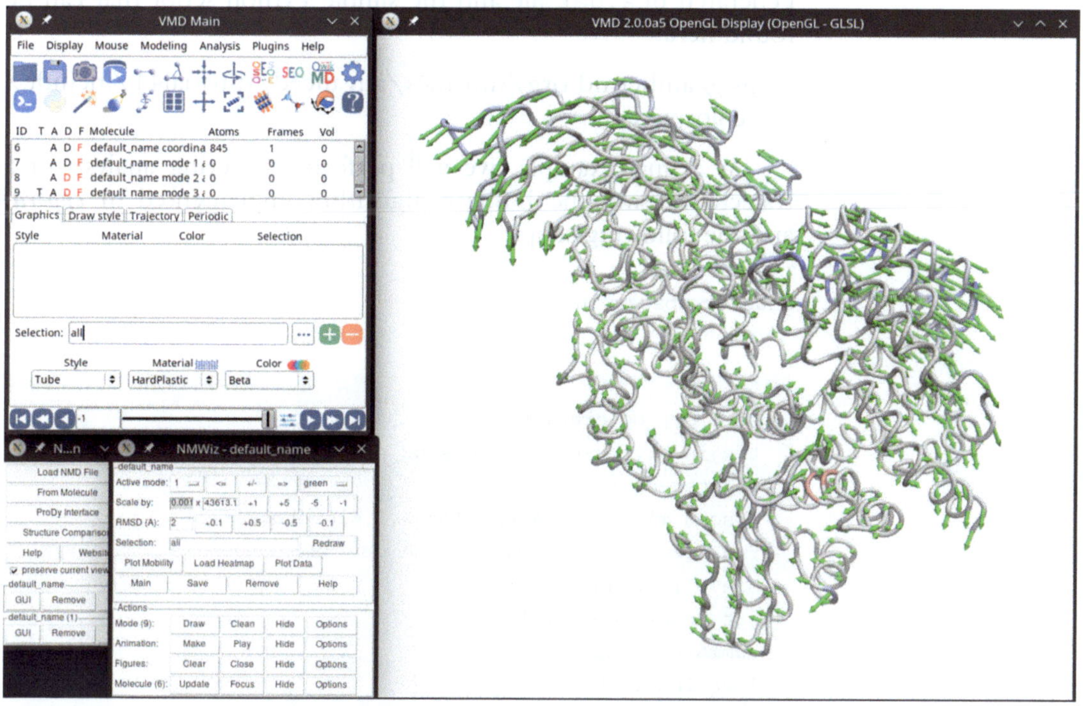

Fig. 6 Visualization of the first principal component (mode 1) obtained from PCA of a 100-ns MD simulation of open ERAP1 using the NMWiz module of VMD. The file `oc-erap1.nmd` obtained from the `pca_anal.in` script is used with standard options, which can be modified from the NMWiz options window

More detailed information about the clustering algorithms implemented in *cpptraj* can be found here:

https://amberhub.chpc.utah.edu/clustering-a-protein-trajectory/

4 Notes

1. Protonation of the terminal amine (+1 charge) and of the phosphinic oxygen of the inhibitor is required, considering their proximity to negatively charged glutamic acid residues of the active site, which create an acidic environment at the N-terminus region of the ligand. The terminal amide group of the ligand is neutral and thus should not be protonated, in contrast to the PDB-deposited model (PBD ID: 6q4r).
2. In case different X-ray structures than those described here will be used as templates, then the corresponding sequence in the alignment (pir) file should be modified accordingly. The Uniprot sequence of ERAP1 isoform 1 (Q9NZ08–1) employed herein corresponds to allotype 2 according to Hutchinson et al. [39]. If one wants to model other ERAP1 allotypes, then the sequence at the last section of the alignment file should be modified according to Table 1 in ref. [39] (e.g. for allotype 1, R127 should be changed to P127).
3. With the most recent version of Amber24 (AmberTools25), a newly developed charge model, ABCG2, can be employed in conjunction with the second-generation General Amber Force Field (GAFF2) [51]. The new BCC parameter set and the GAFF2, which have been shown to perform better, can be used with the command:

   antechamber -i HJ5.mol2 -fi mol2 -o HJ5.prepc -fo prepc /

   -nc 1 -c abcg2 -at gaff2

   To add the missing force field parameters according to the GAFF2, the following command should be used:

   parmchk2 -i HJ5.prepc -f prepc -o HJ5.frcmod -s 1
4. In order to use the most recent Amber force field *ff19SB* [52] in combination with the 4-point, rigid "optimal" point charge (OPC) water model [53], the corresponding commands in the LEaP input file `prep-c1-erap1.cmd` should be changed into:

```
source leaprc.protein.ff19SB
source leaprc.water.opc
source leaprc.gaff2
...
solvateoct mol OPCBOX 10.0 iso
```

It should be noted that the use of the 4-point water model will result in at least 30% overhead in computation time over the 3-point TIP3P model employed here. This combination can be applied to versions of Amber18 and later.

5. The file size of a 100-ns, netcdf-formatted MD trajectory (10,000 frames) for a 125,400-atom system (oc-erap1) is 15 GB. Thus simulations are carried out in chunks of 200 ns each, so as to refrain from generating files of over 30 GB. Disk free space and read/write speed considerations should be planned accordingly before executing the simulations.
6. Special attention is required when loading trajectories of different sampling timescales, as in the case of our protocol, where the sampling rate during production runs has been lowered to one frame per 10 ps, starting from one frame per 5 ps during the relaxation period. In such a case, the [offset] variable in the trajin command of cpptraj should be used.

Conflict of Interest

The authors have no conflicts of interest to declare that are relevant to the content of this chapter.

References

1. Tsujimoto M, Hattori A (2005) The oxytocinase subfamily of M1 aminopeptidases. Biochim Biophys Acta 1751(1):9–18. https://doi.org/10.1016/j.bbapap.2004.09.011
2. Evnouchidou I, Papakyriakou A, Stratikos E (2009) A new role for Zn (II) aminopeptidases: antigenic peptide generation and destruction. Curr Pharm Des 15(31):3656–3670. https://doi.org/10.2174/138161209789271816
3. Saveanu L, Carroll O, Weimershaus M, Guermonprez P, Firat E, Lindo V, Greer F, Davoust J, Kratzer R, Keller SR, Niedermann G, van Endert P (2009) IRAP identifies an endosomal compartment required for MHC class I cross-presentation. Science 325(5937):213–217. https://doi.org/10.1126/science.1172845
4. Kochan G, Krojer T, Harvey D, Fischer R, Chen L, Vollmar M, von Delft F, Kavanagh KL, Brown MA, Bowness P, Wordsworth P, Kessler BM, Oppermann U (2011) Crystal structures of the endoplasmic reticulum aminopeptidase-1 (ERAP1) reveal the molecular basis for N-terminal peptide trimming. Proc Natl Acad Sci USA 108(19):7745–7750. https://doi.org/10.1073/pnas.1101262108
5. Nguyen TT, Chang SC, Evnouchidou I, York IA, Zikos C, Rock KL, Goldberg AL, Stratikos E, Stern LJ (2011) Structural basis for antigenic peptide precursor processing by the endoplasmic reticulum aminopeptidase ERAP1. Nat Struct Mol Biol 18(5):604–613. https://doi.org/10.1038/nsmb.2021
6. Mpakali A, Saridakis E, Harlos K, Zhao Y, Papakyriakou A, Kokkala P, Georgiadis D, Stratikos E (2015) Crystal structure of insulin-regulated aminopeptidase with bound substrate analogue provides insight on antigenic epitope precursor recognition and processing. J Immunol 195(6):2842–2851. https://doi.org/10.4049/jimmunol.1501103
7. Hermans SJ, Ascher DB, Hancock NC, Holien JK, Michell BJ, Chai SY, Morton CJ, Parker MW (2015) Crystal structure of human insulin-regulated aminopeptidase with specificity for cyclic peptides. Protein Sci Publ Protein Soc 24(2):190–199. https://doi.org/10.1002/pro.2604
8. Mpakali A, Saridakis E, Harlos K, Zhao Y, Kokkala P, Georgiadis D, Giastas P,

Papakyriakou A, Stratikos E (2017) Ligand-induced conformational change of insulin-regulated aminopeptidase: insights on catalytic mechanism and active site plasticity. J Med Chem 60(7):2963–2972. https://doi.org/10.1021/acs.jmedchem.6b01890

9. Giastas P, Neu M, Rowland P, Stratikos E (2019) High-resolution crystal structure of endoplasmic reticulum aminopeptidase 1 with bound Phosphinic transition-state analogue inhibitor. ACS Med Chem Lett 10(5): 708–713. https://doi.org/10.1021/acsmedchemlett.9b00002
10. Maben Z, Arya R, Georgiadis D, Stratikos E, Stern LJ (2021) Conformational dynamics linked to domain closure and substrate binding explain the ERAP1 allosteric regulation mechanism. Nat Commun 12(1):5302. https://doi.org/10.1038/s41467-021-25564-w
11. Vourloumis D, Mavridis I, Athanasoulis A, Temponeras I, Koumantou D, Giastas P, Mpakali A, Magrioti V, Leib J, van Endert P, Stratikos E, Papakyriakou A (2022) Discovery of selective Nanomolar inhibitors for insulin-regulated aminopeptidase based on alpha-Hydroxy-beta-amino acid derivatives of Bestatin. J Med Chem 65(14):10098–10117. https://doi.org/10.1021/acs.jmedchem.2c00904
12. Birtley JR, Saridakis E, Stratikos E, Mavridis IM (2012) The crystal structure of human endoplasmic reticulum aminopeptidase 2 reveals the atomic basis for distinct roles in antigen processing. Biochemistry 51(1): 286–295. https://doi.org/10.1021/bi201230p
13. Evnouchidou I, Birtley J, Seregin S, Papakyriakou A, Zervoudi E, Samiotaki M, Panayotou G, Giastas P, Petrakis O, Georgiadis D, Amalfitano A, Saridakis E, Mavridis IM, Stratikos E (2012) A common single nucleotide polymorphism in endoplasmic reticulum aminopeptidase 2 induces a specificity switch that leads to altered antigen processing. J Immunol 189(5):2383–2392. https://doi.org/10.4049/jimmunol.1200918
14. Mpakali A, Giastas P, Mathioudakis N, Mavridis IM, Saridakis E, Stratikos E (2015) Structural basis for antigenic peptide recognition and processing by endoplasmic reticulum (ER) aminopeptidase 2. J Biol Chem 290(43):26021–26032. https://doi.org/10.1074/jbc.M115.685909
15. Mpakali A, Giastas P, Deprez-Poulain R, Papakyriakou A, Koumantou D, Gealageas R, Tsoukalidou S, Vourloumis D, Mavridis IM, Stratikos E, Saridakis E (2017) Crystal structures of ERAP2 complexed with inhibitors reveal pharmacophore requirements for optimizing inhibitor potency. ACS Med Chem Lett 8(3):333–337. https://doi.org/10.1021/acsmedchemlett.6b00505
16. Camberlein V, Fleau C, Sierocki P, Li L, Gealageas R, Bosc D, Guillaume V, Warenghem S, Leroux F, Rosell M, Cheng K, Medve L, Prigent M, Decanter M, Piveteau C, Biela A, Eveque M, Dumont J, Mpakali A, Giastas P, Herledan A, Couturier C, Haupenthal J, Lesire L, Hirsch AKH, Deprez B, Stratikos E, Bouvier M, Deprez-Poulain R (2022) Discovery of the first selective Nanomolar inhibitors of ERAP2 by kinetic target-guided synthesis. Angew Chem Int Ed Engl 61(39):e202203560. https://doi.org/10.1002/anie.202203560
17. Mpakali A, Georgiadis D, Stratikos E, Giastas P (2022) Inhibitor-dependent usage of the S1' specificity pocket of ER aminopeptidase 2. ACS Med Chem Lett 13(2):218–224. https://doi.org/10.1021/acsmedchemlett.1c00582
18. York IA, Chang SC, Saric T, Keys JA, Favreau JM, Goldberg AL, Rock KL (2002) The ER aminopeptidase ERAP1 enhances or limits antigen presentation by trimming epitopes to 8-9 residues. Nat Immunol 3(12):1177–1184. https://doi.org/10.1038/ni860
19. Chang SC, Momburg F, Bhutani N, Goldberg AL (2005) The ER aminopeptidase, ERAP1, trims precursors to lengths of MHC class I peptides by a "molecular ruler" mechanism. Proc Natl Acad Sci USA 102(47): 17107–17112. https://doi.org/10.1073/pnas.0500721102
20. Giastas P, Mpakali A, Papakyriakou A, Lelis A, Kokkala P, Neu M, Rowland P, Liddle J, Georgiadis D, Stratikos E (2019) Mechanism for antigenic peptide selection by endoplasmic reticulum aminopeptidase 1. Proc Natl Acad Sci USA 116(52):26709–26716. https://doi.org/10.1073/pnas.1912070116
21. Papakyriakou A, Stratikos E (2017) The role of conformational dynamics in antigen trimming by intracellular aminopeptidases. Front Immunol 8:946. https://doi.org/10.3389/fimmu.2017.00946
22. Stamogiannos A, Koumantou D, Papakyriakou A, Stratikos E (2015) Effects of polymorphic variation on the mechanism of endoplasmic reticulum aminopeptidase 1. Mol Immunol 67(2 Pt B):426–435. https://doi.org/10.1016/j.molimm.2015.07.010
23. Stamogiannos A, Maben Z, Papakyriakou A, Mpakali A, Kokkala P, Georgiadis D, Stern LJ, Stratikos E (2017) Critical role of Interdomain interactions in the conformational change and catalytic mechanism of endoplasmic reticulum

aminopeptidase 1. Biochemistry 56(10): 1546–1558. https://doi.org/10.1021/acs.biochem.6b01170

24. Evnouchidou I, Weimershaus M, Saveanu L, van Endert P (2014) ERAP1-ERAP2 dimerization increases peptide-trimming efficiency. J Immunol 193(2):901–908. https://doi.org/10.4049/jimmunol.1302855
25. Papakyriakou A, Mpakali A, Stratikos E (2022) Can ERAP1 and ERAP2 form functional heterodimers? A structural dynamics investigation. Front Immunol 13:863529. https://doi.org/10.3389/fimmu.2022.863529
26. Salomon-Ferrer R, Case DA, Walker RC (2013) An overview of the Amber biomolecular simulation package. WIREs Comput Mol Sci 3(2):198–210. https://doi.org/10.1002/wcms.1121
27. Hug S (2013) Classical molecular dynamics in a nutshell. Methods Mol Biol 924:127–152. https://doi.org/10.1007/978-1-62703-017-5_6
28. Monticelli L, Tieleman DP (2013) Force fields for classical molecular dynamics. Methods Mol Biol 924:197–213. https://doi.org/10.1007/978-1-62703-017-5_8
29. Case DA, Cheatham TE III, Darden T, Gohlke H, Luo R, Merz KM Jr, Onufriev A, Simmerling C, Wang B, Woods RJ (2005) The Amber biomolecular simulation programs. J Comput Chem 26(16):1668–1688. https://doi.org/10.1002/jcc.20290
30. Salomon-Ferrer R, Gotz AW, Poole D, Le Grand S, Walker RC (2013) Routine microsecond molecular dynamics simulations with AMBER on GPUs. 2. Explicit solvent particle mesh Ewald. J Chem Theory Comput 9(9): 3878–3888. https://doi.org/10.1021/ct400314y
31. Case DA, Aktulga HM, Belfon K, Cerutti DS, Cisneros GA, Cruzeiro VWD, Forouzesh N, Giese TJ, Gotz AW, Gohlke H, Izadi S, Kasavajhala K, Kaymak MC, King E, Kurtzman T, Lee TS, Li P, Liu J, Luchko T, Luo R, Manathunga M, Machado MR, Nguyen HM, O'Hearn KA, Onufriev AV, Pan F, Pantano S, Qi R, Rahnamoun A, Risheh A, Schott-Verdugo S, Shajan A, Swails J, Wang J, Wei H, Wu X, Wu Y, Zhang S, Zhao S, Zhu Q, Cheatham TE 3rd, Roe DR, Roitberg A, Simmerling C, York DM, Nagan MC, Merz KM Jr (2023) AmberTools. J Chem Inf Model 63(20):6183–6191. https://doi.org/10.1021/acs.jcim.3c01153
32. Fiser A, Do RK, Sali A (2000) Modeling of loops in protein structures. Protein science: a publication of the Protein Society 9(9): 1753–1773. https://doi.org/10.1110/ps.9.9.1753
33. Humphrey W, Dalke A, Schulten K (1996) VMD: visual molecular dynamics. J Mol Graph Model 14(1):33–38. https://doi.org/10.1016/0263-7855(96)00018-5
34. Gordon JC, Myers JB, Folta T, Shoja V, Heath LS, Onufriev A (2005) H++: a server for estimating pKas and adding missing hydrogens to macromolecules. Nucleic Acids Res 33(Web Server issue):W368–W371. https://doi.org/10.1093/nar/gki464
35. Mpakali A, Georgaki G, Buson A, Findlay AD, Foot JS, Mauvais FX, van Endert P, Giastas P, Hamprecht DW, Stratikos E (2024) Stabilization of the open conformation omicronf insulin-regulated aminopeptidase by a novel substrate-selective small-molecule inhibitor. Protein science: a publication of the Protein Society 33(9):e5151. https://doi.org/10.1002/pro.5151
36. Chen L, Lin YL, Peng G, Li F (2012) Structural basis for multifunctional roles of mammalian aminopeptidase N. Proc Natl Acad Sci USA 109(44):17966–17971. https://doi.org/10.1073/pnas.1210123109
37. Kyrieleis OJ, Goettig P, Kiefersauer R, Huber R, Brandstetter H (2005) Crystal structures of the tricorn interacting factor F3 from Thermoplasma acidophilum, a zinc aminopeptidase in three different conformations. J Mol Biol 349(4):787–800. https://doi.org/10.1016/j.jmb.2005.03.070
38. Georgiadis D, Skoulikas N, Papakyriakou A, Stratikos E (2022) Phosphinic peptides as tool compounds for the study of pharmacologically relevant Zn-metalloproteases. ACS Pharmacol Transl Sci 5(12):1228–1253. https://doi.org/10.1021/acsptsci.2c00183
39. Hutchinson JP, Temponeras I, Kuiper J, Cortes A, Korczynska J, Kitchen S, Stratikos E (2021) Common allotypes of ER aminopeptidase 1 have substrate-dependent and highly variable enzymatic properties. J Biol Chem 296:100443. https://doi.org/10.1016/j.jbc.2021.100443
40. Jakalian A, Jack DB, Bayly CI (2002) Fast, efficient generation of high-quality atomic charges. AM1-BCC model: II. Parameterization and validation. J Comput Chem 23(16):1623–1641. https://doi.org/10.1002/jcc.10128
41. Wang J, Wolf RM, Caldwell JW, Kollman PA, Case DA (2004) Development and testing of a general amber force field. J Comput Chem

25(9):1157–1174. https://doi.org/10.1002/jcc.20035

42. Hoops SC, Anderson KW, Merz KM (1991) Force-field Design for Metalloproteins. J Am Chem Soc 113(22):8262–8270. https://doi.org/10.1021/Ja00022a010
43. Maier JA, Martinez C, Kasavajhala K, Wickstrom L, Hauser KE, Simmerling C (2015) ff14SB: improving the accuracy of protein side chain and backbone parameters from ff99SB. J Chem Theory Comput 11(8):3696–3713. https://doi.org/10.1021/acs.jctc.5b00255
44. Jorgensen WL, Chandrasekhar J, Madura JD, Impey RW, Klein ML (1983) Comparison of simple potential functions for simulating liquid water. J Chem Phys 79(2):926–935. https://doi.org/10.1063/1.445869
45. Joung IS, Cheatham TE III (2008) Determination of alkali and halide monovalent ion parameters for use in explicitly solvated biomolecular simulations. J Phys Chem B 112(30):9020–9041. https://doi.org/10.1021/jp8001614
46. Hopkins CW, Le Grand S, Walker RC, Roitberg AE (2015) Long-time-step molecular dynamics through hydrogen mass repartitioning. J Chem Theory Comput 11(4):1864–1874. https://doi.org/10.1021/ct5010406
47. Pierce LC, Salomon-Ferrer R, Augusto FOC, McCammon JA, Walker RC (2012) Routine access to millisecond time scale events with accelerated molecular dynamics. J Chem Theory Comput 8(9):2997–3002. https://doi.org/10.1021/ct300284c
48. Gedeon PC, Thomas JR, Madura JD (2015) Accelerated molecular dynamics and protein conformational change: a theoretical and practical guide using a membrane embedded model neurotransmitter transporter. Methods Mol Biol 1215:253–287. https://doi.org/10.1007/978-1-4939-1465-4_12
49. Roe DR, Cheatham TE III (2013) PTRAJ and CPPTRAJ: software for processing and analysis of molecular dynamics trajectory data. J Chem Theory Comput 9(7):3084–3095. https://doi.org/10.1021/ct400341p
50. Bakan A, Meireles LM, Bahar I (2011) ProDy: protein dynamics inferred from theory and experiments. Bioinformatics 27(11):1575–1577. https://doi.org/10.1093/bioinformatics/btr168
51. He X, Man VH, Yang W, Lee TS, Wang J (2020) A fast and high-quality charge model for the next generation general AMBER force field. J Chem Phys 153(11):114502. https://doi.org/10.1063/5.0019056
52. Tian C, Kasavajhala K, Belfon KAA, Raguette L, Huang H, Migues AN, Bickel J, Wang Y, Pincay J, Wu Q, Simmerling C (2020) ff19SB: amino-acid-specific protein backbone parameters trained against quantum mechanics energy surfaces in solution. J Chem Theory Comput 16(1):528–552. https://doi.org/10.1021/acs.jctc.9b00591
53. Izadi S, Anandakrishnan R, Onufriev AV (2014) Building water models: a different approach. J Phys Chem Lett 5(21):3863–3871. https://doi.org/10.1021/jz501780a

Chapter 11

Guidelines for the Design of Custom Affinity-Based Probes for Metalloproteases

Marie Launay, Lomane Berthy, Sandra Llamas Rizo, Myrto Athina Balachouti, Dimitris Georgiadis, and Laurent Devel

Abstract

Although metalloproteases (MPs) have been studied for decades, their precise roles in various pathological processes remain to be fully elucidated. These enzymes often function within complex networks, and their activation states can vary over time and depending on the pathological context. In this dynamic landscape, there is a critical need for robust analytical tools capable of accurately identifying which MPs are present in their active forms and how their activation patterns evolve. Among the available technologies for profiling active MPs, affinity-based probes (A*f*BPs) have emerged as powerful chemical tools. These probes enable both the tracking and unambiguous identification of active MPs in complex biological systems. Here we attempt to provide guidelines for the design of A*f*BPs that could show selectivity for specific MP subclasses. We present the different alternatives for each structural element of a custom A*f*BP for MPs and the parameters that need to be considered in order to generate an effective A*f*BP for a target enzyme.

Key words Metalloproteinases, Affinity-based probes, Photoactivatable warhead, Cleavable linker

1 Introduction

Metalloproteases (MPs) constitute an essential class of zinc-dependent enzymes that, in their active form, catalyze the hydrolysis of peptide bonds within proteins. In humans, these enzymes account for approximately 1% of total protein-coding genes and comprise several subgroups of endopeptidases, notably including matrix metalloproteinases (MMPs), ADAMs (A Disintegrin and Metalloproteinases), and ADAMTS (ADAM with Thrombospondin motifs), as well as metallocarboxypeptidases, and metalloaminopeptidases. Among these, MMPs represent the most extensively studied subgroup due to their pivotal roles in various physiological processes [1]. Dysregulated MMP expression and activity have been associated with numerous pathological conditions, including cancer, inflammatory diseases, viral infections, and sepsis

Efstratios Stratikos (ed.), *Antigen Processing and Presentation: Methods and Protocols*, Methods in Molecular Biology, vol. 3007, https://doi.org/10.1007/978-1-0716-5092-9_11,

[1]. ADAM proteases primarily function as ectodomain sheddases, modulating cell signaling pathways and intercellular interactions through controlled proteolytic cleavage of membrane-bound proteins [2, 3]. Concerning the role of MMPs/ADAMs in immunomodulation, it is well documented that members of these families ensure that the immune response is adequately propagated and terminated to avoid chronic inflammation and tissue damage [4]. The ADAMTS family mainly participates in extracellular matrix turnover, especially during inflammation, as well as in blood coagulation processes [3, 5]. Metallocarboxypeptidases and metalloaminopeptidases significantly contribute to protein maturation and peptide hormone regulation. Specifically, Endoplasmic Reticulum Aminopeptidase 1 (ERAP1), Endoplasmic Reticulum Aminopeptidase 2 (ERAP2), and Insulin-Regulated Aminopeptidase (IRAP) play critical roles in antigen processing. ERAP1 and ERAP2, predominantly localized within the endoplasmic reticulum, trim precursor peptides by removing N-terminal amino acids, generating peptides of optimal length for presentation by Major Histocompatibility Complex class I (MHC-I) molecules—a process crucial for adaptive immune responses [6, 7]. While MMPs, ADAMs, and ADAMTS are expressed as inactive zymogens requiring proteolytic removal of their inhibitory pro-domains to become active, ERAP-type enzymes are synthesized as fully folded active enzymes. The proteolytic activity of these latter is regulated internally through conformational flexibility and substrate-dependent activation mechanisms [6, 7]. Overall, human metalloproteases significantly influence the pathogenesis of numerous diseases, making them promising targets for diagnostic applications—where their activation status may serve as biomarkers indicative of specific pathological stages—and for therapeutic intervention strategies.

Determining the activation status of metalloproteases (MPs) can be achieved using chemical probes typically designed on substrate- or ligand-based scaffolds [1]. Substrate-based probes, such as fluorogenic substrates, offer real-time detection and high sensitivity but often lack selectivity, particularly when several MPs are simultaneously activated. Moreover, these probes do not allow definitive identification of specific MPs. On the other hand, affinity-based probes (A*f*BPs), a subclass of ligand-based probes, bind to the active sites of MPs and subsequently react with neighboring nucleophilic functionalities, resulting in stable MP-A*f*BP covalent adducts suitable for analysis. Since the seminal study by Cravatt's group [8, 9], this A*f*BP profiling strategy has predominantly focused on MMPs, remaining largely underexplored for other MP families.

In this article, we aim to provide a practical guide for the design of A*f*BPs depending on the MP subgroup targeted. We will present the various modular elements required for both specific MP binding and covalent modification and we will list pros and cons based on current literature. These include i) the targeting ligand that specifically binds the active site of the MP, ii) the reactive warhead that forms a covalent bond with amino acid residues within the MP, and iii) the reporter tag for subsequent detection and analysis of the covalent adduct.

2 Methods and Guidelines

2.1 Select the Binding Motif

1. A suitable binding motif must be selected. This is a critical step since it will determine the specificity of the interaction between the AfBP and the metalloprotease.
2. Three key parameters must be considered (Fig. 1): i) the type of MP being targeted—whether it is an **endopeptidase** (e.g., MMPs), **aminopeptidase** (e.g., ERAPs) or **carboxypeptidase** (e.g., CPA), ii) the orientation of binding within the active site which generates specific enzyme/ligand interactions—whether the ligand spans the primed, the unprimed or both sides of the active cleft, and iii) the positioning of the zinc-binding group (ZBG) within the structure of the ligand—whether is terminal or internal.
3. Despite the wide variety of ZBGs employed to target the catalytic zinc ion within the active site of metalloproteases (MPs), most reported inhibitors are primarily based on hydroxamate or phosphinate scaffolds. These ZBGs provide a solid foundation for designing custom A*f*BPs for MPs and are generally recommended as a starting point. (*see* **Note 1**).
4. In this context, the advantages and limitations of each ZBG type for different classes of MPs are summarized in Table 1. This comparative analysis aims to assist researchers in selecting the most suitable ligand during AfBP design.

2.2 Select the Reactive Warhead

1. The next step is to select a suitable reactive warhead. This structural element is particularly important in the case of MPs because, unlike cysteine and serine proteases—which contain intrinsic catalytic nucleophiles (a cysteine and a serine residue, respectively)—these proteases lack a canonical nucleophile in their active site.
2. Two main types of reactive warhead can be considered.
3. A **photoactivatable warhead**: a reactive group (e.g. benzophenone, aryl azide, diazirine) which upon photoirradiation becomes sufficiently reactive to covalently crosslink adjacent amino acid side chains or protein backbone (Table 2, entries a to c).
4. *Advantages*: (a) Well established, as this covalent crosslinking approach has been widely used for MP-targeting A*f*BPs over

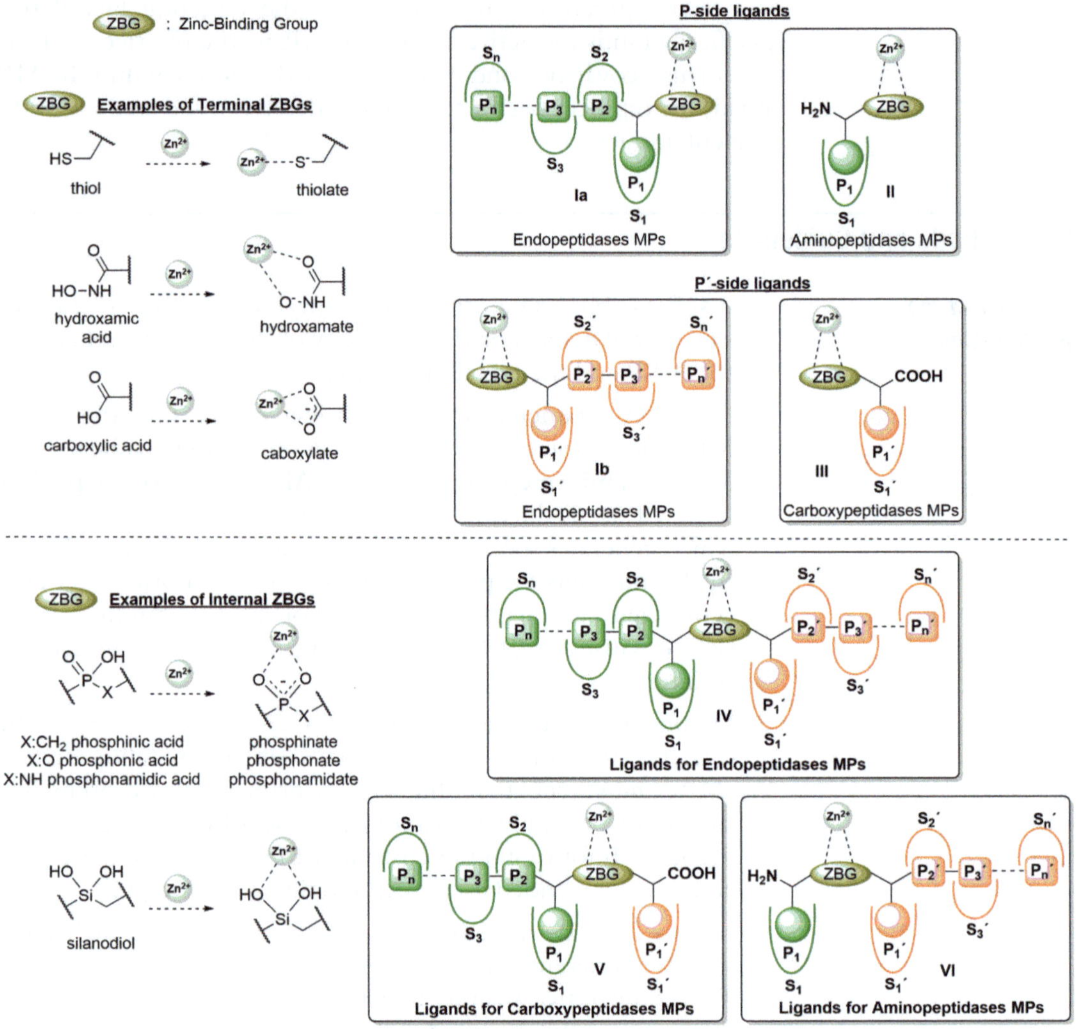

Fig. 1 Generic structures of different types of ligands depending on the type of ZBG being used, the orientation of ligand within the active site and the type of MP being targeted

the past two decades. (b) Preferred method when limited structural information of possible cross-linking sites of targeted MPs are available. (c) Building blocks suitable for solid phase peptide synthesis bearing photoactivatable tags are commercially available.

5. *Drawbacks*: (a) Low crosslinking yields are typically attained. (b) The photoactivatable group must be poorly exposed to solvent to achieve an efficient crosslink. (c) A photoirradiation equipment is necessary.

6. An **electrophilic cleavable linker**: a reactive cleavable linker (e.g. acyl imidazoles [10], *N*-acyl-*N*-alkyl sulfonamides [10–12] which enables proximity-driven reactions with MP nucleophiles (Table 2, entries d and e).

Table 1
Comparison between hydroxamate and phosphinate ligands for targeting endopeptidases, carboxypeptidases and aminopeptidases

	Hydroxamates	Phosphinates
Endopeptidases	Pros: Widely used for MMPs, Strong ZBGs, Both P- or P′-side ligands are accessible Synthetically straightforward Cons: Limited availability of selective ligands P/P′-side ligands cannot be accessed	Pros: Widely used for MMPs, Weaker ZBGs than hydroxamates Nanomolar affinities are achievable P/P′-side ligands are accessible Selective ligands are available Cons: More synthetically demanding
Carboxypeptidases	Pros: Cons: Limited use, not recommended. Only short P′-side ligands of type III (Fig. 1) are accessible	Pros: Efficient ZBGs, nanomolar affinities are achievable Extension to P-side is possible, enriching design options Selective ligands are available Cons: -
Aminopeptidases	Pros: Hydroxamate ligands have been used to target aminopeptidases by unselective probes of type Ib. Cons: Limited availability of selective ligands Moderate affinities as compared to phosphinates Only short P-side ligands of type II (Fig. 1) are accessible	Pros: Nanomolar affinities are achievable Extension to P′-side is possible, enriching design options Typically stronger Zn-ligands than hydroxamates Cons: More synthetically demanding Underexplored as probe ligands, compared to alternatives of higher specificity, such as bestatin [19]

7. *Advantages*: (a) The photoirradiation step is omitted, offering operational simplicity and avoiding unwanted photodecomposition by-products. (b) Compatible with all biological matrices and in vivo applications. (c) Post labelling, this proximity-driven approach may leave the catalytic site partially or fully accessible, a feature that is clearly advantageous for monitoring active MPs in complex biological systems (*see* **Note 2**).

8. *Drawbacks*: (a) This approach is relatively novel for MPs and has been limited so far to labelling of MMPs. (b) It lacks the temporal control afforded by photo-driven labelling methods. (c) The positioning of the cleavable linker within the AfBP structure may critically affect labelling breadth (*see* **Note 3**). (d) More challenging synthetically.

Table 2
Reactive warhead in photoactivatable (entries a to c) and cleavable electrophile (entries d and e) series

	Reactive warhead	Irradiation wavelength *Reactive species*	Position within the AfBP structure	MPs targeted and residue(s) cross-linked, when identified
a	Benzophenone 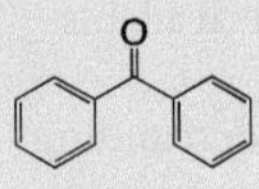	350–365 nm 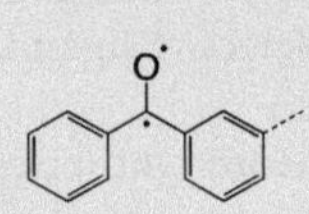 *Triplet diradical*	P_2'/P_3' P_3'/P_4' P_{10}' Primed site (*see* **Note 8**)	MMPs: 1/2/3/7/8/9/12/13 [8, 9, 15, 18, 20] ADAM: 10/17 [9] ADAMTS: 4 [9] Aminopeptidases (e.g. ALAP, PSAP) [8, 9] Dipeptidyl peptidase [9] MMPs: 3/7/14 [21] Endoplasmatic reticulum aminopeptidase (ERAP1) (Leu^{677}/Leu^{686}/Leu^{838}) [22] *P. falciparum* aminopeptidases (PfA-M1 and pf-LAP) [19]
b	Diazirine 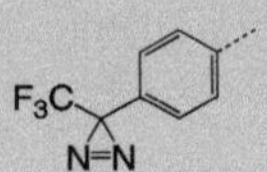	350 nm 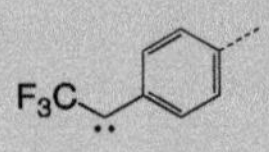 *Carbene*	P_4 P_1' P_2'	MMPs: 9 [23] Carboxypeptidase: CPA [23] MMPs:1/2/3/7/8/9/10/12/13/14 [24–26] ADAM-9/10/17 [24] MMPs: 9/10 [24, 27] ADAM: 9/10/17 [24, 27]
c	Aryl azide 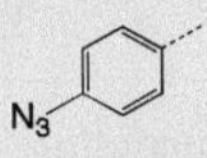	254–400 nm 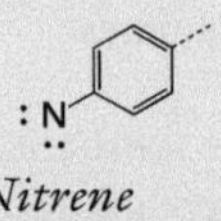 *Nitrene*	P_1'	MMPs: 2/3(His^{241})/8/9/10/11/12 (Lys^{241}) /13/14 [28–31]
d	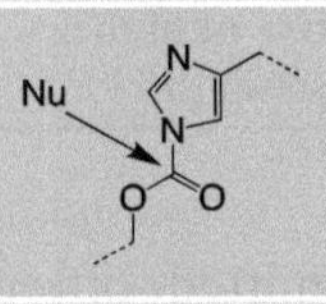	Not applicable	P_3'	MMPs: 3/10/12 (Lys^{177}) [10]
e	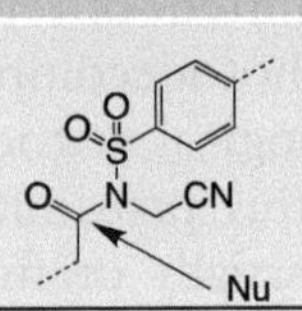	Not applicable	P_3' P_3	MMPs: 3/9(Lys^{214})/12 (Lys^{177})/ 13 (Lys^{212}) [10] MMPs: 2/3/8/9/10/12/13 (N-terminus end) [11]

9. It is essential to emphasize that once the most appropriate reactive warhead has been selected, its optimal positioning within the ligand structure must be determined to ensure efficient labelling.

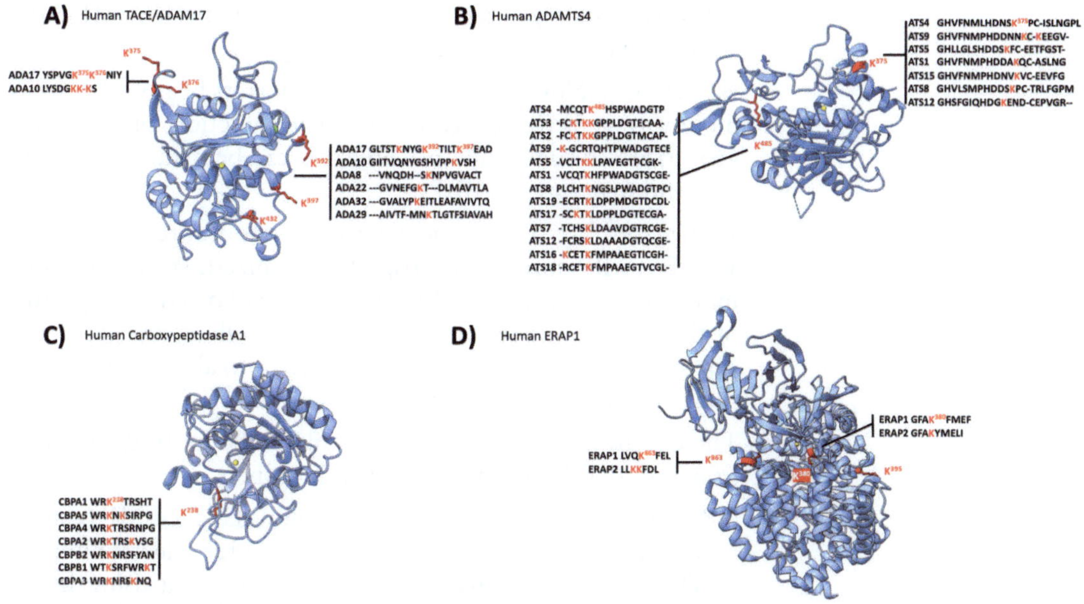

Fig. 2 Targetable lysine residues within the catalytic cleft of human TACE/ADAM17 (**a**), human ADAMTS4 (**b**), human carboxypeptidase A1 (**c**) and human ERAP1 (D). The figures were generated from X-ray structure using ChimeraX®. Key: catalytic domain (cartoon, Cornflower blue), zinc ion (yellow ball); calcium ion (green ball); Lysine (red stick). MPs have been aligned using uniprot (https:/www.uniprot.org)

10. When the warhead is a photoactivatable group, it should ideally be oriented toward hydrophobic subsites of the metalloprotease (MP).
11. In contrast, if a cleavable linker is used, it should be positioned to face nucleophilic amino acid residues, such as amino function of lysine side chain.
12. Available X-ray crystal structures of MPs from the Protein Data Bank (PDB) are valuable resources for identifying such topological features within the active site cleft (*see* **Note 4**). For example, lysine residues that are accessible to electrophilic cleavable linkers can often be located near the catalytic cleft through visual inspection of X-ray structures using ChimeraX® (Fig. 2).
13. Additionally, sequence alignment via UniProt® and advanced structure prediction tools such as AlphaFold [13] and RoseTTAFold [14] can be used to complement experimental data, facilitating the identification of regions suitable for covalent labelling.

2.3 Select the Analytical Handle

1. The choice of analytical handles incorporated within the structure of A*f*BPs is directly guided by the intended modalities of analysis.

2. A **fluorescent reporter** is used to visualize covalent MP-AfBP adducts via in-gel fluorescence imaging.
3. *Advantages*: Numerous fluorescent reporters bearing a reactive moiety (e.g. activated carboxylic acid function or an amino group) are commercially available and can be readily integrated into the structure of AfBPs.
4. *Drawbacks*: (a) If the spacer between the fluorescent moiety and the A*f*BP core is not properly optimized, steric hindrance may interfere with target MP binding. (b) The chemical nature of the fluorescent reporter must be carefully considered, as it can increase nonspecific binding of A*f*BPs to abundant proteins such as serum albumin (*see* **Note 5**).
5. The **biotin tag** serves as an analytical handle for both visualization and identification of MP–AfBP adducts.
6. *Advantages*: (a) Within a single workflow, labelled MPs can be: i) visualized through in-gel imaging after electrophoresis, PVDF membrane transfer, and western blotting using horseradish peroxidase (HRP), and ii) strictly identified through streptavidin-based enrichment using magnetic beads, followed by on-bead trypsin digestion and mass spectrometry (*see* **Note 6**), (b) The biotin moiety can also be leveraged in the development of multiplex bead array assays ([15]; *see* **Note** 7).
7. *Drawbacks*: The linker between AfBP core and biotin must be carefully optimized to minimize steric clashes during both MP labelling and streptavidin-based enrichment, as these can compromise proper target MP visualization and identification.
8. A **clickable handle** (e.g., alkyne or azido group) allows for the modular addition of analytical tags via a simple biorthogonal click chemistry.
9. *Advantages*: (a) High versatility enables attachment of diverse tags using a single reactive handle. (b) Reduced steric hindrance compared to fluorescent or biotin tags, enhancing MP targeting, (c) Potentially lowers nonspecific binding to abundant proteins (*see* **Note 5**).
10. *Drawbacks*: (a) Requires an additional chemical step (click reaction) to install the analytical tag. (b) Limited accessibility of the handle conjugated to MP core may impair derivatization efficiency. (c) The clickable handle, particularly the terminal alkyne, may be unstable in complex biological matrices.

3 Notes

1. For a selective orthosteric ligand, it is advisable to first draw information from the huge amount of SAR data available in the literature in order to identify key selectivity determinants. For

instance, in the case of M_1 aminopeptidases, it has been suggested that variability in S_1 subsite is a key determinant for enzyme specificity [16]. On the other hand, in the case of MMPs, the S_1' subsite of their active site (the so-called S_1 specificity loop) must be targeted to tune ligand selectivity [1, 17].

2. Our group showed that human MMP-12 labelled at its S_3' subsite with a small Cy3® fluorescent reporter, retained significant proteolytic activity compared to its unlabeled form [10].
3. Targeting the S_3' subsite of MMPs with a NASA cleavable linker (Table 2, entry e) led to efficient labelling of a restricted set of MMPs (MMP-3, −9, −12, and − 13) [10], while targeting the S_3 region with the same warhead allowed broad labelling across nearly all tested MMPs (MMP-2, -3, -8, -9, -10, -12, and -13) [11]. This difference is primarily due to the limited presence of reactive lysine in the MMPs S_3' region, whereas both canonical lysine and the N-terminal amino group are present in their S_3 region and can participate in proximity-driven reactions.
4. Extensive structural data are available for human MPs in the Protein Data Bank (PDB), including over 150 structures for MMPs, around 50 for ADAMs (e.g., ADAM10 and ADAM17) and ADAMTSs, and about 60 for carboxypeptidases (e.g., CPA1 and CPA2), and 50 for aminopeptidases.
5. Our group recently demonstrated that an *Af*BP transferring a simple azido tag resulted in lower nonspecific labelling in complex proteomes compared to its Cy3®-tagged counterpart [11]. This clearly highlights that part of the nonspecific reactivity associated with *Af*BPs can be attributed to the fluorescent reporter itself.
6. This approach has been successfully used to identify several MPs across various human cell lines [9] and in the culture medium of bovine cartilage explants [18]. Our group recently employed this dual analysis—combining in-gel visualization and mass spectrometry-based identification—to study murine MMP-12 secreted by engineered human embryonic kidney (HEK) cells [12].
7. Biotin-labelled MPs were utilized in the development of a novel multiplex bead array assay [15]. In this system, the biotin-labelled MPs bind to fluorescent streptavidin, producing a signal proportional to the quantity of active MP. This method has been applied to the multiplex detection of MMPs in peritoneal fluid aspirates, menstrual effluent, and uterine biopsies.

8. From the P_1' position of bestatin, used as a targeting ligand, a long C-terminus polyethylene glycol spacer (PEG_3) was decorated with the photoactivatable group. As a result, the precise positioning of this group within the MP core is difficult to determine.

Acknowledgement

This work was supported by the French National Research Agency (ANR-24-CE18-2934, ML and ANR-24-CE44-0275, LB) and the European Commission in the context of the Marie Skłodowska–Curie Action European Training Network CAPSTONE (954992 – CAPSTONE – H2020 – MSCA–ITN – 2020, SLR).

Conflict of Interest The authors have no conflicts of interest to declare that are relevant to the content of this chapter.

References

1. de Almeida LGN, Thode H, Eslambolchi Y, Chopra S, Young D, Gill S, Devel L, Dufour A (2022) Matrix metalloproteinases: from molecular mechanisms to physiology, pathophysiology, and pharmacology. Pharmacol Rev 74(3):712–768. https://doi.org/10.1124/pharmrev.121.000349
2. Meyer-Schwesinger C, Seipold L, Saftig P (2022) Ectodomain shedding by ADAM proteases as a central regulator in kidney physiology and disease. Biochimica et Biophysica Acta (BBA) 1869(3):119165. https://doi.org/10.1016/j.bbamcr.2021.119165
3. Wang Z, Li W, Chen S, Tang XX (2023) Role of ADAM and ADAMTS proteases in pathological tissue remodeling. Cell Death Discov 9(1):447. https://doi.org/10.1038/s41420-023-01744-z
4. Khokha R, Murthy A, Weiss A (2013) Metalloproteinases and their natural inhibitors in inflammation and immunity. Nat Rev Immunol 13:649. https://doi.org/10.1038/nri3499
5. Santamaria S, de Groot R (2020) ADAMTS proteases in cardiovascular physiology and disease. Open Biol 10(12):200333. https://doi.org/10.1098/rsob.200333
6. de Castro JAL, Stratikos E (2019) Intracellular antigen processing by ERAP2: molecular mechanism and roles in health and disease. Hum Immunol 80(5):310–317. https://doi.org/10.1016/j.humimm.2018.11.001
7. York IA, Chang S-C, Saric T, Keys JA, Favreau JM, Goldberg AL, Rock KL (2002) The ER aminopeptidase ERAP1 enhances or limits antigen presentation by trimming epitopes to 8-9 residues. Nat Immunol 3(12):1177–1184
8. Saghatelian A, Jessani N, Joseph A, Humphrey M, Cravatt BF (2004) Activity-based probes for the proteomic profiling of metalloproteases. Proc Natl Acad Sci USA 101(27):10000–10005. https://doi.org/10.1073/pnas.0402784101
9. Sieber SA, Niessen S, Hoover HS, Cravatt BF (2006) Proteomic profiling of metalloprotease activities with cocktails of active-site probes. Nat Chem Biol 2(5):274–281. https://doi.org/10.1038/nchembio781
10. Kaminska M, Bruyat P, Malgorn C, Doladilhe M, Cassar-Lajeunesse E, Fruchart Gaillard C, De Souza M, Beau F, Thai R, Correia I, Galat A, Georgiadis D, Lequin O, Dive V, Bregant S, Devel L (2021) Ligand-directed modification of active matrix metalloproteases: activity-based probes with no Photolabile group. Angew Chem Int Ed 60(33):18272–18279. https://doi.org/10.1002/anie.202106117
11. Devel L, Malgorn C, Tohon R-W, Launay M, Patiniotis K, Sejalon-Cipolla M, Beau F, Thai R, Bruyat P, Bonino A, Bregant S, Subra G, Cantel S, Georgiadis D (2024) Covalent labeling of matrix metalloproteases with

affinity-based probes containing tuned reactive N-acyl-N-alkyl sulfonamide cleavable linkers. Chembiochem 25(23):e202400441. https://doi.org/10.1002/cbic.202400441

12. Malgorn C, Becher F, Bruyat P, Fruchart-Gaillard C, Beau F, Bregant S, Devel L (2024) A new affinity-based probe to profile MMP active forms. In: Santamaria S (ed) Proteases and cancer: methods and protocols. Springer US, New York, NY, pp 29–39. https://doi.org/10.1007/978-1-0716-3589-6_3
13. Jumper J, Evans R, Pritzel A, Green T, Figurnov M, Ronneberger O, Tunyasuvunakool K, Bates R, Žídek A, Potapenko A, Bridgland A, Meyer C, Kohl SAA, Ballard AJ, Cowie A, Romera-Paredes B, Nikolov S, Jain R, Adler J, Back T, Petersen S, Reiman D, Clancy E, Zielinski M, Steinegger M, Pacholska M, Berghammer T, Bodenstein S, Silver D, Vinyals O, Senior AW, Kavukcuoglu K, Kohli P, Hassabis D (2021) Highly accurate protein structure prediction with AlphaFold. Nature 596(7873):583–589. https://doi.org/10.1038/s41586-021-03819-2
14. Wang J, Lisanza S, Juergens D, Tischer D, Watson JL, Castro KM, Ragotte R, Saragovi A, Milles LF, Baek M, Anishchenko I, Yang W, Hicks DR, Expòsit M, Schlichthaerle T, Chun J-H, Dauparas J, Bennett N, Wicky BIM, Muenks A, DiMaio F, Correia B, Ovchinnikov S, Baker D (2022) Scaffolding protein functional sites using deep learning. Science 377(6604):387–394. https://doi.org/10.1126/science.abn2100
15. Ahrens CC, Chiswick EL, Ravindra KC, Miller MA, Ramseier JY, Isaacson KB, Lauffenburger DA, Griffith LG (2019) Development and application of the metalloprotease activity multiplexed bead-based immunoassay (MAMBI). Biochemistry 58(38):3938–3942. https://doi.org/10.1021/acs.biochem.9b00584
16. Dalal S, Ragheb DRT, Schubot FD, Klemba M (2013) A naturally variable residue in the S1 subsite of M1 family aminopeptidases modulates catalytic properties and promotes functional specialization. J Biol Chem 288(36): 26004–26012. https://doi.org/10.1074/jbc.M113.465625
17. Baidya SK, Banerjee S, Adhikari N, Jha T (2022) Selective inhibitors of medium-size S1′ pocket matrix metalloproteinases: a stepping stone of future drug discovery. J Med Chem 65(16):10709–10754. https://doi.org/10.1021/acs.jmedchem.1c01855
18. Ravindra KC, Ahrens CC, Wang Y, Ramseier JY, Wishnok JS, Griffith LG, Grodzinsky AJ, Tannenbaum SR (2018) Chemoproteomics of matrix metalloproteases in a model of cartilage degeneration suggests functional biomarkers associated with posttraumatic osteoarthritis. J Biol Chem 293(29):11459–11469. https://doi.org/10.1074/jbc.M117.818542
19. Harbut MB, Velmourougane G, Dalal S, Reiss G, Whisstock JC, Onder O, Brisson D, McGowan S, Klemba M, Greenbaum DC (2011) Bestatin-based chemical biology strategy reveals distinct roles for malaria M1- and M17-family aminopeptidases. Proc Natl Acad Sci U S A 108(34):E526–E534. https://doi.org/10.1073/pnas.1105601108
20. Keow JY, Pond ED, Cisar JS, Cravatt BF, Crawford BD (2012) Activity-based labeling of matrix metalloproteinases in living vertebrate embryos. PLoS One 7(8):e43434. https://doi.org/10.1371/journal.pone.0043434
21. Wang J, Uttamchandani M, Li J, Hu M, Yao SQ (2006) "Click" synthesis of small molecule probes for activity-based fingerprinting of matrix metalloproteases. Chem Commun (Cambridge, U K) 36:3783–3785. https://doi.org/10.1039/B609446E
22. Maben Z, Arya R, Georgiadis D, Stratikos E, Stern LJ (2021) Conformational dynamics linked to domain closure and substrate binding explain the ERAP1 allosteric regulation mechanism. Nat Commun 12(1):5302. https://doi.org/10.1038/s41467-021-25564-w
23. Chan EWS, Chattopadhaya S, Panicker RC, Huang X, Yao SQ (2004) Developing photoactive affinity probes for proteomic profiling: Hydroxamate-based probes for metalloproteases. J Am Chem Soc 126(44): 14435–14446. https://doi.org/10.1021/ja047044i
24. Geurink PP, Klein T, Prèly L, Paal K, Leeuwenburgh MA, van der Marel GA, Kauffman HF, Overkleeft HS, Bischoff R (2010) Design of Peptide Hydroxamate-Based Photoreactive Activity-Based Probes of zinc-dependent metalloproteases. Eur J Org Chem 2010(11): 2100–2112. https://doi.org/10.1002/ejoc.200901385
25. Nury C, Bregant S, Czarny B, Berthon F, Cassar-Lajeunesse E, Dive V (2013) Detection of endogenous matrix metalloprotease-12 active form with a novel broad spectrum activity-based probe. J Biol Chem 288(8): 5636–5644. https://doi.org/10.1074/jbc.M112.419499
26. Nury C, Czarny B, Cassar-Lajeunesse E, Georgiadis D, Bregant S, Dive V (2013) A pan photoaffinity probe for detecting active forms of matrix metalloproteinases.

Chembiochem 14(1):107–114. https://doi.org/10.1002/cbic.201200583

27. Leeuwenburgh MA, Geurink PP, Klein T, Kauffman HF, van der Marel GA, Bischoff R, Overkleeft HS (2006) Solid-phase synthesis of Succinylhydroxamate peptides: functionalized matrix metalloproteinase inhibitors. Org Lett 8(8):1705–1708. https://doi.org/10.1021/ol060409e
28. Bregant S, Huillet C, Devel L, Dabert-Gay AS, Beau F, Thai R, Czarny B, Yiotakis A, Dive V (2009) Detection of matrix metalloproteinase active forms in complex proteomes: evaluation of affinity versus photoaffinity capture. J Proteome Res 8(5):2484–2494. https://doi.org/10.1021/pr801069c
29. Dabert-Gay AS, Czarny B, Devel L, Beau F, Lajeunesse E, Bregant S, Thai R, Yiotakis A, Dive V (2008) Molecular determinants of matrix metalloproteinase-12 covalent modification by a photoaffinity probe: insights into activity-based probe development and conformational variability of matrix metalloproteinases. J Biol Chem 283(45):31058–31067. https://doi.org/10.1074/jbc.M805795200
30. Dabert-Gay AS, Czarny B, Lajeunesse E, Thai R, Nagase H, Dive V (2009) Covalent modification of matrix metalloproteinases by a photoaffinity probe: influence of nucleophilicity and flexibility of the residue in position 241. Bioconjug Chem 20(2):367–375. https://doi.org/10.1021/bc800478b
31. David A, Steer D, Bregant S, Devel L, Makaritis A, Beau F, Yiotakis A, Dive V (2007) Cross-linking yield variation of a potent matrix metalloproteinase photoaffinity probe and consequences for functional proteomics. Angew Chem Int Ed Engl 46(18):3275–3277. https://doi.org/10.1002/anie.200604408

Chapter 12

Method for Producing Recombinant Soluble T-Cell Receptors

You Min Ahn, Stephanie Gras, and Demetra S. M. Chatzileontiadou

Abstract

T cells are key players of the cellular immune system, able to detect and fight pathogens' invasion. T cells recognize pathogenic peptides presented by proteins called Human Leukocyte Antigens (HLA) in humans. The molecular mechanism of T cells' interaction with peptide-HLA (pHLA) complexes, via their T cell receptors (TCRs), has been revealed using crystal structures. Indeed, X-ray crystallography, using the soluble version of both the TCR and pHLA, has shed light onto this intricate interaction essential to T cell activation, critical for host survivor. In this chapter, we describe the method to produce soluble TCRs.

Key words T cell receptor, Bacterial expression, Inclusion Body, Purification, Chromatography

1 Introduction

To obtain folded TCR and pHLA complexes, the different chains that compose each protein must be expressed in bacterial cells; the α- and β-chain for the TCR, and the HLA heavy chain and β2 microglobulin (β2m) for HLA-class I (HLA-I) molecules. Once expressed, they can be recovered from inclusion bodies from *Escherichia coli (E. coli)* using a mild solubilization process. We have previously published the methodology for producing soluble and properly folded pHLA complexes [1], and therefore, this has not been described in this chapter. Here we describe the method for producing soluble and properly folded TCR molecules.

The first step includes the bacterial transformation of the TCR α- and β-chain DNA constructs [2, 3]. Here, chemical transformation of competent BL21 (*RIL*) *E. coli* cells with the pET-30a (+) vector (or equivalent, suited for protein expression in bacteria) encoding the α- or β-chain of the TCR is being used. The constructs used for the expression of the TCR α- and β-chain contain only the soluble fraction of the TCR, without the transmembrane domain. This results in an α-chain with a molecular weight of

Efstratios Stratikos (ed.), *Antigen Processing and Presentation: Methods and Protocols*, Methods in Molecular Biology, vol. 3007,
https://doi.org/10.1007/978-1-0716-5092-9_12,

~23–25 kDa, and of ~27 kDa for the β-chain. As the transmembrane domain is removed to produce a soluble TCR protein, there is no natural disulfide bond forming between the two chains. To ensure, and help, with the correct refolding of the TCR α- and β-chain, a single point mutation in both chains can be introduced to replace Threonine 48 and Serine 57 [4], respectively, by a Cysteine that will form a disulfide bond between the two TCR chains.

Following the transformation, large-scale expression of the two TCR chains is performed. High-level expression of recombinant proteins in *E. coli* can lead to protein instability and aggregation, also known as inclusion bodies (IBs) [5, 6]. For the correct folding of the TCR and HLA proteins in vivo glycosylation is required; however, *E. coli* cells do not have the cellular machinery to glycosylate proteins, and therefore the TCR and HLA chains will be expressed unfolded, insoluble, and within the IBs. IBs require extensive processing involving isolation from the cell, solubilization and refolding. The latter involves dilution of the IBs of the TCR α- and β-chain into a refolding solution, followed by dialysis [7, 8] prior to purification.

The TCR purification involves a diethylaminoethyl cellulose, or DEAE resin, a weak anion exchanger, followed by hydrophobic interaction chromatography (HIC) that is used for the binding and desorption of hydrophobic proteins [9]. The last step is a strong anion exchange chromatography column for high-resolution (HiTrap Q HP), small-scale protein purification that separates molecules with slight differences in charge [10].

2 Materials

2.1 Chemical Bacterial Transformation

1. Chemically competent cells (*E. coli* BL21 [DE3 or equivalent]).
2. DNA plasmid (here *pET-30a(+) vector*) that encodes the TCR α-chain or β-chain.
3. Luria-Bertani (LB) agar, 1 L: Mix 1% (w/v) Tryptone, 0.5% (w/v) yeast extract, 1% (w/v) NaCl, 1.5% (w/v) agar, in ddH_2O (*see* **Note 1**).
4. LB broth, 1 L: Mix 1% (w/v) Tryptone, 0.5% (w/v) yeast extract, 1% (w/v) NaCl, in ddH_2O (*see* **Note 2**).
5. Resistance antibiotic (Kanamycin for *pET-30a(+) vector)*.

2.2 Protein Expression

1. LB broth (Sect. 2.1) with corresponding antibiotic (34 μg/mL of Kanamycin for *pET-30a(+) vector)*.
2. Isopropyl-β-D-thiogalactopyranoside (IPTG): prepare 1 M IPTG stock solution by dissolving 2.38 g of IPTG in 10 mL

of MilliQ H_2O (MQ). Filter the solution using a 0.22 μm filter, aliquot into sterile containers and store at −20 °C.

2.3 Inclusion Bodies Preparation

1. Fresh Lysozyme solution (10 mg/ml) in 10 mM Tris-HCl pH 8.0. Aqueous solutions should retain activity for at least one month when stored between 2 and 8 °C.
2. Enzymatic Lysis Buffer for 6 × 800 mL bacterial culture, 200 mL: 0.1 mg/mL Lysozyme, 50 μg/mL DNase I, 1% Triton X-100, 4 mM $MgCl_2$ in ddH_2O. Make fresh each time.
3. Inclusion Bodies Wash Buffer 1, 600 mL: 50 mM Tris-HCl pH 8.0, 0.5% (v/v) Triton X-100, 100 mM NaCl, 1 mM dithiothreitol (DTT), 0.2 mM phenylmethylsulfonyl fluoride (PMSF) in ddH_2O. Chill the buffer solution at 4 °C prior to use.
4. Inclusion Bodies Wash Buffer 2, 150 mL: 50 mM Tris-HCl pH 8.0, 1 mM Na-EDTA in ddH_2O. Chill the buffer solution at 4 °C prior to use.
5. Guanidine-HCl Buffer (1 mL/L of culture), 20 mL: 20 mM Tris-HCl pH 8.0, 6 M guanidine-HCl, 0.5 mM NaEDTA, 1 mM DTT, 0.2 mM PMSF in ddH_2O. Make fresh each time.
6. 15% sodium dodecyl sulfate polyacrylamide gel electrophoresis (SDS-PAGE).
7. Bovine serum albumin (BSA) standard at 1 mg/mL.
8. Reduced loading dye: 10% SDS, 500 mM DTT, 50% glycerol, 250 mM Tris-HCl pH 7.0, 0.5% bromophenol blue dye, pH 6.8.
9. 8 M Urea, 20 μL.

2.4 Refolding and Dialysis

1. Refolding buffer, 500 mL: 5 M Urea, 0.4 M L-Arginine HCl, 0.1 M Tris-HCl pH 8.0, 2 mM Na-EDTA, 0.16% (w/v) Glutathione reduced, 0.03% (w/v) Glutathione non-reduced, 0.5 mM PMSF in ddH_2O. Make fresh each time.
2. Dialysis membranes, 10 kDa cut-off.
3. Dialysis buffer, 30 L: 10 mM Tris-HCl pH 8.0. Store in 3 separate buckets at 4 °C.

2.5 TCR Purification

1. Diethylaminoethyl cellulose (DEAE-C) chromatography resin: Make a 50% w/v of DEAE-C slurry in distilled water and add 5 mL or the mix into an econo-column and allow settling for ~30 min. Measure the settled volume of the resin. This is the Column Volume (CV) to be used for measuring the washing solutions. Keep the column at ~25 °C.
2. DEAE Buffer B, 0.5 L: 10 mM Tris-HCl pH 8.0, 1 M NaCl in ddH_2O. Filter before use.

3. DEAE Buffer A, 0.5 L: 10 mM Tris-HCl pH 8.0 in ddH_2O. Filter before use.
4. DEAE elution buffer, 0.5 L: 10 mM Tris-HCl pH 8.0, 400 mM NaCl. Filter before use.
5. 15% SDS-PAGE.
6. 10 kDa cut-off centrifugal concentrators.
7. 3 M NH_4SO_4, 1330 μL.
8. 0.22 μm syringe filters.
9. HiTrap Phenyl HP (HIC) column.
10. HIC Buffer A, 0.5 L: 10 mM Tris-HCl pH 8.0, 0.8 M NH_4SO_4. Filter and degas before use.
11. HIC Buffer B, 0.5 L: 10 mM Tris-HCl pH 8.0. Filter and degas before use.
12. HiTrap Q HP column.
13. HiTrap Q HP Buffer A, 0.5 L: 10 mM Tris-HCl pH 8.0. Filter and degas before use.
14. HiTrap Q HP Buffer B, 0.5 L: 10 mM Tris-HCl pH 8.0, 1 M NaCl. Filter and degas before use.

3 Methods

3.1 Chemical Bacterial Transformation of the TCR α-Chain and β-Chain DNA

1. Thaw chemically competent cells [11] on ice (avoid directly holding the cells by the bottom to prevent warming up the cells). Here *E. coli* BL21 (DE3 or equivalent) strain has been mainly used.
2. Add 0.5–1 μL (~100 ng of DNA) of plasmid to the competent cells (50 μL) (*see* **Note 3**).
3. Incubate cells:DNA mixture on ice for ~20 min.
4. Heat shock competent cells at 42 °C for 45 seconds (water bath or heat block).
5. Add 1 mL of sterile LB broth to the cells:DNA and incubate at 37 °C in the shaking incubator at 180 revolution per minute (rpm) for 1 h. Meanwhile, prepare LB agar plates with the corresponding antibiotics that will select for both the competent cells and the DNA plasmid. For example, Kanamycin for pET30 plasmid selection and Chloramphenicol for *E. coli* BL21 (DE3) selection.
6. Plate 50, 100 and 200 μL of cells onto LB agar plates with corresponding antibiotics and place the plates in a stationary incubator overnight at 37 °C.

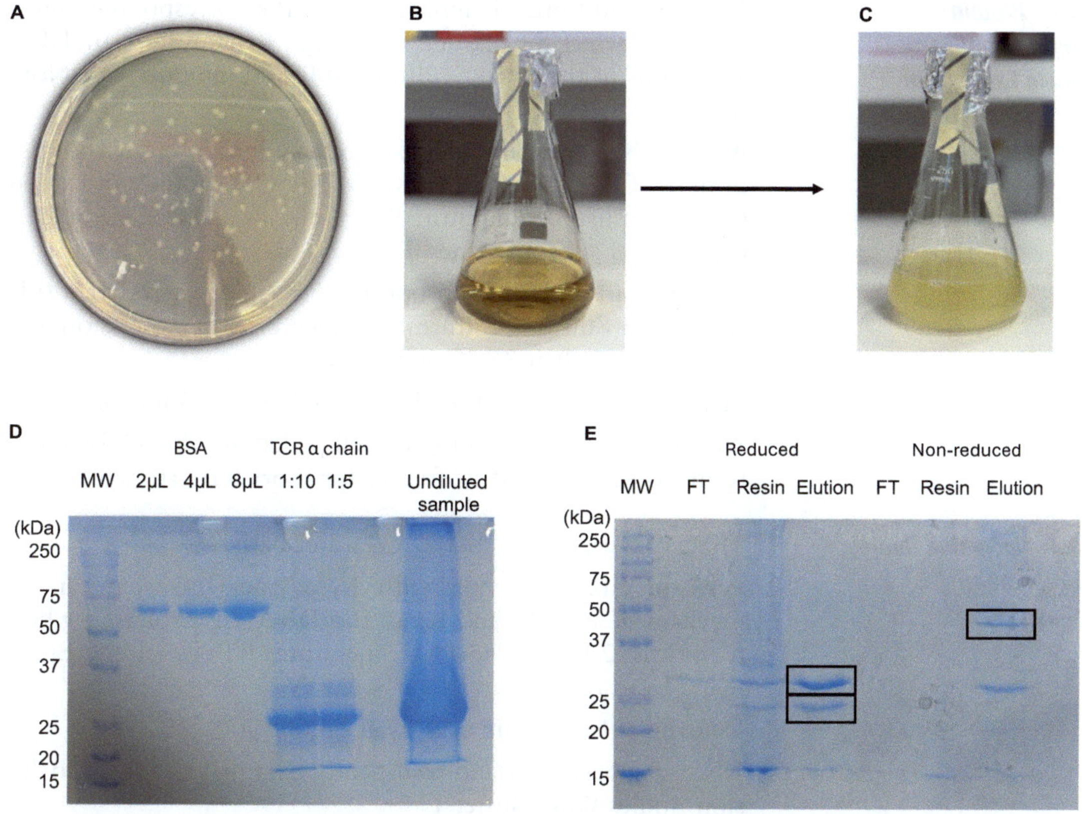

Fig. 1 Transforming and producing properly folded soluble TCRs (**a**). Bacterial transformation of BL21 (RIL) *E. coli* cells with pET30a(+) vector encoded with one TCR chain DNA on an agar plate. (**b**, **c**). Starter cultures inoculated with a bacterial colony with corresponding antibiotics at the beginning of the incubation (**b**) and at the end of the overnight incubation (**c**). (**d**) 15% SDS-PAGE for the quantification of the TCR α-chain after performing IB preparation. Lane 1 is the protein molecular weight marker (MW); lanes 2–4 are the BSA standards; lanes 5 and 6 are the 1:10 and 1:5 IB dilutions, respectively; lane 8 is the undiluted IB sample. (**e**). 15% SDS-PAGE after DEAE purification of the TCR, assessing the refolded TCR in both reduced and non-reduced conditions. Lane 1 is the MW; lanes 2–4 are the reduced samples of DEAE flow-through (FT), DEAE resin and DEAE elution, respectively; lanes 5–7 are the non-reduced samples of DEAE FT, DEAE resin and DEAE elution, respectively

7. After incubation, single colonies should appear (Fig. 1a), which means that bacterial cell transformation has been induced (*see* **Note 4**).

3.2 Protein Expression: Starter Culture

1. Inoculate 150 mL of sterile LB with a colony from the agar plate (step above), with the corresponding antibiotics for both the competent cells and DNA plasmid (Fig. 1b) (*see* **Note 5**).
2. Incubate overnight in a shaker incubator at 37 °C at 180 rpm. The resulting starter culture should have a high optical density of 0.8–1.2 at 600 nm (OD 600 nm) (Fig. 1c).

3.3 Protein Expression: Large-Scale Expression

1. Inoculate 800 mL of sterile LB with the corresponding antibiotics to select for plasmid only (e.g., Kanamycin if pET30), and 8 mL of the starter culture (100-fold dilution) (*see* **Note 6**).
2. Allow cells to grow in the shaker incubator at 37 °C at 180 rpm until the OD 600 nm reaches 0.6. The OD should not exceed 0.6 and must be monitored carefully as the cell growth is exponential after it reaches 0.2.
3. Add 800 μL of 0.1 M isopropyl-β-D-thiogalactopyranoside (IPTG) to 800 mL culture to induce protein expression and incubate for 4 h at 37 °C at 180 rpm.
4. Harvest cells by centrifugation at 5000 g for 15 min at 4 °C.
5. Discard supernatant and store cell pellet at −80 °C for long-term storage or − 20 °C freezer for short-term storage.

3.4 Inclusion Bodies Preparation

1. Thaw bacterial pellet in a 37 °C water bath.
2. Perform enzymatic lysis with Lysozyme (this step can be mechanical as well) and break the DNA on a rotating wheel for at least 20 min at room temperature (RT) by adding Enzymatic Lysis Buffer.
3. Centrifuge for 15 min at 4 °C at 15,000 *g*.
4. Discard supernatant and resuspend pellet in 150 mL of Inclusion Bodies Wash Buffer 1.
5. Homogenize and centrifuge at 10,000 *g* at 4 °C for 10 min.
6. Repeat **steps 4** and **5** at least three times or until cell pellet is clean of dark debris and supernatant is clear. The pellet should be white.
7. Discard supernatant and resuspend pellet in 150 mL of Inclusion Bodies Wash Buffer 2.
8. Homogenize and centrifuge at 10,000 *g* at 4 °C for 10 min.
9. Discard supernatant and resuspend pellet in Guanidine-HCl Buffer (1 mL/L of culture).
10. Mix on a rotating wheel overnight at 4 °C.
11. Centrifuge for 30 min at 30,000 g at 4 °C.
12. Harvest supernatant and add 0.2 mM PMSF.
13. Quantification by 15% SDS-PAGE with 1 mg/ml BSA standard, in reduced loading dye. Load wells to look reflect the following example:

    Protein Marker (MW).

    2 μL of BSA

    4 μL of BSA

    8 μL of BSA

1 μL of sample diluted 10 times in 8 M Urea

1 μL of sample diluted 5 times in 8 M Urea.

Undiluted sample (if the sample appears yellow and clear).

14. Look for the size and thickness of the band and estimate the concentration of the sample (Fig. 1D). For the TCR α- and β-chains, aliquot them out to 50 mg/tube. Store IBs in −80 °C or proceed to the next step.

3.5 Refolding of the TCR α-Chain and β-Chain

Add 50 mg TCR α-chain and 50 mg TCR β-chain into the Refolding buffer and stir for 3 days at 4 °C.

3.6 Dialysis

Place refolding solution (0.5 L) into dialysis membranes (10 kDa cut-off), and dialyze over 10 L of 10 mM Tris-HCl pH 8.0 for 12 h (3× times) at 4 °C.

3.7 Purification of the Refolded TCR: Ion Exchange Chromatography–Diethylaminoethyl Cellulose (DEAE-C)

1. Set the flow rate at 1 mL/min and wash the DEAE column with 5 column volumes (CV) of DEAE Buffer B.
2. Equilibrate the DEAE column with 5 CV of DEAE Buffer A.
3. Filter the dialyzed refolding solution prior to loading onto the column at RT.
4. Elute bound proteins with 20 mL (for 0.5 L refold) of DEAE elution buffer. This should result in the elution of the TCR based on the charge of the folded protein.
5. Visualize the purity of the TCR and check for the presence of homodimers on a 15% SDS-PAGE (*see* **Note 7**). Use the following as an example to load the wells of the gel (Fig. 1E):

   Protein marker (MW).

   DEAE flow-through (FT) (reduced*).

   DEAE resin (reduced).

   Eluted protein (reduced).

   DEAE FT (non-reduced*).

   DEAE resin (non-reduced).

   Eluted protein (non-reduced).

   * Reduced: loading buffer that contains DTT; non-reduced: loading buffer without DTT.

6. The DEAE column can be washed with 5 CV of DEAE Buffer B followed by 5 CV of DEAE Buffer A and stored in 20% EtOH to be reused at RT.

3.8 Purification of the Refolded TCR: Hydrophobic Interaction Chromatography (HIC)–HiTrap Phenyl HP

1. Prepare the sample by doing the following:
   - Concentrate the DEAE eluted sample to 3 mL using a 10 kDa cut-off centrifugal concentrator.
   - Put 1 mL each into an Eppendorf tube (total of 3 tubes).
   - Add 443 μL of 3 M NH_4SO_4 to each Eppendorf tube and mix the solution.
   - Centrifuge at 18,000 g for 10 min at RT.
   - Keep the supernatant and filter using a 0.22 μm syringe filter before injecting into the Fast Liquid Protein Chromatography (FPLC) machine.
2. Connect the HiTrap Phenyl HP column on the FPLC with HIC Buffer A.
3. Check the pressure limit that can be set up on the machine which is 0.5 MPa for (HiTrap Phenyl HP, Cytiva).
4. Set the flow rate at 1 mL/min and wash the column with 5 CV of HIC Buffer B.
5. Equilibrate the column with 5 CV of HIC Buffer A.
6. Start fractionation before injecting the protein of interest. Collect 1 mL fractions.
7. Inject the protein sample and run at a flow rate of 3 mL/min. Let it run with HIC Buffer A for 10 mL.
8. Once 10 mL has passed, set the gradient from 0 to 100% HIC Buffer B over 20 min at a flow rate of 1 mL/min.
9. Once 20 min have passed, wash the column with 3 CV of 100% HIC Buffer B (Fig. 2a).
10. Run 15% SDS-PAGE of the fractions of interest (Fig. 2b).
11. Pool fractions needed based on the purification profile and gel analysis. Using a 10 kDa cut-off centrifugal concentrator (10 min each time at 3200 g at 4 °C), buffer exchange over 10 mM Tris-HCl pH 8.0 until the concentration of NH_4SO_4 is very low (i.e. ~10 times less the initial concentration, 0.08 M). Finally concentrate your protein sample down to 250 μL before proceeding to the last purification step.

3.9 Purification of the Refolded TCR: Ion Exchange Chromatography–HiTrap Q HP

1. Connect the HiTrap Q HP column on the FPLC with HiTrap Q HP Buffer A (*see* **Note 8**).
2. Check the pressure limit that can be set up on the FPLC, which is 0.5 MPa for HiTrap Phenyl HP, Cytiva.
3. Set the flow rate at 1 mL/min and wash the column with 5 CV of HiTrap Q HP Buffer B.
4. Equilibrate the column with 5 CV of HiTrap Q HP Buffer A.

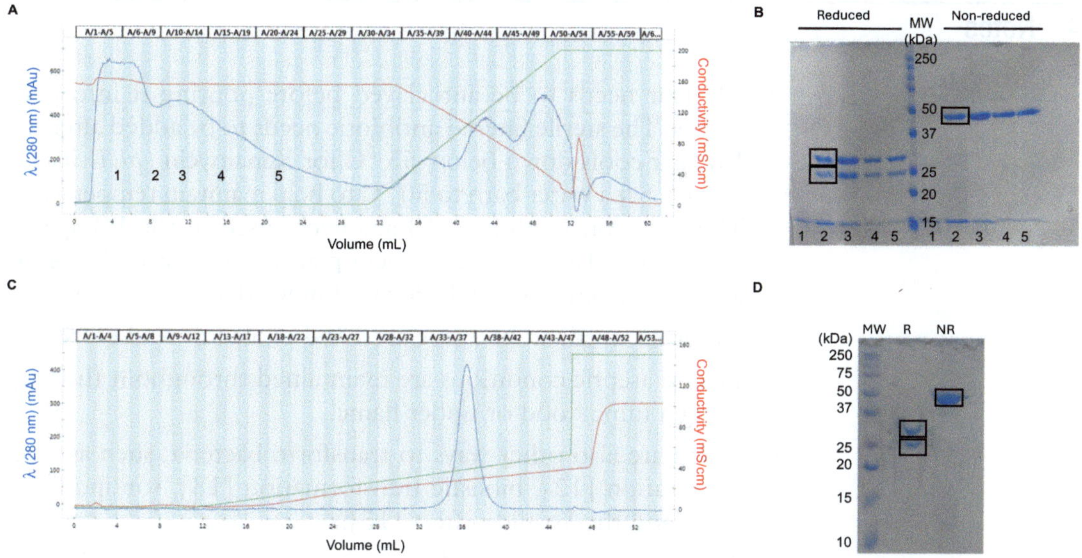

Fig. 2 Purification of properly folded TCRs. (**a**) TCR purification FPLC profile with a hydrophobic column. (**b**). The fractions 1–5 of the TCR elution peak ran on 15% SDS-PAGE in both reduced (lanes 1–5) and non-reduced conditions (lanes 7–12), lane 6 is the MW. (**c**)>. TCR purification FPLC profile with an anion exchange column. (**a**, **c**). The blue trace represents the protein absorbance at 280 nm (mAU). The red trace represents the changes in conductivity (mS/cm), which occur with changing salt concentration (green line). The top bar of the graph represents the fractions collected. (**d**). 15% SDS-PAGE was performed for the eluted fractions within the single peak shown on the panel, pooled and ran in reduced (R) and non-reduced (NR) conditions. Lane 1 is the MW; lane 2 is the reduced sample of TCR from fraction 1 showing two bands for the α- and β-chain of the TCR, while lane 3 is the non-reduced sample showing a single band at the cumulative MW of both the α- and β-chain (25 and 27 kDa, respectively) of the TCR (MW >50 kDa)

5. Start fractionation before protein injection and ensure the protein of interest is filtered before injecting. Collect 1 mL fractions.
6. Inject <250 μL of the protein sample, using a 500 μL loop, at a flow rate of 3 mL/min for 10 mL of HiTrap Q HP Buffer A.
7. Set a gradient from 0 to 35% HiTrap Q HP Buffer B over 35 min.
8. TCR complexes usually elute at ~25% HiTrap Q HP Buffer B; however, this can vary between different proteins (Fig. 2c).
9. Run the fractions of interest on an SDS-PAGE using both non-reduced and reduced conditions (Fig. 2d).
10. Pool fractions required based on purification profile and gel analysis.
11. Concentrate the protein of interest using a 10 kDa cut-off centrifugal concentrator and measure the protein concentration using a UV spectrophotometer at 280 nm (*see* **Note 9**).

4 Notes

1. LB agar needs to be autoclaved before pouring it into petri dishes. The sterile filtered antibiotic needs to be added after the LB agar cools down below 55 °C for Kanamycin (*pET-30a(+) vector*). Set a water bath to 50 °C to help maintain temperature. The final concentration of Kanamycin used in the LB agar is 34 μg/mL. LB agar/Kanamycin plates can be stored at 4 °C and away from direct light for no longer than two weeks.
2. LB broth needs to be autoclaved and cooled down before use.
3. Ensure aseptic conditions are maintained throughout this step using a fume hood or under flame.
4. There are also other ways to transform bacteria, such as electroporation [12], biolistic transformation [13], sonoporation [14], and the Yoshida effect [15].
5. A starter culture is used, prior to large-scale expression, to control the environment and time that the bacterial cells will take to multiply and reach the optimum density for protein expression. If it is the first time expressing a particular protein, starting with a small-scale expression, of for example 50 mL of LB, is recommended and can be directly adapted from the method described here.
6. 3–6 flasks of 800 mL per TCR chain is recommended to produce enough material to refold a TCR. The yield of protein expressed varies with the TCR sequence, but usually 50 mg/L would be considered a good yield.
7. The SDS-PAGE of the DEAE elution and resin can indicate if there is degradation of one of the chains, or both, and if one is in excess compared to the other, which could lead to a change of ratio during the refolding. The TCR α-chain should be ~25 kDa and β-chain ~27 kDa, and the total molecular weight of a TCR is ~52 kDa but may vary depending on the TCR sequence. Ensure there is a clear single ~52 kDa band in the non-reduced samples and two clear and singular bands (25 and 27 kDa) in the reduced samples (Fig. 1e). The α- and β-chain can homodimerize. If homodimers are present in the samples, the next step to further purify the refolded TCR is to perform hydrophobic interaction chromatography (HIC), which is then followed by ion exchange chromatography with HiTrapQ HP. If no homodimers are present in the samples, the next step is to perform ion exchange chromatography with HiTrap Q HP.
8. If you have performed a HIC interaction chromatography, your protein sample is ready to be loaded into the HiTrap Q HP column. If HIC was not needed, then you need to

concentrate the DEAE eluted protein using a centrifugal concentrator with a 10 kDa cut-off and buffer exchange via centrifugation for 10 min at 3200 g at 4 °C over 10 mM Tris-HCl pH 8.0, to remove NaCl, before loading the sample onto the HiTrap Q HP column.

9. The Beer-Lambert law is used to calculate the concentration of the protein based on the absorbance at 280 nm: A = ecl, where A = Absorbance; e = Molar absorption coefficient, M-1 cm-1; c = Molar concentration, M; l = optical path length, cm. The protein sample needs to be concentrated to the desired concentration for subsequent experiments.

Acknowledgments

The authors want to thank all their colleagues that contributed to the development of the methods to produce both soluble pHLA and TCR to enable the analysis of the interaction between these molecules. DSMC was supported by an AINSE ECR grant from ANSTO, and SG is supported by an NHMRC Leadership Investigator Grant (#2034677).

Conflict of Interest The authors have no conflicts of interest to declare that are relevant to the content of this chapter.

References

1. Chatzileontiadou DSM, Szeto C, Jayasinghe D, Gras S (2021) Protein purification and crystallization of HLA-A *02:01 in complex with SARS-CoV-2 peptides. STAR Protoc 2(3):100635. https://doi.org/10.1016/j.xpro.2021.100635
2. Yoshida N, Sato M (2009) Plasmid uptake by bacteria: a comparison of methods and efficiencies. Appl Microbiol Biotechnol 83(5): 791–798. https://doi.org/10.1007/s00253-009-2042-4
3. Hanahan D (1983) Studies on transformation of Escherichia coli with plasmids. J Mol Biol 166(4):557–580. https://doi.org/10.1016/s0022-2836(83)80284-8
4. Boulter JM, Glick M, Todorov PT, Baston E, Sami M, Rizkallah P, Jakobsen BK (2003) Stable, soluble T-cell receptor molecules for crystallization and therapeutics. Protein Eng 16(9):707–711. https://doi.org/10.1093/protein/gzg087
5. Rosano GL, Ceccarelli EA (2014) Recombinant protein expression in Escherichia coli: advances and challenges. Front Microbiol 5: 172. https://doi.org/10.3389/fmicb.2014.00172
6. Burgess RR (2009) Refolding solubilized inclusion body proteins. In: Methods Enzymol, vol 463, pp 259–282. https://doi.org/10.1016/S0076-6879(09)63017-2
7. Singh A, Upadhyay V, Upadhyay AK, Singh SM, Panda AK (2015) Protein recovery from inclusion bodies of Escherichia coli using mild solubilization process. Microb Cell Factories 14:41. https://doi.org/10.1186/s12934-015-0222-8
8. Tsumoto K, Ejima D, Kumagai I, Arakawa T (2003) Practical considerations in refolding proteins from inclusion bodies. Protein Expr Purif 28(1):1–8. https://doi.org/10.1016/s1046-5928(02)00641-1
9. McCue JT (2009) Theory and use of hydrophobic interaction chromatography in protein purification applications. Methods Enzymol 463:405–414. https://doi.org/10.1016/S0076-6879(09)63025-1

10. Duong-Ly KC, Gabelli SB (2014) Using ion exchange chromatography to purify a recombinantly expressed protein. Methods Enzymol 541:95–103. https://doi.org/10.1016/B978-0-12-420119-4.00008-2
11. Daegelen P, Studier FW, Lenski RE, Cure S, Kim JF (2009) Tracing ancestors and relatives of Escherichia coli B, and the derivation of B strains REL606 and BL21(DE3). J Mol Biol 394(4):634–643. https://doi.org/10.1016/j.jmb.2009.09.022
12. Sheng Y, Mancino V, Birren B (1995) Transformation of Escherichia coli with large DNA molecules by electroporation. Nucleic Acids Res 23(11):1990–1996. https://doi.org/10.1093/nar/23.11.1990
13. Shark KB, Smith FD, Harpending PR, Rasmussen JL, Sanford JC (1991) Biolistic transformation of a procaryote, bacillus megaterium. Appl Environ Microbiol 57(2):480–485. https://doi.org/10.1128/aem.57.2.480-485.1991
14. Bao S, Thrall BD, Miller DL (1997) Transfection of a reporter plasmid into cultured cells by sonoporation in vitro. Ultrasound Med Biol 23(6):953–959. https://doi.org/10.1016/s0301-5629(97)00025-2
15. Yoshida N (2007) Discovery and application of the Yoshida effect: nano-sized acicular materials enable penetration of bacterial cells by sliding friction force. Recent Pat Biotechnol 1(3):194–201. https://doi.org/10.2174/187220807782330147

Chapter 13

Method for Performing T Cell Receptor-Peptide-Human Leukocyte Antigen Co-complexation and Structural Analysis

Demetra S. M. Chatzileontiadou and Stephanie Gras

Abstract

An important component of the adaptive immune system is the T cells that detect and clear pathogen-infected cells. T cells have surface receptors (TCRs) that recognize pathogen-derived peptides bound by immune proteins called Human Leukocyte Antigens (HLA) in humans. The intricate interaction between the TCR and the peptide-HLA (pHLA) complex drives the T cell activation that, in turn, dictates infection's outcome. To understand this key interaction, X-ray crystallography has been used to determine the atomic structures of both the TCR and the pHLA after complex formation. In this chapter, we describe the method to form a TCR-pHLA complex, crystallize it and analyze the resulting crystal structure.

Key words T cell receptor, Human leukocyte antigen, Peptide, Purification, X-ray, Crystal structure

1 Introduction

The first step for the co-complexation of the TCR with the pHLA is to mix the purified TCR with the purified pHLA in 1: 1 molar ratio at room temperature (RT) and then proceed to the purification of the co-complex. Size-exclusion chromatography (SEC) is a chromatographic method in which molecules in solution are separated by their size, and in some cases molecular weight. Typically, when an aqueous solution is used to transport the sample through the column, the technique is known as gel-filtration chromatography. The chromatography column is packed with fine, porous beads which are composed of dextran polymers (Sephadex), agarose (Sepharose), or polyacrylamide (Sephacryl or BioGel P). SEC works by trapping smaller molecules in the pores of the adsorbent "stationary phase" [1].

To form protein crystals, a high concentration of protein is mixed with crystallizing agents, also known as crystallants, which are usually a combination of buffers, precipitants, polymers and salts. The protein-crystallant mixture is placed as a droplet in either

Efstratios Stratikos (ed.), *Antigen Processing and Presentation: Methods and Protocols*, Methods in Molecular Biology, vol. 3007, https://doi.org/10.1007/978-1-0716-5092-9_13,

a sitting or hanging position next to a reservoir of identical crystallants and the set-up is sealed with a cover slip and petroleum jelly to create a vacuum [2]. Protein is mixed with crystallant at a 1:1 ratio and a droplet is placed in one of the two arrangements; the well is covered with a glass cover slip sealed with petroleum jelly in the hanging drop arrangement whereas a Crystal-Clear plate Sealing Tape can be alternatively used in the sitting drop arrangement. In both cases, as the concentration of crystallants is higher in the reservoir solution than in the protein-crystallant drop, solvent is gradually evaporated from the protein-crystallant drop through vapor diffusion [3]. This causes protein concentration to increase slowly and eventually reach a supersaturated state. Under particular crystallizing conditions in the supersaturated state, nucleation can occur where protein molecules aggregate into a crystalline form [3]. The nucleation event decreases the concentration of protein to a low-supersaturated state where protein molecules will adhere to the crystal surface and contribute to the growth of the crystal. This will continue to decrease the concentration of the protein to its solubility limit and cease crystal growth. Alternatively, if the protein reaches a supersaturated state without nucleation occurring, protein concentration continues to increase and protein precipitant will form in the well [3].

Optimization steps may be needed, depending on the quality of the crystals (e.g., size, diffraction quality, formation of monocrystal). There are multiple ways to optimize the initial conditions, including the adjustment of the pH or/ and the concentration of either the precipitant or the protein. Alternatively, the Additive Screen HT and microseeds from the drop that gave crystals can be used [4]. Once a high-resolution crystal structure of the TCR-pHLA complex has been obtained [5], the molecular interaction can be analyzed. There are several parameters to consider such as the overall docking and docking angle of the TCR; TCR footprint on the surface of the pHLA; complementarity-determining region (CDR) loops contributions to the interaction; comparison with other TCR-pHLA complexes. For this section, the methods are described using CCP4 [6] and PyMol (version 1.20; copyright, Schrodinger, LLC), but other tools can be used with similar results.

2 Materials

2.1 Co-complexation of the TCR and pHLA

1. Tris Buffered Salt buffer (TBS), 0.5 L: 10 mM Tris-HCl, 150 mM NaCl, pH 8.0. Filter before use.
2. Superdex® 200 Increase 10/300 GL column.
3. 15% sodium dodecyl sulfate polyacrylamide gel electrophoresis SDS-PAGE.

4. 10 kDa cut-off centrifugal concentrators.
5. 0.22 μm syringe filters.

2.2 TCR-pHLA Co-crystallization

1. Sitting drop 96-well plates for robot set-up or 48-well plates for hand-tray set-up.
2. Hampton Research commercial screening kits, such as: PEG/Ion HT Crystallization reagent kit, Morpheus, Salt RX, Crystal Screen HT, Index, Wizard Screen, Additive Screen HT.
3. Crystal Clear plate Sealing Tape.

2.3 Structural Analysis

1. Software: CCP4, PyMol.

3 Methods

3.1 Co-complexation of the TCR and pHLA

1. Prepare the sample by filtering the protein using 0.22 μm syringe filter before injecting it into the fast protein liquid chromatography (FPLC) machine.
2. Connect the column on the FPLC with the TBS buffer. The column we use for the co-complexation is Superdex® 200 Increase 10/300 GL.
3. Check the pressure limit that can be set up on the machine, which is 1.5 MPa for Superdex® 200 Increase 10/300 GL.
4. Set the flow rate at 0.5 mL/min and equilibrate the column with 5 CV of TBS.
5. Start fractionation before injecting the protein of interest. Collect 0.2 mL fractions. Inject the protein sample and run at a flow rate of 0.5 mL/min (the volume of the sample should be 0.5 mL in TBS).
6. The volume that the protein elutes can vary between different proteins. Usually for TCR-pHLA complexes, the elution volume is at 14 mL (Fig. 1a).
7. Run 15% SDS-PAGE of the fractions of interest to identify where the co-complex elutes (Fig. 1b).
8. Pool fractions needed based on the purification profile and gel analysis. Concentrate the TCR-pHLA complex samples using a 10 kDa cut-off concentrator to the desired concentration, usually at 5–10 mg/mL for the initial crystallization screen (*see* **Note 1**).

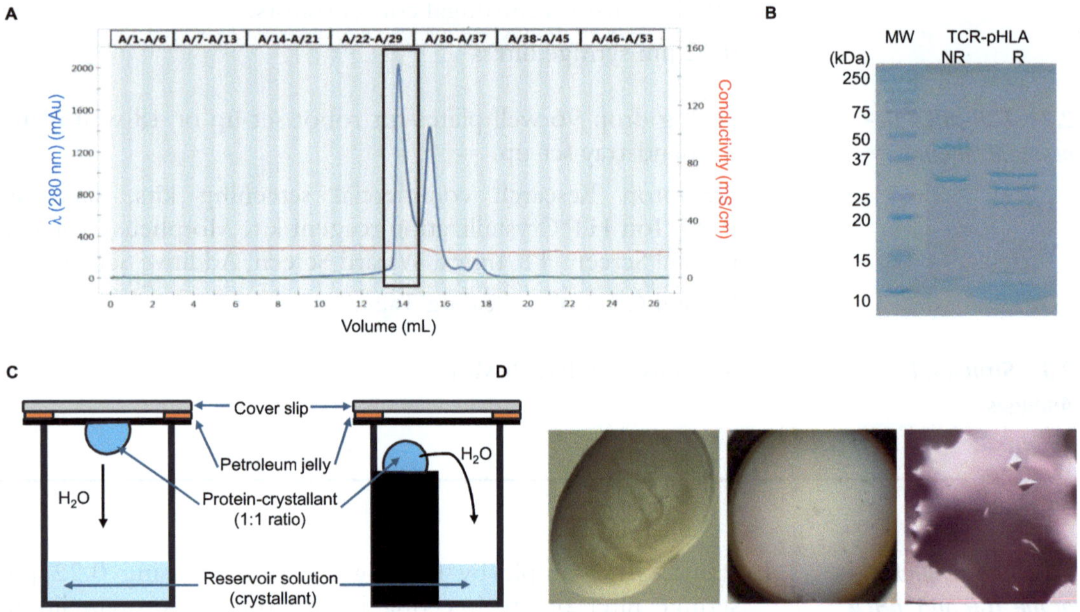

Fig. 1 Purification and crystallization of the TCR-pHLA co-complex (**a**) Gel filtration chromatograph of the TCR-pHLA complex. The blue trace represents the protein absorbance at 280 nm (mAU). Co-complexed TCR-pHLA elutes first, at around 14 mL (highlighted in black rectangle), whereas uncomplexed TCR and/or pHLA elute later, at around 15–16 mL. The top bar of the graph shows the collected fractions. (**b**) 15% SDS-PAGE performed for the gel filtration. The eluted first peak (black rectangle) for which the fractions were pooled was run on the SDS-PAGE, under non-reduced (NR) and reduced (R) conditions. The first lane of the gel is the protein molecular weight marker (MW) in kDa. (**c**) Crystallization via vapor diffusion using hanging drop and sitting drop methods. (**d**) Outcomes of a protein crystallization experiment; left, amorphous precipitation (protein and/or precipitant are in high concentration); middle, clear drop (under saturated drop); right, protein crystals

3.2 TCR-pHLA Co-crystallization: Broad Crystallization Screening

For the initial screen, robot or hand trays can be set up (*see* **Note 2**).

1. Robot: Sitting drop 96-well plates (Fig. 1c) are being set up using reagent formulations from Hampton Research commercial screening kits. The kits usually used for TCR-pHLA complexes are the PEG/Ion HT Crystallization reagent kit, Morpheus, Salt RX, Crystal Screen HT, Index, Wizard Screen. The Mosquito robot is used for the plate setup. 50 μL of each of the 96 conditions are put in each well and the protein: reservoir drop ratio is 1:1 (50 nL protein: 50 nL reservoir). The initial concentration of the protein complex used varies between 5 and 10 mg/mL in TBS.
2. Hand trays: Sitting drop 48-well plates (Fig. 1c) can be set up using reagent formulations from Hampton Research commercial screening kits. Pipette 100 μL of crystallization solution into the deep well. 1 μL of crystallization solution from each deep is put well onto the well, followed by 1 μL of the protein sample.

3. In both cases, the plate is sealed with the Crystal Clear plate Sealing Tape and stored in a 20 °C incubator monitored frequently for the presence of crystals (Fig. 1d).

3.3 TCR-pHLA Co-crystallization: Crystallization Optimization

1. Set up sitting drop 48-well plates using the original condition that produced crystals.
2. Pipette 1 μL of the crystallization solution from each deep well onto the well, followed by 1 μL of the protein.
3. Add 0.2 μL of the additive included in the Additive Screen HT to your drop.
4. Add crystal micro seeds if initial crystals have been obtained. Crush up the seeds finely and add 0.2 μL of crushed seeds into each drop.
5. Seal the plate with Crystal Clear plate Sealing Tape and store at 20 °C. Wait until crystals appear.

3.4 Structural Analysis: TCR Docking

To calculate the TCR docking angle (*see* **Note 3**):

1. Determine the center of mass (COM) for the TCR Variable α- and β-chains, which can be done by different methods. Under the CCP4 software [6] you can use Arealmol to calculate the coordinates (x,y,z) of the COM of each chain (e.g., COM (x,y, z): 34.41 3.89 60.10). The COM coordinates can then be added to the PDB file of the complex as a new water molecule and displayed in PyMol as a large sphere (set sphere_scale, 0.15 or larger) (Fig. 2a).
2. Orientate the pHLA structure in PyMol (or other visualization tools) as a top view of the cleft with the α2-helix at the bottom and both helices parallel (Fig. 2a). Remove all the display for the TCR and only keep the COM as a sphere.
3. Draw a line parallel to the HLA helices, which should run on the top of the peptide backbone, then draw a second line between the 2 COM of the TCR. The angle between the two lines is the docking angle of the TCR (Fig. 2b) (*see* **Note 4**).

3.5 Structural Analysis: TCR Footprint

To calculate the TCR footprint angle (*see* **Note 5**):

1. Calculate the TCR interaction with the peptide and the HLA using, for example, Contact in CCP4 [6]. Four files should be generated, two for each TCR chain that represent TCR-peptide and TCR-HLA interaction with a cut-off at 4 Å.
2. A list of interactions and the atoms involved in these interactions will be generated as shown in Fig. 2C.
3. From this file a list of contacts can be generated that will summarize all the contacts made by the TCR. To draw the

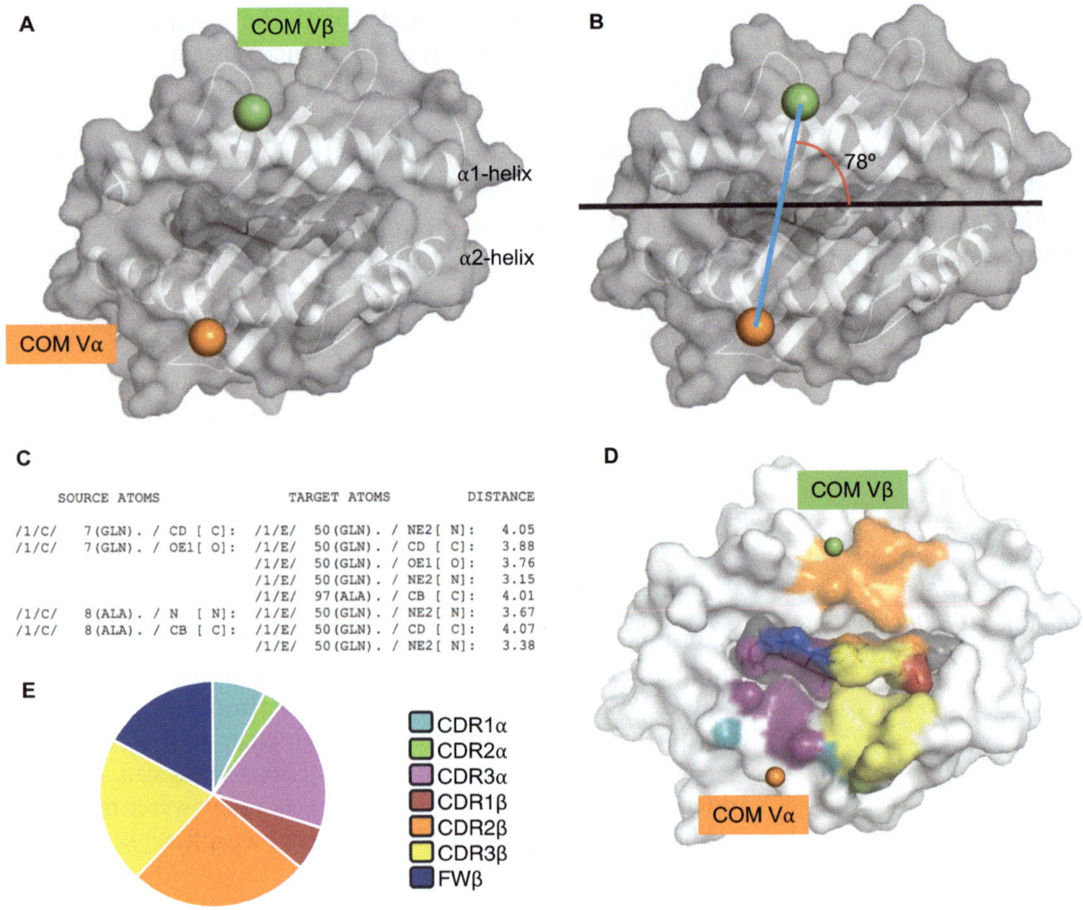

Fig. 2 Structural analysis of the TCR interaction with the pHLA complex. (**a**) Surface representation of the HLA (white) and the peptide (grey) in a top view, with the HLA helices shown as a cartoon representation. The center of mass (COM) of both the α- and β-chain of the TCR are represented by orange and green spheres, respectively. (**b**). The TCR docking angle is represented by the red curved line formed between the black line (parallel to the HLA helices and running over the peptide backbone) and the cyan line (in between the TCR COMs represented as per panel **a**). (**c**) Example of the output file from Contact in CCP4 suite to calculate the atomic contacts between the peptide (chain C) and the TCR β-chain (chain E). (**d**) TCR footprint on the surface of the peptide (grey surface) and HLA (white surface) with each TCR fragment colored in teal, green and purple for the CDR1/2/3α, respectively, and in firebrick, orange, yellow and dark blue for the CDR1/2/3β and the framework (FW) of the β-chain, respectively. The COMs are displayed as per panel **A**. (**e**) Pie chart representation of the TCR buried surface area contribution of each TCR segment in contact with the peptide and HLA, colored as per panel **D**. The TCR-pHLA structure used is of the KS1 TCR in complex with the HLA-B*57:01-TW10 (PDB code: 8F5A) [7]

footprint map, a text file called ".pml" can be generated with the contact list added as shown in the example below:

Example: color orange,/a1a2//A/ARG'69/NE (*see* **Note 6**).

4. Once all the contacts are entered in the ".pml" file, it can be opened in PyMol using the command "run script" or "open

file." The resulting figure will show each TCR contact with the peptide and HLA, and the COM of the TCR can also be added to the footprint representation(Fig. 2d) (*see* **Note 7**)

3.6 Structural Analysis: CDR Loops Contribution

To calculate the CDR loops contribution (*see* **Note 8**):

1. Calculate the individual buried surface area (BSA) upon TCR binding to the pHLA with AreaIMol in CCP4 [6] with the option "pairwise area difference between molecules" and add for Molecule #1 the TCR chains, and for Molecule #2 the peptide and the HLA heavy chain. As β2 microglobulin (β2m) does not participate in the interaction, there is no need to include it.
2. An output file will be created. At the end of the file there is the "analysis of area differences" that describes the surface buried for each residue as a negative value. Select the residues that belong, for example, to the CDR2β loop, add the values and this is the BSA of the CDR2β loop in $Å^2$. Repeat for all TCR segments, including framework (FW) for each chain.
3. The overall TCR BSA is given under "total area difference" and should be equal to the BSA of the α- and β-chains. If the total TCR BSA is 800 $Å^2$, and the CDR2β BSA is 200 $Å^2$, it would be a 25% contribution to the TCR interaction from the CDR2β loop. The total of each TCR segment contribution should be equal to 100% (*see* **Note 9**).

4 Notes

1. The protein complex sample can be stored at 4 °C short-term or at −20 °C long-term. However, it is highly recommended to be used straight away to set up crystal trays or within a week after the purification. Avoid freeze/ thawing protein samples multiple times; this increases degradation. Try to use freshly purified protein for crystallization. After long-term storage, it is recommended to re-purify the complex by running the above step again.
2. This process can be applied to a variety of different proteins/ peptides/ protein complexes; however, many factors such as the crystallization condition, protein concentration, use of seeds and additives, can vary significantly. Furthermore, the purity of the protein sample should be evaluated prior to crystallization. For the initial crystallization screen, the protein sample should be at least 95% pure on SDS-PAGE. It is recommended, if possible, to also evaluate the homogeneity and monodispersity of the protein sample as both factors can significantly influence the crystallization outcome [4].

3. The canonical docking mode of a TCR onto the surface of the pHLA complex is diagonal and central. This means that the TCR variable segments of the α- and β-chains are docked above the α2- and α1- helix of the HLA molecule, respectively, forming a ~ 60° angle with the peptide.
4. If comparing multiple TCR-pHLA structures, it is advised to first align them together so the view of the HLA will be the same and the angles directly comparable.
5. A way to visualize the extent and nature of the TCR contacts onto the surface of the pHLA complex is to draw what is called a "footprint" map of the interaction. Each TCR atom that interacts with the peptide and/or HLA is attributed a color depending on the TCR segment it belongs to (for example, CDR2β is shown in orange). The peptide or HLA atoms contacted by the TCR will be colored accordingly. The colored area represents the "footprint" of the TCR onto the surface of the pHLA, and the larger the surface one color covers, the higher the contribution of the TCR segment corresponding to this color.
6. This will color Arg69 in orange. If orange is the key color for the CDR2β loop of the TCR this means that this loop is interacting with the HLA Arg69 residue.
7. To distinguish the peptide and HLA surface clearly, it is recommended to create two different objects in PyMol for the peptide and the HLA separately before generating the surface.
8. To determine if some specific TRBV or TRAV or motif biases observed in the TCR repertoire play an important role, the individual contribution of each TCR segment can be calculated. This indicates the % of the contribution of each TCR segment relative to the total TCR surface interacting with the pHLA complex.
9. The % BSA of each TCR segment can be represented as a pie chart next to the footprint figure (Fig. 2e), and both should show the same contribution. It is another way to visualize the importance of each TCR part to the interaction with the pHLA complex.

Acknowledgments

The authors want to thank all their colleagues that contributed to the development of the methods to produce both soluble pHLA and TCR to enable the analysis of the interaction between these molecules. DSMC was supported by an AINSE ECR grant from ANSTO, SG is supported by an NHMRC Leadership Investigator Grant (#2034677).

Conflict of Interest The authors have no conflicts of interest to declare that are relevant to the content of this chapter.

References

1. Lubomirsky E, Khodabandeh A, Preis J, Susewind M, Hofe T, Hilder EF, Arrua RD (2021) Polymeric stationary phases for size exclusion chromatography: a review. Anal Chim Acta 1151:338244. https://doi.org/10.1016/j.aca.2021.338244
2. Dessau MA, Modis Y (2011) Protein crystallization for X-ray crystallography. J Vis Exp (47). https://doi.org/10.3791/2285
3. Chayen NE, Saridakis E (2008) Protein crystallization: from purified protein to diffraction-quality crystal. Nat Methods 5(2):147–153. https://doi.org/10.1038/nmeth.f.203
4. Chatzileontiadou DSM, Szeto C, Jayasinghe D, Gras S (2021) Protein purification and crystallization of HLA-A *02:01 in complex with SARS-CoV-2 peptides. STAR Protoc 2(3):100635. https://doi.org/10.1016/j.xpro.2021.100635
5. Szeto C, Lobos CA, Nguyen AT, Gras S (2020) TCR recognition of peptide-MHC-I: rule makers and breakers. Int J Mol Sci 22(1). https://doi.org/10.3390/ijms22010068
6. Collaborative Computational Project N (1994) The CCP4 suite: programs for protein crystallography. Acta Crystallogr D Biol Crystallogr 50 (Pt 5):760–763. https://doi.org/10.1107/S0907444994003112
7. Chatzileontiadou DSM, Lobos CA, Robson H, Almedia CA, Szeto C, Castley A, D'Orsogna LJ, Gras S (2024) Public T cell clonotypes are selected in HLA-B(*)57:01(+)/HIV(+) patients independently of the viral load. Cell Rep 43(8): 114555. https://doi.org/10.1016/j.celrep.2024.114555

Chapter 14

Method for Determining the Affinity of the T Cell Receptor for the Peptide-Human Leukocyte Antigen Complex

Lawton D. Murdolo, Stephanie Gras, and Demetra S. M. Chatzileontiadou

Abstract

T cells are key players of the cellular immune system, able to detect and fight pathogens' invasion. T cells recognize pathogen-derived peptides presented by molecules called Human Leukocyte Antigens (HLA) in humans. The interaction between the T cells and peptide-HLA (pHLA) complexes is driven by the surface T cell receptors (TCRs). This interaction is critical and the first step of T cell activation. The affinity of the TCR for the pHLA is one of the key drivers of T cell activation. In this chapter, we describe the method to determine the affinity of the TCR for the pHLA via surface plasmon resonance.

Key words T cell receptor, HLA, Peptide, Surface plasmon resonance, Affinity measurements

1 Introduction

There are several methods to determine the binding affinity of macromolecules. These include isothermal titration calorimetry (ITC) [1], fluorescence spectroscopy [2], bio-layer interferometry (BLI) [3], and surface plasmon resonance (SPR) [4]. In addition to 3-dimensional (D) kinetics, because both the TCR and HLA molecules are membrane-bound proteins, 2D kinetic is also an important parameter to link the TCR recognition to the T cell function [5, 6].

The overall affinity of the TCR for the pHLA is moderate, in the μM range [7], making SPR a well-suited method that requires small amounts of protein and does not require protein labeling. As a result, SPR has been preferentially used to study TCR-pHLA interactions since the 1990s [8, 9] and has become the primary technique to determine the 3D binding kinetics of soluble TCRs to pHLA complexes.

SPR is a physical phenomenon that occurs when atoms within a thin metal film become excited by light directed through a prism at a specific angle of incidence. This light is then reflected (total

Efstratios Stratikos (ed.), *Antigen Processing and Presentation: Methods and Protocols*, Methods in Molecular Biology, vol. 3007, https://doi.org/10.1007/978-1-0716-5092-9_14,

internal reflection) except at a certain angle of incidence, known as the SPR angle. At this angle, light is absorbed and converted into an evanescent wave, which excites electrons on the metal surface, generating surface plasmons. The SPR angle can change depending on the refractive indices of the film or, more specifically, the materials attaching or detaching from the film. This was first described by Liedberg et al. (1983) [10] in the context of immunosensing. A gold layer is positioned at the interface between the glass and the sample solution flowing through the microfluidic chip. The first molecule (ligand) can be immobilized to the gold surface in different ways. One immobilization method involves pre-immobilized streptavidin (SA) on the sensor chip, which requires the ligand protein to be biotinylated before binding. Another method involves amine coupling, via Lysine residues, using a carboxyl moiety within a dextran matrix on the surface of the sensor chip (CM5 chip). Those two immobilization methods are the most used for TCR kinetic assays, and either the TCR or the pHLA molecules can be immobilized on the SRP sensor chip (the other molecule will be the analyte). The choice is usually made based on the protein yield of the two molecules, as the immobilized molecule will be required in a smaller quantity than the analyte, but also based on the stability of the TCR or pHLA. Here the pHLA molecule is immobilized on the SPR sensor chip.

Described below is the protocol for using the GE Healthcare Streptavidin (SA) or Dextran (CM5) sensor chip and the Biacore™ T200 SPR machine to determine the binding affinity of TCRs to pHLA molecules. The use of a CM5 chip allows coupling the protein, without the requirement of biotinylation. There are four flow cells on each chip (SA or CM5) on a Biacore™ T200 instrument; therefore, up to 4 different pHLA molecules can be immobilized on the chip. If the SPR machine is different, there might be more than 4 flow cells.

2 Materials

2.1 SA Chip Buffers

1. SA running buffer, 500 mL: 0.5% Bovine serum albumin (BSA), 0.002% P20 (Cytiva), 100 mM Tris-HCl pH 8.0, 50 mM NaCl in Milli Q water (MQ). Make fresh each time. Filter and degas for at least 30 min before use.
2. 1 L MQ. Filter and degas for at least 30 min before use.

2.2 SA Chip Activation

1. SA chip (Cytiva).
2. Activation buffer, 50 mL: 50 mM NaOH, 1 M NaCl in MQ. Filter before use.
3. BIAnormalizing solution: 70% Glycerol (Cytiva).

2.3 Protein Immobilization and TCR Test Injections on SA Chip

1. 2 μg of the biotinylated protein that will be immobilized in 70 μL of SA running buffer.
2. 1 mg of biotin in 1 mL of MQ. Store in −20 °C.
3. Prepare 35 μL of analytes (TCRs of interest) in SA running buffer at 10 μM and if needed (if there is no binding at 10 μM) at 100 μM, 200 μM and 400 μM to test binding (*see* **Note 1**).

2.4 Affinity Measurement Experiment

132 μL of the TCR in SA running buffer at 200 μM is needed for the full experiment.

2.5 CM5 Chip Buffers

1. CM5 immobilization buffer, 500 mL: 10 mM HEPES pH 6.8, 0.005% P20 (Cytiva), 150 mM NaCl in MQ. Make fresh each time. Filter and degas for at least 30 min before use (*see* **Note 2**).
2. 1 L MQ. Filter and degas for at least 30 min before use.

2.6 CM5 Chip Activation

1. CM5 chip (Cytiva).
2. 75 μL of 0.05 M N-hydroxysuccinimide (NHS) and 75 μL of 0.2 M N-ethyl-N′-(3-dimethylaminopropyl)carbodiimide hydrochloride (EDC) (Amine Coupling Kit, by Cytiva, *see* **Note 3**).

2.7 Protein Immobilization on CM5 Chip

1. 4 μL of pHLA protein that will be immobilized at 1 mg/mL in CM5 immobilization buffer.
2. 0.6 M Sodium acetate, pH 4.2, 20 mL in MQ. Filter before use.
3. Ethanolamine (Cytiva).
4. Prepare 35 μL of analytes (TCRs of interest) in SA running buffer at 10 μM and if needed (if there is no binding at 10 μM) at 100 μM, 200 μM and 400 μM to test binding (*see* **Note 1**).

3 Methods

3.1 SA Chip Activation

1. Using the Biacore™ T200 control software, eject the chip that might be inside (maintenance chip), and insert a new SA chip.
2. Enter the lot number and expiry date of the SA chip. Without introducing air into the system, connect the SA running buffer (filtered and degassed) to Line A (left side) (*see* **Note 4**).
3. Insert Line B (right side) into the 1 L MQ (filtered and degassed) and check that the waste line is placed into a large empty bottle.

4. Prime the SA sensor chip using the control software (Prime machine). This step takes ~6 min.
5. Start a Manual Run (Run → Manual Run) and confirm that the correct rack has been placed inside the machine.
6. Set the flow path to all flow cells (2-1, 3-1, 4-1).
7. Equilibrate all four flow cells using a flow rate of 10 μL/min, until baseline (flat line). This step takes ~40 min.
8. Eject the rack (Tools → Eject rack), place 60 μL of activation buffer into one position, and re-insert.
9. Activate the SA chip (*see* **Note 5**) using three separate injections of activation buffer:
   - Set flow rate to 5 μL/min.
   - Set the contact time of the activation buffer to 240 seconds.
   - Repeat two more times.
10. Stop the run once complete (Stop run → End Run).
11. Place 500 μL of BIAnormalizing solution into the rack.
12. Normalize the surface using (Tools → More Tools → Normalize). This step takes 9 min.

3.2 Protein Immobilization on SA Chip

1. Start a new manual run, set the flow rate at 10 μL/min, select the appropriate rack and run SA running buffer on all flow cells for 10 min to remove excess glycerol from the previous step (Fig. 1a, block #1).
2. While this normalizing step is occurring, prepare your protein samples:
   - 2 μg of biotinylated pHLA molecule in a total of 70 μL of SA running buffer.
   - Per SA chip:
     - 1 negative control (NC) pHLA (that the TCR will not bind to).
     - 1 positive control pHLA (that the TCR will bind to, if required).
     - 1 or 2 pHLA molecules of interest to be tested.

   Make sure each pHLA sample is placed in the correct position on the rack.
3. Change the flow path to flow cell 1 only. Change the flow rate to 5 μL/min and inject the NC into flow cell 1 for 60 s (33 μL).
   - Biotinylated pHLA molecules will bind to the SA chip, increasing the response units on the sensorgram (Fig. 1a, block #2).
   - Try to reach a response change of 2000 response units (RU) before moving to the next protein.

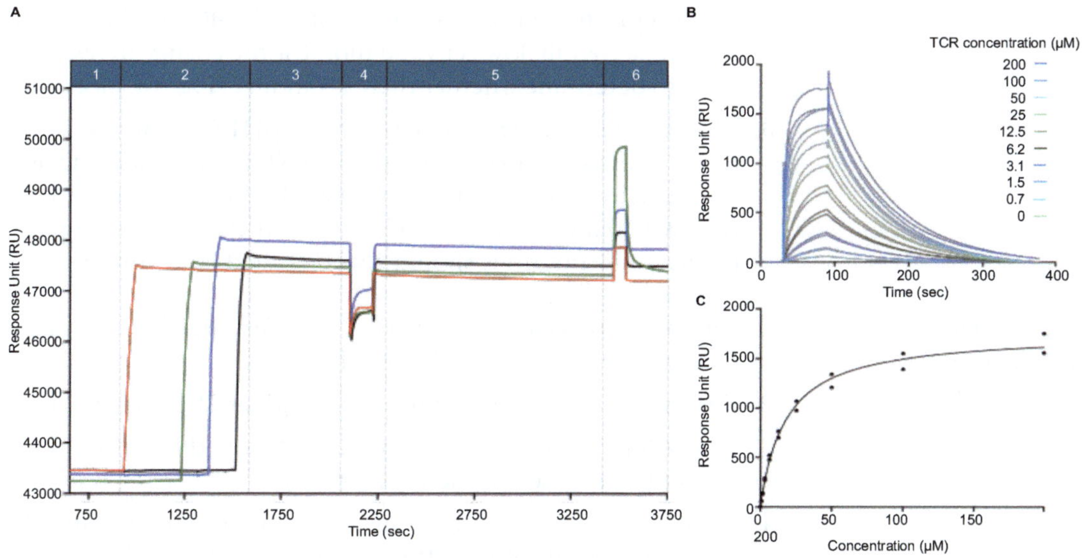

Fig. 1 Surface plasmon resonance between a TCR (analyte) and a pHLA (ligand). (**a**) Immobilization of the ligands (pHLA) and testing of TCR binding. The negative control is immobilized in flow cell 1 (red line) and three different pHLAs in separate flow cells: 2 (green line), 3 (blue line), and 4 (black line). The 6 different stages of the run are represented by the blue blocks. Block 1: flowing running buffer into all activated flow cells until they reach baseline. Block 2: injection and immobilization of each target pHLA into each flow cell separately. Block 3: stabilization of the sensorgram after immobilization and measurement of the RU compared to Block 1. Block 4: injection of biotin for the SA chip (or ethanolamine if using a CM5 chip) onto all four flow cells to deactivate the chip. Block 5: baseline all flow cells and re-compare response units to Block 1. Block 6: Test injection of 10 μM TCR as analyte onto all flow cells and determination of the change in RU compared to the negative control (e.g., 2-1) to ensure TCR binding before starting the experiment. (**b**) Sensorgram curves indicating association, baseline, and dissociation of the target TCR injected at different concentrations (colored curves), to a pHLA from a selected flow cell. (**c**) Steady-state binding curve and the K_d value generated

- If immobilization is low (<1000 RU), reinject more pHLA onto the flow cell.

4. Change the flow path to flow cell 2 and inject the second protein until a response of 2000 RU is reached.
   - Repeat for each protein of interest, ensuring the injection is only occurring on each flow path separately and that the RU reached for each target protein are similar.
5. To deactivate the SA chip, inject biotin at 1 mg/mL into all flow cells for 120 seconds (38 μL) (Fig. 1a, block #4).
6. Baseline and check the response values after biotin.

3.3 TCR Test Injections

1. Prepare the TCR protein of interest for testing its binding to the pHLA molecules using two concentrations (10 μM, 100 μM), using the SA running buffer to dilute the sample.

2. Inject 5 μL (60 s injection, 33 μL) of the TCR (10 μM concentration) across all flow cells and monitor the change in response units to determine whether there is binding (Fig. 1a, block #6).
3. The response on the flow cell with the pHLA of interest needs to be subtracted from flow cell 1 (NC), to obtain the RU resulting from TCR binding. If the RU value is positive, then the TCR binds the pHLA of interest. If the value is negative higher concentrations should also be tested (*see* **Note 6**).

3.4 Affinity Measurement Experiment

1. Set up a method using Wizard (refer to Biacore™ or equivalent training) or open an existing method.
2. For the TCR of interest, prepare 132 μL at 200 μM in a tube.
3. Take 66 μL and place it in the rack (this is the highest TCR concentration used in the experiment).
4. Perform 8 serial dilutions in SA running buffer with the remaining 66 μL of the 200 μM TCR solution and add them to the rack.
5. The tenth sample will only contain the SA running buffer.
6. The tubes on the rack need to be placed in order from the highest to the lowest concentration.
7. Add 120 μL of BIAnormalizing solution onto the rack and a separate tube of 66 μL of SA running buffer as startup.
8. Set flow paths as 2-1, 3-1, 4-1.
9. Check that all tubes are in the correct positions before starting the experiment.
10. Prime and normalize before starting.

3.5 Results

1. Open the Biacore™ T200 evaluation software and open the experiment.
2. Sensorgrams for each flow cell minus the negative control flow cell (e.g., 2-1) will appear automatically (Fig. 1b).
3. Click "Surface Bound" from the Kinetics/Affinity drop-down menu.
4. Select the correct TCR (Sample) and click Next → Affinity → Next, which will display the steady-state binding.
5. Click "Fit"; this will calculate a non-linear steady-state binding curve and provide a K_d value (Fig. 1c).

3.6 Protein Immobilization on CM5 Sensor Chip

1. Connect the buffer pump inlet to your CM5 immobilization buffer, prime the biosensor, and start the sensorgram using a flow rate of 5 μL/min to all flow cells until baseline (steps same as for SA chip).
2. Thaw and mix 75 μL of NHS and 75 μL of 0.2 M EDC.

3. Activate the CM5 chip by injecting the EDC/NHS sample at 5 μL/min for 15 min (*see* **Note 5**).
4. While the CM5 chip is activating, prepare your protein samples in the following mixture:
   - 4 μL of pHLA at 1 mg/mL.
   - 4 μL of 0.6 M Sodium acetate, pH 4.2.
   - 32 μL of MQ water.
5. Once the EDC/NHS injection is complete, make sure you change the flow path to select only one flow cell at the time (e.g., flow cell 1). Inject 5 μL (60 s injection) of the prepared protein sample (NC) into flow channel 1. Try to reach a response change of 2000 RU before moving to the next protein. If immobilization is still low after several injections, try using a different pH of sodium acetate (between pH 4.0 and 5.5) if the protein of interest is stable at a lower pH.
6. Change the flow path to another flow channel (e.g., flow cell 2) and inject the second protein until a response of 2000 RU is reached.
7. Repeat for each flow channel and each protein of interest.
8. Inject 50 μL of ethanolamine (600s injection) into all activated flow cells to avoid future covalent coupling.
9. Change the buffer to SA running buffer and baseline with the new buffer before running the binding testing.

3.7 TCR Test Injections and Affinity Measurement Experiment

Follow the same steps as per Subheadings 3.3, 3.4, and 3.5 (*see* **Note 7**). The analyte will be the pHLA if the TCR has been immobilized on the CM5 sensor chip, or the TCR if the pHLA has been immobilized.

4 Notes

1. The biosensor is very sensitive and will pick up buffer mismatches, so it is important to have your TCR at a high concentration to ensure that the dilution will be optimum for buffer match with the SA running buffer.
2. As the immobilization method uses random coupling of the protein via its amine group, do not use Tris as a buffer during the activation, as Tris-HCl will bind firmly to the activated surface of the CM5 chip, not allowing for the protein of interest to be coupled.
3. Prepare NHS and EDC according to Cytiva Amine Coupling Kit instructions. Make aliquots of 75 μL and store in −20 °C.

Both NHS and EDC need to be thawed only once. Do not reuse them once thawed.

4. Set aside 50 mL of SA running buffer to be used for protein sample dilution.
5. Once the SA or the CM5 chip is activated, the protein molecules need to be immobilized as soon as possible (within 1 h).
6. If the affinity of the TCR is >100 μM, you might need to test higher TCR concentrations to observe binding (i.e., 200 μM, 400 μM). Commonly $CD8^+$ TCRs have an affinity of 1–10 μM for viral peptides presented by HLA-I [7].
7. The buffer used in these steps for the CM5 chip is the same as that used for the SA chip, the SA running buffer.

Acknowledgments

The authors want to thank all their colleagues that contributed to the development of the methods to produce both soluble pHLA and TCR to enable the analysis of the interaction between these molecules. DSMC was supported by an AINSE ECR grant from ANSTO, and SG is supported by an NHMRC Leadership Investigator Grant (#2034677).

Conflict of Interest The authors have no conflicts of interest to declare that are relevant to the content of this chapter.

References

1. Freire E (2004) Isothermal titration calorimetry: controlling binding forces in lead optimization. Drug Discov Today Technol 1(3): 295–299. https://doi.org/10.1016/j.ddtec.2004.11.016
2. Yammine A, Gao J, Kwan AH (2019) Tryptophan fluorescence quenching assays for measuring protein-ligand binding affinities: principles and a practical guide. Bio Protoc 9(11):e3253. https://doi.org/10.21769/BioProtoc.3253
3. Muller-Esparza H, Osorio-Valeriano M, Steube N, Thanbichler M, Randau L (2020) Bio-layer interferometry analysis of the target binding activity of CRISPR-Cas effector complexes. Front Mol Biosci 7:98. https://doi.org/10.3389/fmolb.2020.00098
4. de Mol NJ, Fischer MJ (2010) Surface plasmon resonance: a general introduction. Methods Mol Biol 627:1–14. https://doi.org/10.1007/978-1-60761-670-2_1
5. Huang J, Zarnitsyna VI, Liu B, Edwards LJ, Jiang N, Evavold BD, Zhu C (2010) The kinetics of two-dimensional TCR and pMHC interactions determine T-cell responsiveness. Nature 464(7290):932–936. https://doi.org/10.1038/nature08944
6. Kolawole EM, Lamb TJ, Evavold BD (2020) Relationship of 2D affinity to T cell functional outcomes. Int J Mol Sci 21(21). https://doi.org/10.3390/ijms21217969
7. Szeto C, Lobos CA, Nguyen AT, Gras S (2020) TCR recognition of peptide-MHC-I: rule makers and breakers. Int J Mol Sci 22(1). https://doi.org/10.3390/ijms22010068
8. Corr M, Slanetz AE, Boyd LF, Jelonek MT, Khilko S, al-Ramadi BK, Kim YS, Maher SE, Bothwell AL, Margulies DH (1994) T cell receptor-MHC class I peptide interactions: affinity, kinetics, and specificity. Science 265(5174):946–949. https://doi.org/10.1126/science.8052850

9. Matsui K, Boniface JJ, Steffner P, Reay PA, Davis MM (1994) Kinetics of T-cell receptor binding to peptide/I-Ek complexes: correlation of the dissociation rate with T-cell responsiveness. Proc Natl Acad Sci USA 91(26): 12862–12866. https://doi.org/10.1073/pnas.91.26.12862

10. Liedberg B, Nylander C, Lunström I (1983) Surface plasmon resonance for gas detection and biosensing. Sensors Actuators 4:299–304. https://doi.org/10.1016/0250-6874(83)85036-7

Chapter 15

Methods for Probing NK Cell Surveillance of Antigen Presentation

Kamila Król and Doriana Fruci

Abstract

Natural killer (NK) cells are critical components of the innate immune system, capable of rapidly identifying and eliminating virus-infected and tumor cells without prior antigen sensitization. This rapid cytotoxic response is regulated by a balance between activating and inhibitory receptor signals, a mechanism known as "missing self." Inhibitory receptors recognize HLA class I molecules, preventing unwanted activation against healthy cells, while activating receptors detect stress-induced ligands on abnormal cells, triggering the release of perforin and granzyme B.

Recent evidence shows that inhibitory receptor-HLA class I interactions are peptide-dependent, highlighting a complexity similar to T cell receptor recognition. Alterations in the antigen processing and presentation pathway in cancer cells can modify the immunopeptidome, enhancing NK-mediated cytotoxicity.

To assess NK cell function and receptor-ligand interactions, degranulation and killing assays are widely employed. These methods are applicable to both primary NK cells and established NK cell lines, which differ in receptor expression and cytotoxic capabilities. Strategic selection of NK models and target systems allows for in-depth studies of NK cell biology and supports the development of NK cell-based immunotherapies.

Key words NK cells, HLA class I, Cytotoxicity, Killing assay, Degranulation assay, Antigen probing, Antigen presentation, Immunotherapy, Flow cytometry

1 Introduction

Natural killer (NK) cells play a central role in immune surveillance by rapidly recognizing and eliminating virus-infected and cancer cells, without prior antigen sensitization. Their function relies on the "missing self" mechanism, which integrates activating and inhibitory receptor signals to distinguish healthy cells from abnormal ones [1].

NK cell activity depends on the balance of these opposing signals through receptor engagement with ligands on target cells [2]. Inhibitory receptors such as killer cell immunoglobulin-like

Efstratios Stratikos (ed.), *Antigen Processing and Presentation: Methods and Protocols*, Methods in Molecular Biology, vol. 3007, https://doi.org/10.1007/978-1-0716-5092-9_15,

receptors (KIRs) and NKG2A recognize human leukocyte antigen (HLA) class I molecules: KIRs bind classical HLA-A, -B, and -C [3], while NKG2A interacts with non-classical HLA-E [4]. These interactions suppress NK cell activation, helping to preserve self-tolerance and prevent damage to healthy tissues. Conversely, activating receptors recognize stress-induced or aberrant ligands, tipping the balance toward activation. When activation prevails, NK cells release cytotoxic granules containing perforin and granzyme B, which induce apoptosis by forming membrane pores and activating caspase pathways [5, 6]. Unlike cytotoxic T lymphocytes, NK cells mount this response rapidly, without prior sensitization.

Recent findings have shown that inhibitory NK cell receptors interact with HLA class I molecules in a peptide-dependent manner, similar to T cell receptor-HLA interactions [7–9]. Alterations in the antigen processing and presentation (APP) pathway in cancer cells can modify the surface peptide repertoire (immunopeptidome), enhancing their recognition and elimination by immune cells [10–12]. Notably, NK cells respond significantly faster than T cells, killing target cells within approximately 4 h of coculture, compared to 24–48 h for T lymphocytes [13, 14], underscoring their role as a rapid first line of defense [15].

To study how an alteration of the APP pathway may affect NK cell responses, both degranulation and killing assays are widely used. Degranulation is typically assessed by flow cytometry using markers such as CD107a (LAMP-1), which is transiently expressed on the surface during granule release [16, 17]. Combined with antibody panels and optimized gating strategies, this assay allows detailed investigation of receptor-ligand interactions in various experimental settings. Killing assay which quantify surviving target cells, provide insights into mechanisms of susceptibility or resistance to NK-mediated cytotoxicity.

Both primary NK cells from healthy donor peripheral blood and established NK cell lines, such as YTS, NKL, and NK-92, are widely employed. Each cell line has a distinct receptor profile that influences its functional behavior. For instance, NKL cells, with high expression of the inhibitory receptor NKG2A, are well-suited for studying HLA-E interactions [4, 18]. In contrast, YTS cells require costimulatory molecules like CD80 and CD86 on target cells for effective killing [19], making them more specific but also more limited in application [20–22]. The careful selection of NK cell sources and target cells enables comprehensive exploration of receptor-ligand dynamics and supports the development of NK cell-based immunotherapies.

2 Materials

2.1 NK Cell Isolation from Healthy Donor Blood and Culture

1. Starting material: Human peripheral blood (recommended volume: 15–30 mL).
2. NK Cell Enrichment: RosetteSep™ Human NK Cell Enrichment Cocktail (Stem Cell Technologies).
3. Density Gradient Medium: Pancoll (Pan-Biotech).
4. Buffers and Media:
   - Phosphate-buffered saline (PBS) 1×, sterile-filtered (Euroclone).
   - NK MACS medium human with supplements (Miltenyi Biotec).
5. Cytokines and Supplements:
   - Human Interleukin-2 (IL-2) (R&D Systems).
   - Human serum (Sigma).
   - Penicillin-Streptomycin 100x (Euroclone).
6. Cell Counting Reagents: Turk's solution (Sigma).

2.2 NK Cell Degranulation Assay

1. Cells: NK cells, target cells of interest, and K562 cells (CCL-243) as a positive control.
2. Culture Medium: RPMI medium (Euroclone).
3. Supplements: FBS (Gibco).
4. Viability Dye: 0.4% Trypan blue solution (Thermo Fisher).
5. Protein Transport Inhibitor: BD GolgiStop™ Protein Transport Inhibitor (Containing Monensin) (BD Biosciences).
6. Fixation Buffer: BD Cytofix Fixation Buffer (BD Biosciences).
7. Flow cytometry antibodies panel (Table 1).

2.3 Killing Assay Using NK Cell Lines

1. Cells: NK cells (e.g., NKL, YTS, NK-92), target cells of interest, and either K562 (CCL-243) or 721.221 cells as positive controls.
2. Culture Medium: RPMI medium (Euroclone).
3. Supplements: FBS (Gibco).
4. Cytokines: Human IL-2 (R&D Systems), for pre-activation of NK cells (R&D Systems).
5. Buffers: PBS 1x sterile-filtered (Euroclone).
6. Antibiotics: Penicillin-Streptomycin 100x (Euroclone).
7. Viability Dye: 0.4% Trypan blue solution (Thermo Fisher).
8. Cell Tracking and Counting:

Table 1
List of antibodies

Marker	Clone	Fluorochrome	Supplier	Cataloge no.	Dilution
CD56	B159	PE-Cy7	BD bioscience	557747	1:40
CD107a	H4A3	BV421	BD bioscience	562623	1:50
KIR2DL1/S1	EB6B	PE-Cy5.5	Beckman coulter	A66898	1:25
KIR2DL2/L3/S2	GL183	PE	Beckman coulter	IM2278U	1:25
KIR3DL1	DX9	APC	R&D systems	FAB1225A-100	1:25
NKG2A	a131411	AF700	R&D systems	FAB1059N-100	1:40

- CellTracker Fluorescent Probes, e.g., Deep Red (ThermoFisher).
- CountBright™ Absolute Counting Beads (ThermoFisher).

9. Live/dead Staining: Propidium iodide/7-AAD (for distinguishing live and dead cells).
10. Flow Cytometry Tubes: Sterile 5 ml round-bottom tubes with caps (Corning).

3 Methods

3.1 Isolation of NK Cells from Healthy Donor Blood and Culture

NK cells can be isolated from peripheral blood of healthy donors using the RosetteSep™ Human NK Cell Enrichment Cocktail, which enables rapid and selective enrichment.

1. Add RosetteSep cocktail to the whole blood sample at a concentration of 50 μL/mL of blood.
2. Incubate for 20 min at room temperature (RT).
3. Dilute the whole blood 1:1 with sterile PBS (e.g., mix 15 mL whole blood with 15 mL PBS).
4. Fill a 50 mL Falcon tube with an appropriate volume of Pancoll (e.g., 15 ml of Pancoll for 30 mL of diluted blood).
5. Carefully layer the diluted blood over the Pancoll without mixing the layers.
6. Centrifuge at 1200× *g* for 20 min without a brake.
7. Carefully collect the NK cell layer, visible as a distinct ring between the layers.
8. Transfer the NK cells to a new 50 mL Falcon tube.
9. Wash the cells by filling the tube with warm PBS to remove residual Pancoll.
10. Centrifuge at 300× *g* for 10 min.

11. Carefully remove the supernatant to avoid losing the cell pellet.
12. Repeat the wash step once more.
13. Carefully remove the supernatant first with a serological pipette, and then more accurately with a 1 mL pipette.
14. Resuspend the cells in freshly prepared, warmed NK-cell complete medium composed of 50 mL NK MACS supplemented with 1% NK Medium Supplement, 1% Penicillin-Streptomycin, 5% heat-inactivated human serum, and 500 IU/mL human IL-2. Start with 5–10 mL of medium to resuspend the cells.
15. Count the cells using Turk's solution and adjust the medium volume to achieve two million cells/mL.
16. Plate the cells in a 24-well plate at 1 mL per well.
17. The following day, assess NK cell purity by flow cytometry using antibodies targeting main immune cell populations. Samples with >90% NK cells are suitable for functional assays.

3.2 NK Cell Degranulation Assay

The degranulation assay measures NK cell functional activation by detecting the surface expression of degranulation markers such as CD107a, following coculture with target cells.

1. The day before the experiment, ensure both NK cells and target cells are healthy and in log-phase growth, refreshing media and splitting if needed.
2. Plan the experimental setup, including treatments, different effector-to-target (E:T) ratios, and target cell lines. For example, a typical experiment may include 12 different experimental conditions plus the K562 control cells at an E:T ratio of 1:1. All conditions are run in triplicate, along with NK cells alone in duplicate, and a set of unstained NK cells as a control.
3. Calculate the total number of NK cells required (150,000 cells per sample). For 42 samples plus 3 extra to compensate for pipetting loss, approximately 6.75 million cells are needed.
4. Harvest NK cells from culture, count, and resuspend in RPMI complete medium at 40 μL per sample.
5. Plate 40 μL of NK cells into one well of a 96-well U-bottom plate to serve as the unstained control.
6. Add anti-CD107a antibody to the rest of NK cells at 1:50 dilution (e.g., 35.2 μL for the total volume).
7. Plate 40 μL of NK cells per well in a 96-well U-bottom plate.
8. Prepare target cells by counting and adjusting the concentration to 150,000 cells per 50 μL per well (including extra volume to compensate for pipetting loss).
9. Add 50 μL of target cells to each well containing NK cells.

10. Centrifuge the plate gently at 100× *g* for 2 min to facilitate cell contact.
11. Incubate the coculture at 37 °C for 1 h.
12. Add 10 μL of Golgi Stop solution (diluted 1:50 in complete RPMI) to each well, centrifuge at 100× *g* for 2 min to mix.
13. Incubate for an additional 3 h at 37 °C (total 4 h coculture).
14. After incubation, centrifuge the plate at 320× *g* for 5 min.
15. Discard the supernatant by inverting the plate quickly.
16. Stain cells with an antibody cocktail targeting NK cell receptors and markers in FACS buffer (PBS + 2% FBS), adding 20 μL per well (except unstained controls).
17. Incubate for 20 min at 4 °C in the dark.
18. Wash cells with 150 μL FACS buffer, centrifuge at 320× *g* for 5 min, and discard supernatant by flipping the plate.
19. Fix cells with 100 μL Cytofix buffer for at least 20 min in the dark at RT.
20. Adjust sample volumes to at least 200 μL and analyze by flow cytometry. The gating strategy used to detect degranulation of NK cells expressing individual inhibitory receptors is shown in Fig. 1.

3.3 Killing Assay with NK Cell Lines

This assay measures the cytotoxic activity of NK cells against target cells by assessing the loss of live target cells after coculture.

1. Count the NK cells and collect the required number for the experiment. It is advisable to prepare extra cells to ensure sufficient volume.
2. Centrifuge the NK cells in a 50 mL Falcon tube at 300× *g* for 5 min.
3. Stain the NK cells with the selected CellTracker dye (e.g., Deep Red) diluted in serum-free medium. Incubate at RT in the dark for 20 min at the dye's optimal concentration (0.25 μM for Deep Red), according to the manufacturer's instructions.
4. Add four volumes of complete medium containing FBS and incubate for an additional 5 min at RT.
5. Centrifuge the stained NK cells, then resuspend them in complete medium to achieve the desired cell concentration in 100 μL.
6. Count the target cells and collect the required number, preparing extra for volume assurance.
7. Centrifuge the target cells in a 15 mL Falcon tube at 300× *g* for 5 min.

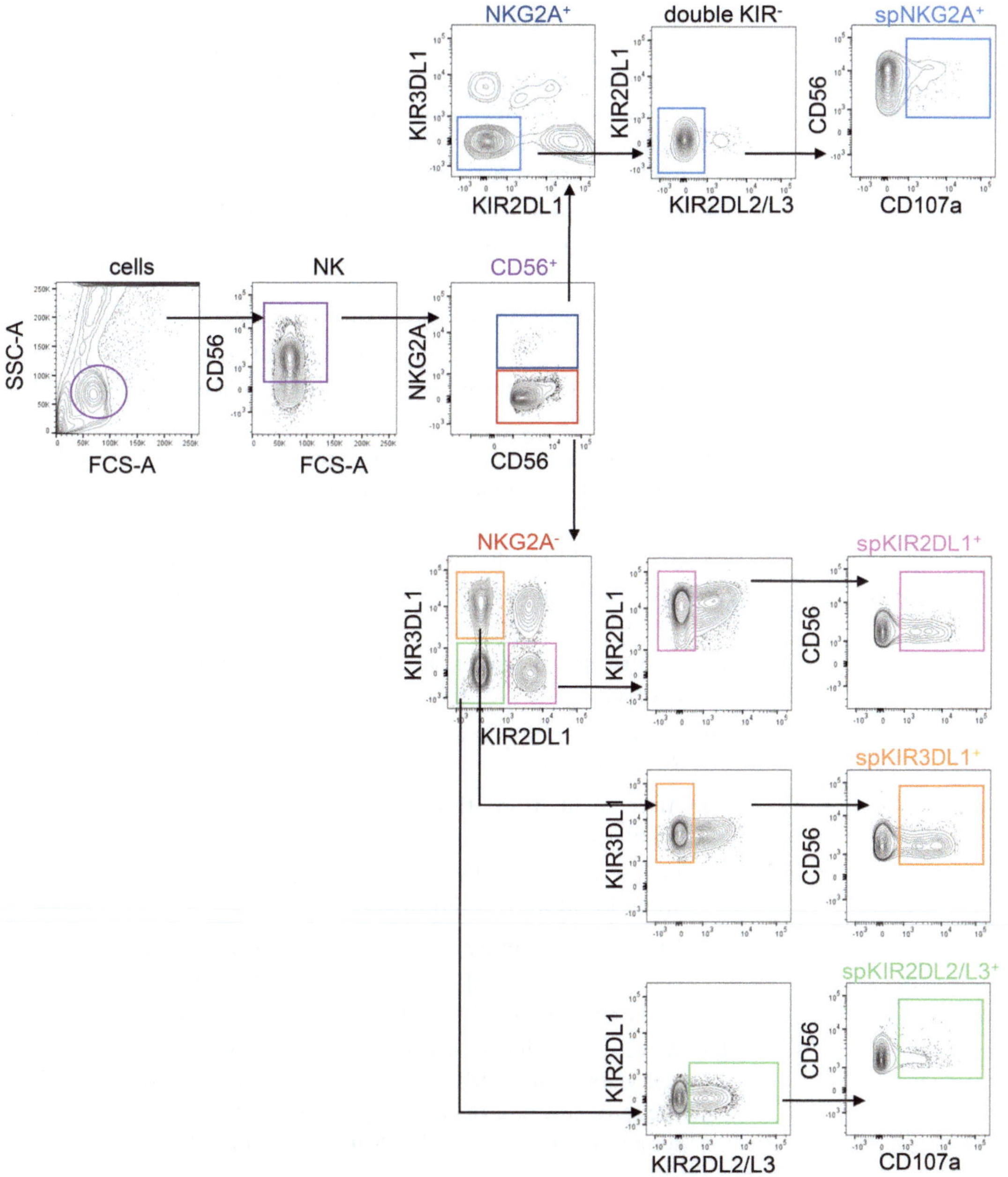

Fig. 1 Gating strategy for flow cytometry analysis of NK cell degranulation subsets. Lymphocytes were initially identified by forward scatter (FSC) versus side scatter (SSC) gating to exclude debris and non-lymphoid cells. Within this lymphocyte gate, CD56$^+$ NK cells (purple) were selected to exclude other immune populations. The CD56$^+$ NK cells were then subdivided based on NKG2A expression into NKG2A$^+$ (dark blue) and NKG2A$^-$ (bordeaux) subsets. The NKG2A$^+$ population was further gated to exclude KIR3DL1$^+$ and KIR2DL1$^+$ cells, isolating the double-negative subset (blue). From this, KIR2DL2/L3-negative cells were selected to focus on the single-positive NKG2A$^+$ (spNKG2A$^+$) subset for degranulation analysis. For analysis of single KIR-expressing NK cells, the NKG2A$^-$ (bordeaux) subset was stratified based on KIR expression into KIR2DL1$^+$ (pink), KIR3DL1$^+$ (orange), and double KIR-negative (green) populations. Each of these was further gated for KIR2DL2/L3 expression to identify single-positive subsets: spKIR2DL1$^+$, spKIR3DL1$^+$, and spKIR2DL2/L3$^+$, respectively, for evaluation of their degranulation activity

8. If your target cells express a fluorescent protein (e.g., GFP or mCherry), no additional staining is necessary. If target cells are not fluorescently distinguishable from NK cells and are similar in size (e.g., leukemia or lymphoma cells), stain them with a distinct dye such as CellTracker Blue CMAC, following the manufacturer's instructions.
9. After staining, centrifuge and resuspend the target cells in complete medium to obtain the desired concentration in 100 μL.
10. Label round-bottom FACS tubes clearly for all conditions.
11. Add 100 μL of complete medium to all control tubes (without coculture) to ensure equal volume across all tubes in the experiment.
12. Mix the target cell suspension gently and pipette 100 μL into each tube designated for coculture and target controls, in triplicate.
13. Mix the NK cell suspension well and add 100 μL to each coculture tube and to effector-only control tubes, also in triplicate.
14. Centrifuge all tubes at 100× *g* for 2 min to promote cell contact.
15. Place the tubes on a sterile rack in a cell culture incubator at 37 °C for 4 h.
16. After incubation, transfer the tubes to ice to stop further cellular activity.
17. Vortex the CountBright absolute counting beads thoroughly and add 50 μL to each tube.
18. Immediately before acquisition, add a live/dead viability dye (e.g., 1 μL of propidium iodide at a final concentration of 5 μg/mL), mix gently the sample.
19. Acquire samples promptly on a flow cytometer to assess target cell viability and NK cell killing efficiency by always acquiring the same number of counting beads between the samples (e.g., set a stopping gate on 10,000 beads) (Fig. 2).

Calculation Formula

To calculate NK cell-mediated killing, use the following formula based on absolute counts of real live target cells (normalized to bead count) (Fig. 3):

1. Cell concentration (cells/μL):

$$\text{cells}/\mu l = \left(\frac{\text{live targets}}{\text{beads nr}} \times \frac{\text{beads conc.}}{\text{sample volume}}\right)$$

2. Absolute number of target cells in sample:

$$\text{absolute count} = \frac{\text{cell}}{\mu l} \times \text{sample volume}$$

3. NK cell-mediated killing (%):

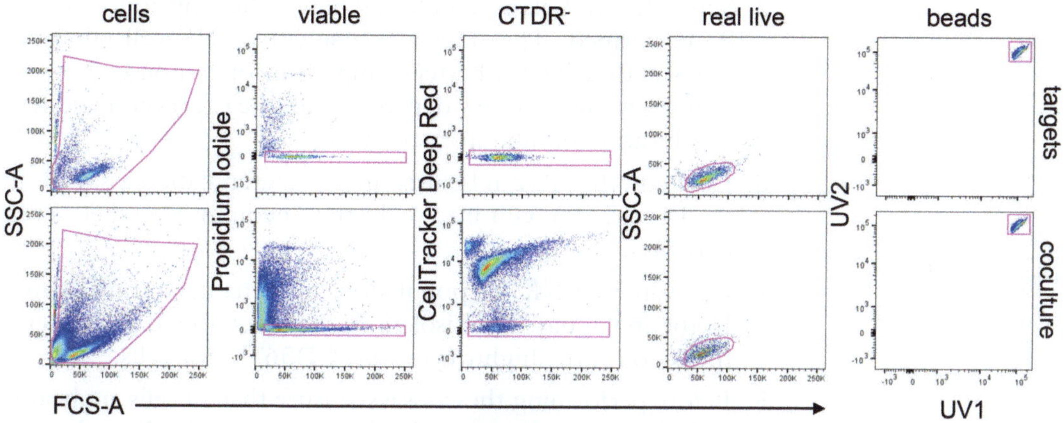

Fig. 2 Gating strategy for killing assay. Cells are first gated on the FSC-A vs. SSC-A plot, followed by selection of viable cells (propidium iodide negative). CTDR^{-} target cells are then gated to exclude CTDR^{+} NK cells. The real live target compact population is further refined using FSC-A vs. SSC-A gating. In parallel, UV1 vs. UV2 beads are gated. The calculation formula considers the real live target and bead numbers

	A	B	C	D	E	F	G	H
1	Killing assay FACS							
2				formula= (cells/beads amount)* (beads concentration/total sample volume) [cells/ul]				
3	Controls (targets alone)	live events	beads	nr of cells/ul	nr of cells in sample volume	mean of live targets alone		
4	target1_1	2591	10037	51,11	12778			
5	target1_2	2557	9996	50,65	12662			
6	target1_3	2523	9987	50,02	12505	12648		
7								
8								
9	Experimental	live events	beads	nr of cells/ul	nr of cells in sample volume	cells in coculture/mean of cells alone	% killing	% mean killing
10	nk_target1_1	1613	10132	31,52	7880,33	0,62	37,70	
11	nk_target1_2	1442	10133	28,18	7044,21	0,56	44,31	
12	nk_target1_3	1554	10133	30,37	7591,34	0,60	39,98	40,66
13								
14	sample volume [ul]	beads conc.	beads volume [ul	total sample volume [ul]				
15	200	49500	50	250				

Fig. 3 Example calculation sheet for the killing assay. The calculation starts by determining the number of real live target cells and beads for each replicate. Cell concentration (cells/μL) is then calculated by normalizing the number of live target cells to the bead count. The absolute count of target cells in each sample is obtained by multiplying the cell concentration by the total sample volume (e.g., 250 μL). Finally, NK cell-mediated killing (%) is calculated by comparing the ratio of the absolute count of live target cells remaining after coculture with the mean absolute count of live target cells in control wells (without NK cells). This ratio is subtracted from 1 and multiplied by 100 to yield the percentage of target cell killing

$$\text{killing } (\%) = \left(1 - \frac{\text{absolute count of live targets in coculture}}{\text{mean of absolute count of live targets alone (no NK)}}\right) \times 100$$

4 Notes

1. It is recommended to perform HLA class I typing on target cells to assess which NK cell inhibitory receptors they may engage.
2. Similarly, HLA class I and KIR typing of healthy donors should be performed. This allows assessment of NK cell education levels and facilitates better comparison of responses between donors, which can vary depending on the expression and level of education of inhibitory receptors.
3. Isolated NK cells can be used in the degranulation assay for up to 14 days in culture, with the medium refreshed every two days.
4. Be aware that NK cell populations may change during culture. In the initial days after isolation, the most abundant subset in the blood is the highly cytotoxic $CD56^{dim}$ NK cells.
5. Before performing these assays, ensure that all cells are healthy and viable. It is best to split them and completely replace the culture medium one day prior to the experiment.
6. The killing assay relies on comparing relative target cell counts between coculture and control tubes containing only target cells. To ensure reproducibility, use the same set of calibrated, high-accuracy pipettes across all tubes.
7. If needed, activate NK cells with IL-2 the day before the experiment, depending on the NK cell type. For example, blood-derived NK cells and the NKL cell line typically require IL-2 activation.
8. The killing assay can be performed using various E:T ratios depending on the experimental needs, though the standard and most commonly used ratio is 10:1.
9. Perform all cell handling and manipulations under a sterile laminar flow hood to maintain aseptic conditions.
10. When preparing cells for coculture, ensure you count enough cells to accommodate technical triplicates for all experimental conditions and target-only controls.
11. For example, if using an E:T ratio of 10:1, coculture 150,000 NK cells with 15,000 target cells per well.
12. Include a positive control for NK cell activity, such as K562 cells (for NKL cells) or 721.221 cells (for YTS cells). These

control target cells lack HLA class I expression and should be efficiently lysed by NK cells, confirming NK cytotoxic function.

13. Always include unstained and single-stained controls with cell tracker dyes for both effector and target cells. These controls are essential for properly setting compensation and gates during flow cytometry analysis.
14. The antibody panel for the degranulation assay can be expanded to include other NK cell receptors that interact with HLA class I molecules, such as activating NKG2C and activating KIRs.
15. The additional staining of target cells (if they do not express any fluorescent protein) in the killing assay is recommended, as it allows confirmation that the final 'real live' population contains only target cells.

Acknowledgement

This work was supported by grants awarded by the European Commission in the context of the Marie Skłodowska-Curie Action European Training Network CAPSTONE (954992 - CAPSTONE—H2020-MSCA-ITN-2020) (D. Fruci), the Associazione Italiana Ricerca sul Cancro (AIRC) IG24345 (D. Fruci), the Italian Ministry of Health with Ricerca Finalizzata No. PE-2011-02351866 (D. Fruci) and Current Research funds (D. Fruci). K. Król is a CAPSTONE-ETN H2020 Early Stage Researcher.

Conflict of Interest The authors have no conflicts of interest to declare that are relevant to the content of this chapter.

References

1. Ljunggren H-G, Kärre K (1990) In search of the 'missing self': MHC molecules and NK cell recognition. Immunol Today 11:237–244. https://doi.org/10.1016/0167-5699(90)90097-S
2. Lanier LL (2005) NK cell recognition. Annu Rev Immunol 23:225–274. https://doi.org/10.1146/annurev.immunol.23.021704.115526
3. Pende D, Falco M, Vitale M, Cantoni C, Vitale C, Munari E, Bertaina A, Moretta F, Del Zotto G, Pietra G, Mingari MC, Locatelli F, Moretta L (2019) Killer Ig-like receptors (KIRs): their role in NK cell modulation and developments leading to their clinical exploitation. Front Immunol 10. https://doi.org/10.3389/fimmu.2019.01179
4. Lee N, Llano M, Carretero M, Ishitani A, Navarro F, López-Botet M, Geraghty DE (1998a) HLA-E is a major ligand for the natural killer inhibitory receptor CD94/NKG2A. Proc Natl Acad Sci 95(9):5199–5204. https://doi.org/10.1073/pnas.95.9.5199
5. Krzewski K, Coligan JE (2012) Human NK cell lytic granules and regulation of their exocytosis. Front Immunol 3:335. https://doi.org/10.3389/fimmu.2012.00335
6. Gwalani LA, Orange JS (2018) Single Degranulations in NK cells can mediate target cell killing. J Immunol 200(9):3231–3243. https://doi.org/10.4049/jimmunol.1701500

7. Cifaldi L, Lo Monaco E, Forloni M, Giorda E, Lorenzi S, Petrini S, Tremante E, Pende D, Locatelli F, Giacomini P, Fruci D (2011) Natural killer cells efficiently reject lymphoma silenced for the endoplasmic reticulum aminopeptidase associated with antigen processing. Cancer Res 71(5):1597–1606. https://doi.org/10.1158/0008-5472.CAN-10-3326
8. Cifaldi L, Romania P, Falco M, Lorenzi S, Meazza R, Petrini S, Andreani M, Pende D, Locatelli F, Fruci D (2015) ERAP1 regulates natural killer cell function by controlling the engagement of inhibitory receptors. Cancer Res 75(5):824–834. https://doi.org/10.1158/0008-5472.CAN-14-1643
9. Das J, Khakoo SI (2015) NK cells: tuned by peptide? Immunol Rev 267(1):214–227. https://doi.org/10.1111/imr.12315
10. Nikopaschou M, Samiotaki M, Stylianaki E-A, Król K, Gragera P, Raja A, Aidinis V, Chroni A, Fruci D, Panayotou G, Stratikos E (2025) ERAP1 activity modulates the Immunopeptidome but also affects the proteome, metabolism, and stress responses in cancer cells. Mol Cell Proteomics 24(5):100964. https://doi.org/10.1016/j.mcpro.2025.100964
11. Tempora P, D'Amico S, Gragera P, Damiani V, Krol K, Scaldaferri V, Pandey K, Chung S, Lucarini V, Giorda E, Scarsella M, Volpe G, Pezzullo M, De Stefanis C, D'Oria V, De Angelis L, Giovannoni R, De Ioris MA, Melaiu O, Purcell AW, Locatelli F, Fruci D (2024) Combining ERAP1 silencing and entinostat therapy to overcome resistance to cancer immunotherapy in neuroblastoma. J Exp Clin Cancer Res 43(1):292. https://doi.org/10.1186/s13046-024-03180-y
12. Tsao H-W, Anderson S, Finn KJ, Perera JJ, Pass LF, Schneider EM, Jiang A, Fetterman R, Chuong CL, Kozuma K, Stickler MM, Creixell M, Klaeger S, Phulphagar KM, Rachimi S, Verzani EK, Olsson N, Dubrot J, Pech MF, Silkworth W, Lane-Reticker SK, Allen PM, Ibrahim K, Knudsen NH, Cheng AY, Long AH, Ebrahimi-Nik H, Kim SY, Du PP, Iracheta-Vellve A, Robitschek EJ, Suermondt JSMT, Davis TGR, Wolfe CH, Atluri T, Olander KE, Rush JS, Sundberg TB, McAllister FE, Abelin JG, Firestone A, Stokoe D, Carr SA, Harding FA, Yates KB, Manguso RT (2024) Targeting the aminopeptidase ERAP enhances antitumor immunity by disrupting the NKG2A-HLA-E inhibitory checkpoint. Immunity 0(0). https://doi.org/10.1016/j.immuni.2024.10.013
13. Lanzavecchia A, Iezzi G, Viola A (1999) From TCR engagement to T cell activation: a kinetic view of T cell behavior. Cell 96(1):1–4. https://doi.org/10.1016/S0092-8674(00)80952-6
14. De Maria A, Bozzano F, Cantoni C, Moretta L (2011) Revisiting human natural killer cell subset function revealed cytolytic CD56dimCD16+ NK cells as rapid producers of abundant IFN-γ on activation. Proc Natl Acad Sci 108(2):728–732. https://doi.org/10.1073/pnas.1012356108
15. Shimasaki N, Jain A, Campana D (2020) NK cells for cancer immunotherapy. Nat Rev Drug Discov 19(3):200–218. https://doi.org/10.1038/s41573-019-0052-1
16. Alter G, Malenfant JM, Altfeld M (2004) CD107a as a functional marker for the identification of natural killer cell activity. J Immunol Methods 294(1–2):15–22. https://doi.org/10.1016/j.jim.2004.08.008
17. Aktas E, Kucuksezer UC, Bilgic S, Erten G, Deniz G (2009) Relationship between CD107a expression and cytotoxic activity. Cell Immunol 254(2):149–154. https://doi.org/10.1016/j.cellimm.2008.08.007
18. Robertson MJ, Cochran KJ, Cameron C, Le JM, Tantravahi R, Ritz J (1996) Characterization of a cell line, NKL, derived from an aggressive human natural killer cell leukemia. Exp Hematol 24(3):406–415
19. Chen X, Allan DSJ, Krzewski K, Ge B, Kopcow H, Strominger JL (2006) CD28-stimulated ERK2 phosphorylation is required for polarization of the microtubule organizing center and granules in YTS NK cells. Proc Natl Acad Sci 103(27):10346–10351. https://doi.org/10.1073/pnas.0604236103
20. Peterson ME, Long EO (2008) Inhibitory receptor signaling via tyrosine phosphorylation of the adaptor Crk. Immunity 29(4):578–588. https://doi.org/10.1016/j.immuni.2008.07.014
21. Saunders PM, Vivian JP, Baschuk N, Beddoe T, Widjaja J, O'Connor GM, Hitchen C, Pymm P, Andrews DM, Gras S, McVicar DW, Rossjohn J, Brooks AG (2015) The interaction of KIR3DL1*001 with HLA class I molecules is dependent upon molecular microarchitecture within the Bw4 epitope. J Immunol 194(2):781–789. https://doi.org/10.4049/jimmunol.1402542
22. Sim MJW, Malaker SA, Khan A, Stowell JM, Shabanowitz J, Peterson ME, Rajagopalan S, Hunt DF, Altmann DM, Long EO, Boyton RJ (2017) Canonical and cross-reactive binding of NK cell inhibitory receptors to HLA-C Allotypes is dictated by peptides bound to HLA-C. Front Immunol 8. https://doi.org/10.3389/fimmu.2017.00193

Chapter 16

Cellular Assays to Monitor Antigen Cross-presentation In Vitro

François-Xavier Mauvais and Peter van Endert

Abstract

Antigen cross-presentation is a functional feature of some specialized antigen-presenting cells, allowing them to internalize and process antigens from extracellular sources and present them via major histocompatibility class I (MHC-I) molecules to $CD8^+$ T lymphocytes. This process is critical for the mounting of protective immunity to tumors and intracellular pathogens. Antigen cross-presentation can be monitored through in vivo and in vitro assays that are key to evaluate the efficiency and the molecular and cellular mechanisms of cross-presentation by a given cell type. This chapter intends to complement published protocols by describing how to use MutuDC, a murine immortalized type 1 conventional dendritic cell-like line, and a "nonprofessional" human antigen-presenting cell, the human embryonic kidney HEK 293T cell line, endowed with the ability to cross-present immune complexes through overexpression of an Fc receptor, for in vitro antigen cross-presentation assays.

Key words Cross-presentation, Dendritic cells, Antigen presentation, MHC-I, Endocytosis

1 Introduction

Antigen cross-presentation is the process by which a professional antigen-presenting cell can internalize exogenous antigens, process them to small (8–10 mer) antigenic peptides suitable for loading onto and presentation by Major Histocompatibility Class-I (MHC-I) molecules. $CD8^+$ T lymphocyte clones expressing a T Cell Receptor (TCR) specific for a given peptide/MHC-I complexes will arrest close to and form an immunological synapse with the antigen-presenting cell, resulting in $CD8^+$ T lymphocyte activation, expansion, and acquisition of an effector/cytotoxic phenotype.

Antigen cross-presentation is critical for the mounting of protective immunity against tumors and intracellular pathogens, including parasites or some intracellular bacteria. Contrasting with the endogenous pathway of MHC-I antigen presentation, which is active in all nucleated cells, cross-presentation is a specific

Efstratios Stratikos (ed.), *Antigen Processing and Presentation: Methods and Protocols*, Methods in Molecular Biology, vol. 3007, https://doi.org/10.1007/978-1-0716-5092-9_16,

feature of professional antigen-presenting cells and notably type 1 conventional dendritic cells (cDC1s). Yet, several other immune and possibly non-immune cells can also be efficient depending on the context [1].

Monitoring antigen-cross presentation in vitro through cellular assays can help address the following questions: (1) Does a specific (in most cases, myeloid) antigen-presenting cell efficiently cross-present antigens? We have recently provided an overview of published reports of cross-presentation by non-DCs and proposed criteria for assessing the in vivo relevance of such observations [1]. (2) Which mechanisms do the antigen-presenting cells examined use to cross-present antigens to cognate $CD8^+$ T lymphocytes? The dependence on a specific proteasome-dependent or proteasome-independent pathway can be tested conveniently in vitro using specific pharmacological inhibitors or by gene-targeting approaches. However, in vivo assays remain mandatory to test the physiological relevance of observations made in in vitro settings.

Several protocols suitable for monitoring cross-presentation using in vitro cellular assays have previously been published for mouse and human cells [2–6] and can be of interest as alternatives to the protocols described hereafter. These procedures may be applied to primary antigen-presenting cells purified from mouse or human tissues [6–12] or produced in vitro by differentiation of bone marrow or blood progenitors [5, 13–16] with appropriate differentiation and growth factors, as well as to various murine or human dendritic cell lines. Approximately 20 murine DC lines with different genetic backgrounds and tissue origins are available and have been used to test DC biology, including cross-presentation [17]. Similarly, human DC lines have been generated from leukemia or lymphoma cells [17], including the myelogenous leukemia cell line KG-1 [18], the monocytic leukemia cell line THP-1 [19], the myeloid leukemia cell line HL-60 [20], or the MUTZ-3 cell line [21] derived from the peripheral blood of a patient with acute myeloid leukemia. These cell lines are permanently available, can be expanded to high yields, are easy to manipulate genetically and require neither an animal facility nor approbation by an ethical review board for use. However, important caveats are associated with murine and human immortalized DC lines, which frequently require incubation periods with specific cytokines or other reagents to induce DC-like phenotypes [17]. For example, the proportion of DC-like cells obtained upon differentiation of human cell lines can be marginal or very low, with the exception of MUTZ-3 cell line [17]. Moreover, none of them can be said to resemble cross-presenting cDC1s closely.

Regarding murine DC-like lines, an immortalized cell line, MutuDC, derived by transgenic expression of the simian virus 40 (SV40) T antigen in mature DCs presently is considered to correspond best to primary cDC1s [22, 23]. MutuDCs have been used to test the role of the Zeb1 and Zbtb10 molecules in cDC1 activation and cytokine secretion [24, 25], to identify the cooperative role of Interleukin-10 (IL-10) and IL-35 in promoting tolerance [26], or to test the effect of inhibitors on antigen cross-presentation [27]. An immortalized cDC2-like cell line has also been produced by the same laboratory [28]. One objective of this chapter is to provide a protocol for cross-presentation assays using MutuDCs that complements published protocols using other less cDC1-like lines.

Specifically, we will provide a detailed protocol designed for examining cross-presentation by MutuDCs of complexes formed between the model antigen ovalbumin and ovalbumin-specific antibodies, which are endocytosed by Fcγ receptors. For the sake of space, we will not provide procedures for testing cross-presentation of other relevant types of antigens, including soluble protein, lectin receptor-targeted antigens or phagocytosed antigens (yeast or apoptotic cells); we refer the reader to our previously published protocols describing preparation and use of such antigens [29].

Considering the lack of an immortalized human cell line resembling cDC1 closely, we will also provide a protocol that can be used to study cross-presentation of immune complexes (but not other types of cross-presented antigens) by a human cell, FcRγIIa-transfected HEK 293 (Human Embryonic Kidney) cells. These cells were originally produced to demonstrate that expression of an Fcγ receptor is sufficient to transform a "nonprofessional" antigen-presenting cell into a cross-presenting cell [30]. Although HEK cells are not related to DCs, the FcRγIIa-transfected line is of interest for in vitro studies of cross-presentation by human MHC-I molecules, either endogenously expressed by the line (HLA-A2, B7 and C7) or conveniently introduced in the cells that are known to be easy to transfect. HEK 293 cells expressing FcRγIIa and the murine MHC-I molecule H-2K^b have been used to test the roles of various molecules in the efficiency of antigen cross-presentation [30] through gene targeting [31–34] or pharmacological approaches [34]. The line provides a unique opportunity to monitor antigen cross-presentation of human and murine antigens.

2 Materials

2.1 Common Consumables and Laboratory Equipment

1. Tissue culture room: vertical laminar flow hood, 37 °C CO_2 incubators for cell culture, refrigerated centrifuges for 1.5, 2, 15 and 50 mL tubes.
2. Sterile plastic-ware: 5, 10 and 25 mL pipettes, 25, 75 and 150 cm^2 tissue culture flasks, 10 to 1000 μL disposable tips, 40 μm cell strainers, reservoirs, 48-well flat-bottomed and

96-well V-bottomed and flat-bottomed culture plates treated for adherent cells, 1.5, 15 and 50 mL conical tubes, 5 mL flow cytometry tubes.

3. 96-well flat-bottomed Maxisorp ELISA plates.

2.2 Antigens

1. Ovalbumin, Low Endotoxin, purified (Worthington Biochemical Corporation).
2. Anti-ovalbumin rabbit polyclonal IgG antibody (Sigma-Aldrich).
3. Anti-ovalbumin rabbit polyclonal IgG antibody (Polyscience).
4. SIIFNFEKL synthetic peptide.
5. TSYKFESV synthetic peptide.

2.3 Mouse Antigen-Presenting Cell Cultures

1. MutuDC cells, an immortalized C57BL/6 murine splenic type 1 conventional dendritic cell (cDC1)-like line, obtained from Dr. Hans Acha-Orbea [22]. *See* **Note 1**.
2. MutuDC complete medium: RPMI-1640 medium, 10% heat-inactivated Fetal Bovine Serum, 2 mM L-Glutamine, 1% Penicillin/Streptomycin, 50 μM β-mercaptoethanol, 25 mM HEPES, 1X Non-Essential Amino Acids, 1 mM Sodium Pyruvate. *See* **Note 2**.
3. Detaching solution for adherent cells: PBS 2 mM EDTA.
4. Fixation solution: PBS 0.0008% Glutaraldehyde.

2.4 Human Antigen-Presenting Cell Cultures

1. Human embryonic kidney 293T FcR-EGFP.K^b cells: HEK293T cells expressing the mouse MHC class I molecule H-2K^b and the human FcγRIIA receptor fused to an eGFP tag (293T FcR-EGFP.K^b), kindly provided by P. Cresswell (Yale University) [30]. *See* **Note 3**.
2. FcR-EGFP.K^b complete medium: DMEM high-glucose Glutamax medium, 10% heat-inactivated Fetal Bovine Serum, 1% Penicillin/Streptomycin.
3. Detaching solution for adherent cells: PBS, 2 mM EDTA.
4. Fixation solution: PBS 0.0008% Glutaraldehyde.

2.5 $CD8^+$ T Lymphocyte Culture

1. Antigen-specific $CD8^+$ T cells: OT-I cells purified from lymph nodes of transgenic OT-I (C57BL/6-Tg$^{(TcraTcrb)1100Mjb/J}$) *Rag2*$^{-/-}$ mice. *See* **Note 4**.
2. OT-I complete medium: RPMI-1640 medium, 10% heat-inactivated Fetal Bovine Serum, 2 mM L-Glutamine, 1% Penicillin/Streptomycin.

2.6 Quantification of IL-2 Production by Activated CD8+ T Cells Using Sandwich ELISA

1. Purified anti-IL2 antibody (clone JES6-1A12). *See* **Note 5**.
2. Biotin-conjugated anti-IL2 antibody (clone JES6-5H4).
3. Horse Radish Peroxidase (HRP)-streptavidin.
4. IL-2 recombinant protein.
5. Phosphate-Buffered Saline, PBS, pH 7.4.
6. ELISA blocking solution: PBS, 10% FBS.
7. ELISA washing solution: PBS, 0.05% Tween20.
8. TMB substrate.
9. HCl 2N solution, prepared by diluting 12N stock solution into deionized sterile water.
10. ELISA plate reader.

2.7 Quantification of CD8+ T Cell Proliferation Using Flow Cytometry

1. Tag-it Violet™ Proliferation and Cell Tracking Dye (Biolegend Cat# 425101).
2. Flow cytometry buffer: PBS, 2% Fetal Bovine Serum, 2 mM EDTA.
3. TruStain FcX™ PLUS (anti-mouse CD16/32) antibody (Biolegend, Cat# 156604).
4. Fluorochrome-conjugated monoclonal antibodies following the manufacturer's instruction. *See* **Note 6**.
   (a) Monoclonal anti-mouse CD8β.2 (Rat, clone 53–5.8), Phycoerythrin conjugated.
   (b) Monoclonal anti-mouse TCR beta chain (Rat, clone H57–597), Alexa Fluor 488 conjugated.
   (c) Monoclonal anti-mouse CD45 Brilliant Violet 605™ (Rat, clone I3/2.3) conjugated.
5. 7-amino-actinomycin D (7-AAD) Viability Staining Solution.
6. Flow cytometer—analyzer.
7. Flow cytometry analysis software.

3 Methods

3.1 Cellular Assays to Monitor Cross-presentation in Human

3.1.1 Antigen-Presenting Cell Culture

1. Collect an ampoule of 293T FcR-EGFP.K^b cells from liquid nitrogen storage. Place the cryovial into a 37 °C water bath. Quickly thaw the cells (<2 min) by gently swirling the vial in the water bath until one or two small ice crystals are visible in the vial.
2. Transfer the vial to a laminar flow hood. Pipette slowly the whole content of the ampoule into a 15 mL sterile conical tube containing 9 mL of 37 °C pre-warmed 293T FcR-EGFP.K^b

complete medium. Mix and remove 20 μL to determine the viable cell count using trypan blue.

3. Centrifuge the 15 mL tube for 5 min at 300×*g*, at room temperature and discard supernatant. Add 37 °C pre-warmed 293T FcR-EGFP.K^b complete medium to obtain a density of 50,000–100,000 viable cells/mL and transfer 500,000–1,000,000 cells into 75 cm^2 culture flasks treated for adherent cells. Incubate for 24 h at 37 °C in a CO_2-supplemented incubator.
4. Examine every day the cells under a phase contrast microscope to determine their confluence. Aspirate culture medium and replace with fresh complete medium every 48 h.
5. Split the cells when the confluence reaches 70–90% for adherent cells. Aspirate culture medium, wash once in pre-warmed PBS, aspirate PBS, add 5 mL of room temperature detaching solution for adherent cells and incubate at 37 °C for 5 min maximum. Detach the cells by gently pipetting, transfer the 5 mL detaching solution for adherent cells into a 15 mL sterile tube containing 10 mL of 293T FcR-EGFP.K^b complete medium. Repeat from **step 3**.

3.1.2 Preparation of Immune Complexes for Internalization by Fcγ Receptor

1. Prepare ovalbumin stock solution in ion-free sterile water at a concentration of 50 mg/mL in 1.5 mL conical tubes. Vortex well until the solution is clear. Centrifuge for 20 min at 20,000×*g* at 4 °C to pellet any protein aggregates.
2. Collect the whole supernatant and transfer it into new 1.5 mL conical tubes. Store at −20 °C for several months any remaining ovalbumin solution.
3. Prepare ovalbumin/anti-ovalbumin immune complexes by incubating 1 mg of purified IgG anti-ovalbumin (Polysciences) with 10 mg of ovalbumin (50 mg/mL in water) in 1 mL PBS for 4–6 h on ice. Centrifuge for 10 min, 4 °C, 14,000×*g*. Collect supernatant into a new 1.5 mL conical tube (constituting soluble immune complexes). Add 1.0 mL of sterile PBS and resuspend the pellets (constituting precipitated immune complexes). *See* **Note 7**. Keep on ice until using. *See* Subheading 3.1.4.

3.1.3 Preparation of Transgenic CD8$^+$ T OT-I Lymphocytes

1. Collect lymph nodes from OT-I *Rag2*$^{-/-}$ mice into 4–10 mL sterile PBS. This should be done 1 h maximum before the end of cell incubation with immune complexes. *See* Subheading 3.1.4.
2. Position the lymph nodes on a 40 μm cell strainer placed on a 50 mL conical tube. Transfer the PBS and lymph nodes from the culture dish to the strainer. Remove the piston/plunger from a sterile 2 mL syringe. Use the flat end of the plunger to

mince the lymph nodes by pressing them several times in gentle circular motions.

3. Add PBS for 30 mL final volume, centrifuge at 300×*g* for 5 min at room temperature.
4. Discard supernatant, add 30 mL of sterile PBS, resuspend cells, and collect a small volume to determine viable cell density using trypan blue staining or equivalent procedure.
5. Centrifuge at 300×*g* for 5 min at room temperature, discard supernatant and add room temperature OT-I complete medium to adjust concentration to 100,000 OT-I cells/ 250 μL. Keep on ice until use for in vitro cross-presentation assays. *See* Subheading 3.1.4.

3.1.4 Cross-presentation Assay Using IL-2 Readout

1. Seed 50,000 293T FcR-EGFP.K^b per well, in 500 μL 293T FcR-EGFP.K^b complete medium, in a 48-well flat-bottomed culture plate treated for adherent cells. Incubate for 2 h at 37 ° C in a CO_2-supplemented incubator. *See* **Note 8**. *See* Table 1 for an example of a typical plate layout.
2. Centrifuge the plate at 300×*g* for 5 min at room temperature, aspirate the culture medium, wash once in PBS, and add the ovalbumin/anti-ovalbumin immune complexes prepared as previously mentioned (*see* Subheading 3.1.2). Perform serial dilutions by diluting 450 μL, 150 μL and 50 μL of precipitated immune complexes in 1 mL (final volume) of 293T FcR-EGFP. K^b incubation medium. Add 500 μL of diluted antigens per well. In parallel, incubate some wells containing 293T

Table 1
Example of a typical plate layout for in vitro assays to test the cross-presentation capacity of ovalbumin/anti-ovalbumin complexes by 293T FcR-EGFP.K^b cells

1	1	1	1	1	1	1	1
1	2	2	3	3	4	4	1
1	2	2	3	3	4	4	1
1	2	2	3	3	4	4	1
1	2	2	3	3	4	4	1
1	1	1	1	1	1	1	1

1 no cells, 500 μL PBS/well; *2* 293T FcR-EGFP.K^b cells incubated with serial dilutions of bovine serum albumin/anti-bovine serum albumin immune complexes (500, 250, 125, 62.5 μg bovine serum albumin with technical duplicates); *3* 293T FcR-EGFP.K^b cells incubated with serial dilutions of ovalbumin/anti-ovalbumin immune complexes (500, 250, 125, 62.5 μg ovalbumin with technical duplicates); *4* as in 3 but cells are treated with fixative solution prior to incubation with ovalbumin/anti-ovalbumin complexes. Typically, optical density readings should be <0.300 for samples 2 and 4, and for sample 3 readings should range from 0.600–0.900 (highest dose) to 0.300–0.400 (lowest doses). Two identical plates should be run in parallel in one experiment to get four technical replicates per experiment. A third plate should include the positive control, e.g., serial dilutions of SIINFEKL from 10^{-13} M to 10^{-18} M (10× dilution factor) and a condition with irrelevant peptide TSYKFESV from 10^{-13} M to 10^{-18} M, all run in technical duplicates

FcR-EGFP.K^b cells with serial dilutions of SIINFEKL or TSYK-FESV synthetic peptides diluted from 10^{-8} to 10^{-18} M in 293T FcR-EGFP.K^b incubation medium (500 μL/well). *See* **Note 9**.

3. After 2 h, centrifuge the plate at 300×*g* for 5 min at room temperature, aspirate the medium containing the immune complexes or synthetic peptides, and wash twice in 293T FcR-EGFP.K^b complete medium, and add 500 μL 293T FcR-EGFP.K^b incubation medium for another 18 h.
4. At the end of incubation time, centrifuge the plate at 300×*g* for 5 min at room temperature, wash twice in pre-warmed sterile PBS, add fixative solution for 60 s (200 μL/well), add 1.8 mL 293T FcR-EGFP.K^b complete medium, centrifuge the plate at 300×*g* for 5 min at room temperature, wash twice in 37 °C pre-warmed 293T FcR-EGFP.K^b complete medium.
5. After aspirating 293T FcR-EGFP.K^b complete medium, add the specific T cells as previously described above (*see* Subheading 3.1.3), in 250 μL of OT-I complete medium and incubate for 24–36 h. At the end of incubation, centrifuge the plate at 4 °C, 300×*g*, 5 min, collect supernatants and transfer 220–230 μL into a 96-well flat-bottomed plate. Store the plate at −20 °C or −80 °C until use for ELISA assays. *See* **Note 10**.
6. Coat Maxisorp ELISA plates overnight with purified anti-IL2 antibody (clone JES6-1A12, 2 μg/mL) diluted in PBS (50 μL/well).
7. Discard coating solution, block plates with ELISA blocking solution (PBS FBS 10%, 200 μL/well) for 10–30 min at room temperature.
8. Discard ELISA blocking solution, add for 60–90 min at room temperature with gentle shaking (300 rpm) culture supernatants or serial dilutions (500 μg, 250 μg, 125 μg, 62 μg, 31 μg, 16 μg, 8 μg, 0 μg) of mouse IL-2 recombinant protein to the wells (100 μL/well). Culture supernatants from cells treated with precipitated immune complexes can be used undiluted or 2× diluted into PBS. Culture supernatants from cells treated with immune complexes should be 5× or 10× diluted into PBS. Each condition, including recombinant IL-2 controls, should be performed in duplicate or triplicate.
9. Discard culture supernatants and recombinant IL-2 protein dilutions, wash twice with ELISA washing solution (300 μL/well), and add biotin-conjugated anti-IL2 antibody (clone JES6-5H4, 2 μg/mL) diluted in PBS (50 μL/well) for 45 min at room temperature with gentle shaking (300 rpm).

10. After 5 washes in ELISA washing solution (300 μL/well), add HRP-streptavidin (1:2000) diluted in PBS for 30 min (50 μL/well) at room temperature with gentle shaking (300 rpm).
11. After 5 washes in ELISA washing solution (300 μL/well), add TMB substrate solution (100 μL/well) for 5–10 min at room temperature to reveal bound cytokine, block reaction by adding HCl 2 N solution (20 μL/well).
12. Read the absorbance at 450 nm using a ELISA plate reader.
13. Correct IL-2 optical density readings for background values, before their conversion into IL-2 concentrations based on the linear curve of mouse recombinant IL-2 standards. *See* **Note 11**.

3.2 Cellular Assays to Monitor Cross-presentation in Mouse

3.2.1 Antigen-Presenting Cell Culture

1. Collect an ampoule of MutuDC cells from liquid nitrogen storage.
2. Thaw MutuDC cells following steps 1–7 of Subheading 3.1.1. but use the MutuDC complete medium instead of 293T FcR-EGFP.K^b complete medium.

3.2.2 Preparation of Immune Complexes for Internalization by Fc Receptor

1. Prepare ovalbumin stock solution in ion-free sterile water at a concentration of 1 mg/mL as described in **steps 1** and **2** from Subheading 3.1.2.
2. Purify antibodies from ovalbumin-specific rabbit polyclonal serum (Sigma-Aldrich) using Protein G Sepharose beads. Dialyze against PBS. Sterilize by filtration, and store at −20 °C in PBS with 10% glycerol at a target concentration of 5–10 mg/mL. *See* **Note 12**.
3. Form ovalbumin/anti-ovalbumin complexes by incubating 20 μg/mL ovalbumin with serial dilutions (from 500 to 31 μg/mL or less) of anti-ovalbumin polyclonal antibodies for 2 h in MutuDC complete medium.

3.2.3 Preparation of Transgenic $CD8^+$ T OT-I Lymphocytes

1. Perform as described above (*see* Subheading 3.1.3, **steps 1** to **4**).
2. Centrifuge cells, discard supernatant and add 37 °C pre-warmed sterile PBS to obtain a cell concentration of $1 - 10 \times 10^6$ cells/mL. Label with 5 μM Tag-it Violet™ Proliferation and Cell Tracking Dye for 20 min according to manufacturer's instructions, with an incubation at 37 °C for dye labeling. *See* **Note 13**.
3. At the end of labeling procedure, resuspend cells and adjust the volume of MutuDC complete medium to obtain a target concentration of 100,000 cells/250 μL.

3.2.4 Cross-presentation Assays

1. Plate 30,000 MutuDCs per well into a 96-well U-bottomed culture plate in 200 μL of MutuDC complete medium and incubate in a 37 °C incubator, 5% CO_2, for 2 to 4 h.
2. Centrifuge the plate at 300×*g* for 5 min at room temperature, aspirate the culture medium, wash once in PBS, and add the ovalbumin/anti-ovalbumin immune complexes in 200 μL of MutuDC complete medium. *See* Subheading 3.1.4. and **Note 9** to determine how to include positive (serial dilutions of SIINFEKL or irrelevant TSYKFESV peptide) and negative controls (MutuDC cells fixed before pulsing with ovalbumin/ anti-ovalbumin immune complexes or MutuDC cells pulsed with bovine serum albumin/anti-bovine serum albumin immune complexes). *See* Subheading 3.2.2 for the preparation of immune complexes. *See* Table 2 for an example of a typical plate layout.
3. Twelve to sixteen hours later, wash MutuDCs twice in complete MutuDC medium. *See* **Note 14**.
4. Add specific $CD8^+$ T cells labeled with Tag-it Violet™ Proliferation and Cell Tracking Dye (200 μL/well), prepared in the above. *See* Subheading 3.2.3.
5. After 3 days of co-culture, wash the 96-well plates twice in PBS at room temperature.
6. Add 200 μL/well of flow cytometry buffer, resuspend cells and transfer into 96-well V-bottomed culture plate.

Table 2
Example of a typical plate layout for in vitro assays to test the cross-presentation capacity of ovalbumin/anti-ovalbumin complexes by MutuDCs

1	1	1	1	1	1	1	1	1	1	1	1
1	2	2	2	2	2	4	4	4	4	4	1
1	2	2	2	2	2	4	4	4	4	4	1
1	2	2	2	2	2	4	4	4	4	4	1
1	3	3	3	3	3	5	5	5	5	6	1
1	3	3	3	3	3	5	5	5	5	6	1
1	3	3	3	3	3	5	5	5	5	6	1
1	1	1	1	1	1	1	1	1	1	1	1

1 no cells, 200 μL PBS/well; *2* MutuDCs incubated with serial dilutions of bovine serum albumin/anti-bovine serum albumin immune complexes (500, 250, 125, 62.5, 31 μg bovine serum albumin with technical triplicates); *3* MutuDCs incubated with serial dilutions of ovalbumin/anti-ovalbumin immune complexes (500, 250, 125, 62.5, 31 μg ovalbumin with technical triplicates); *4* as in 3 but cells are treated with fixation solution prior to incubation with ovalbumin/anti-ovalbumin complexes; *5* serial dilutions of SIINFEKL from 10^{-13} to 10^{-18} M (10× dilution factor, technical duplicates); *6* TSYKFESV 10^{-13} M (technical triplicates)

7. Centrifuge the plate at 450×g for 3 min at 4 °C, discard supernatant, and add 50 μL/well of TruStain FcX™ PLUS (anti-mouse CD16/32) antibody diluted in flow cytometry buffer (1:100).
8. After 10 min, add 50 μL of fluorochrome-conjugated anti-mouse CD8β.2, TCR beta chain and CD45 monoclonal antibodies diluted in flow cytometry (1:00 considering the final volume of 100 μL) for another 30 min on ice.
9. Wash twice in flow cytometry buffer, add 200 μL of flow cytometry buffer, transfer into 5 mL tubes for flow cytometry, and add 800 μL flow cytometry buffer.
10. Five minutes before acquisition, add 5 μL 7-AAD and vortex gently.
11. Acquire data on the flow cytometer/analyzer available locally. Record at least 1000 events in the gate of interest that identifies $CD8^+$ T lymphocytes (7-$AAD^-$$CD45^+$TCR beta $chain^+$$CD8\beta.2^+$Tag-it Violet™ Proliferation and Cell Tracking Dye^+).
12. Analyze exported flow cytometry data files with the flow cytometry software available locally. Doublets and dead cells should be excluded. OT-I cells are identified by gating on $CD45^+$TCR beta $chain^+$$CD8\beta.2^+$Tag-it Violet™ Proliferation and Cell Tracking Dye^+ cells. The activation of OT-I cells can be assessed by calculating the proportion of divided cells or more accurately by using an algorithm calculating the division index. An algorithm is available on some flow cytometry analysis software (FlowJo) or described in the literature [2, 35]. *See* Fig. 1.

4 Notes

1. MutuDC cells can be easily genetically targeted to deplete, reduce or increase candidate gene/protein expression [24–26, 36] and test their contribution to key conventional dendritic cell functions such as cross-presentation.
2. Sets of experiments should be performed using the same batch (same lot and provider) of fetal bovine serum. In addition, any new lot of fetal bovine serum should be tested in a pilot experiment to assess MutuDC viability, growth, and functionality.
3. Similar to MutuDCs, 293T FcR-EGFP.K^b cells can be genetically targeted to deplete, reduce or increase expression of proteins of interest [31, 32, 34] and test their contribution to antigen cross-presentation.

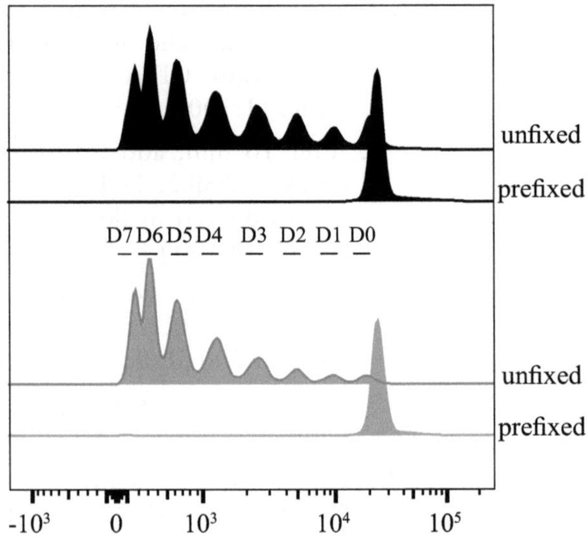

Fig. 1 Flow cytometric analysis of OT-I CD8$^+$ T cell proliferation induced by ovalbumin/anti-ovalbumin immune complexes-pulsed MutuDC. Histograms show the fluorescence intensity of Tag-it Violet™ Proliferation and Cell Tracking Dye after gating on OT-I cells (*see* Subheading 3.2.4) incubated for 3 days with MutuDC pulsed with ovalbumin/anti-ovalbumin immune complexes (500 μg/mL) prefixed or not before addition of immune complexes. The different peaks of divisions are annotated from Division 0 (D0, undivided cells) to Division 7 (D7). Two technical replicates are shown (black and grey histograms)

4. In case *Rag2*$^{-/-}$ OT-I mice are not accessible, the B3Z cell line can be used as a convenient alternative [37]. The B3Z T cell hybridoma recognizes SIINFEKL ovalbumin peptide/H2-Kb complexes and secretes IL-2 upon T cell receptor activation, which can be detected and quantified in culture supernatants. The B3Z hybridoma also displays intracellular accumulation of lacZ protein upon T cell receptor activation, which can be measured by β-galactosidase assay. These cells can be cultured similarly to MutuDCs (*see* Subheading 3.2.1) but in B3Z complete medium: RPMI-1640 medium, 10% heat-inactivated Fetal Bovine Serum, 2 mM L-Glutamine, 1% Penicillin/Streptomycin, 1 mM pyruvate, 50 μM 2-mercaptoethanol. B3Z cells should not be kept in culture for more than 15–20 consecutive days because they may become less reactive to antigen-restricted activation.
5. Commercial ELISA kits may naturally be used but home-made IL-2 ELISA assays can significantly reduce cost when a high number of experiments are planned.
6. Fluorochromes should be adapted to the configuration of the flow cytometer analyzer available locally. Isotype controls should be included in assays as they may not only determine

signal specificity but also improve the delineation of gates. Other antibody clones might alternatively be used but they should be tested in small pilot experiments to ensure they allow identification of the cell populations of interest.

7. Both soluble immune complexes and precipitated immune complexes can be used for antigen cross-presentation assays in vitro. However, precipitated immune complexes can activate, more efficiently, cognate $CD8^+$ T cells [30] and should be preferred.

8. Assay should be done at least in triplicate wells (e.g., at least three technical replicates for each experimental condition) per experiment to estimate and reduce technical variability.

9. The presence of free SIINFEKL peptide should be rigorously controlled in the antigen preparation. Assays should include additional wells, in which antigen-presenting cells are prefixed (fixed before antigen incubation) with fixative solution for 1 min (50 μL/well). After the addition of 150 μL/well of 293T FcR-EGFP.K^b complete medium, plates should be centrifuged at 300×g, room temperature, for 5 min before discarding culture supernatants. After two washes twice with 293T FcR-EGFP.K^b complete medium, proceed with the next steps. Obtention of a high assay background for these control wells strongly suggests the presence of free SIINFEKL peptide and should invalidate the results of the experiment. In this case, a new preparation of ovalbumin should be done. In addition, pulsing antigen-presenting cells with serial dilutions of bovine serum albumin/anti-bovine serum albumin complexes may serve as a complementary negative control.

10. Incubation time should be determined from pilot experiments. Longer incubation times (>36 h) may result in increased OT-I activation (IL-2 secretion) but also in higher background values and should be avoided. Low assay values may be corrected by optimizing the post-fixation (fixation after antigen incubation) procedure, by either decreasing the duration of incubation with fixative solution and/or decreasing the concentration of glutaraldehyde in fixative solution. Any change in the composition of fixation solution or the duration of the fixation step should not allow 293T FcR-EGFP.K^b cells to proliferate as this may strikingly impair OT-I activation.

11. Optical density readings of negative controls should generally not exceed 0.300 arbitrary units. Otherwise, this may indicate the presence of free SIINFEKL peptides in antigen preparation (*see* **Note 9**).

12. Free SIINFEKL peptides in ovalbumin preparation may be removed by separation of small-size peptides from full-length

ovalbumin through the use of the Bio-Spin® 6 Columns as previously described [29].

13. Numerous commercially available dyes may be used alternatively to monitor OT-I proliferation. The final choice should be made according to the experience of the user and the configuration of the flow cytometer/analyzer available locally.
14. Post-fixation (fixation after antigen incubation) is facultative in the procedure described here. In some situations, post-fixation should be performed as described above (*see* **Note 9**), especially when an inhibitor added to the cultures has a reversible effect or when no dose-effect can be observed in assay results. The latter may be solved by either increasing antigen dilution or by post-fixation, which will stop any further production of peptide-MHC complexes during the period of incubation with OT-I cells.

Acknowledgments

This work benefited from a government grant managed by the Agence Nationale de la Recherche under the France 2030 program, under the reference ANR-23-IAHU-0010.

Conflict of Interest The authors have no conflicts of interest to declare that are relevant to the content of this chapter.

References

1. Mauvais F-X, van Endert P (2023) Cross-presentation by the others. Semin Immunol 67:101764. https://doi.org/10.1016/j.smim.2023.101764
2. Clausen BE, Amon L, Backer RA et al (2023) Guidelines for mouse and human DC functional assays. Eur J Immunol 53:e2249925. https://doi.org/10.1002/eji.202249925
3. Alloatti A, Kotsias F, Hoffmann E, Amigorena S (2016) Evaluation of cross-presentation in bone marrow-derived dendritic cells in vitro and splenic dendritic cells ex vivo using antigen-coated beads. Bio Protoc 6. https://doi.org/10.21769/BioProtoc.2015
4. Sadiq BA, Mantel I, Blander JM (2020) A comprehensive experimental guide to studying cross-presentation in dendritic cells in vitro. Curr Protoc Immunol 131:e115. https://doi.org/10.1002/cpim.115
5. Segura E (2016) Cross-presentation assay for human dendritic cells. Methods Mol Biol 1423:189–198. https://doi.org/10.1007/978-1-4939-3606-9_14
6. Lubin R, Gvili R, Hazan I, Yona S (2023) Human dendritic cell enrichment and their activation of T cells. Curr Protoc 3:e873. https://doi.org/10.1002/cpz1.873
7. Amon L, Dudziak D, Backer RA et al (2023) Guidelines for DC preparation and flow cytometry analysis of mouse lymphohematopoietic tissues. Eur J Immunol 53:e2249893. https://doi.org/10.1002/eji.202249893
8. Heger L, Dudziak D, Amon L et al (2023) Guidelines for DC preparation and flow cytometric analysis of human lymphohematopoietic tissues. Eur J Immunol 53:e2249917. https://doi.org/10.1002/eji.202249917
9. Rødahl I, Gotley J, Andersen SB et al (2021) Acquisition of murine splenic myeloid cells for protein and gene expression profiling by advanced flow cytometry and CITE-seq.

STAR Protoc 2:100842. https://doi.org/10.1016/j.xpro.2021.100842

10. De Leeuw E, Bosteels C, Lambrecht BN, Hammad H (2022) Isolation of conventional murine lung dendritic cell subsets. Methods Mol Biol 2506:237–255. https://doi.org/10.1007/978-1-0716-2364-0_17
11. Rocca Y, Voissière A, Valladeau-Guilemond J, Bendriss-Vermare N (2023) Isolation and identification of dendritic cell subsets from human and mouse tumors. Methods Mol Biol 2618:187–197. https://doi.org/10.1007/978-1-0716-2938-3_14
12. Swartz AM, Nair SK (2022) The in vitro differentiation of human CD141+CLEC9A+ dendritic cells from mobilized peripheral blood CD34+ hematopoietic stem cells. Curr Protoc 2:e410. https://doi.org/10.1002/cpz1.410
13. Santa P, Roubertie A, Loizon S et al (2023) Enrichment of large numbers of splenic mouse dendritic cells after injection of Flt3L-producing tumor cells. Methods Mol Biol 2618:173–186. https://doi.org/10.1007/978-1-0716-2938-3_13
14. Song R, Bafit M, Tullett KM et al (2024) A simple and rapid protocol for the isolation of murine bone marrow suitable for the differentiation of dendritic cells. Methods Protoc 7. https://doi.org/10.3390/mps7020020
15. Weimershaus M, van Endert P (2013) Preparation of dendritic cells by in vitro cultures. Methods Mol Biol 960:351–357. https://doi.org/10.1007/978-1-62703-218-6_25
16. Balan S, Dalod M (2016) In vitro generation of human XCR1(+) dendritic cells from CD34(+) hematopoietic progenitors. Methods Mol Biol 1423:19–37. https://doi.org/10.1007/978-1-4939-3606-9_2
17. Hargadon KM (2016) Murine and human model systems for the study of dendritic cell immunobiology. Int Rev Immunol 35:85–115. https://doi.org/10.3109/08830185.2014.952413
18. Ackerman AL, Cresswell P (2003) Regulation of MHC class I transport in human dendritic cells and the dendritic-like cell line KG-1. J Immunol 170:4178–4188. https://doi.org/10.4049/jimmunol.170.8.4178
19. Berges C, Naujokat C, Tinapp S et al (2005) A cell line model for the differentiation of human dendritic cells. Biochem Biophys Res Commun 333:896–907. https://doi.org/10.1016/j.bbrc.2005.05.171
20. Koski GK, Schwartz GN, Weng DE et al (1999) Calcium ionophore-treated myeloid cells acquire many dendritic cell characteristics independent of prior differentiation state, transformation status, or sensitivity to biologic agents. Blood 94:1359–1371
21. Masterson AJ, Sombroek CC, De Gruijl TD et al (2002) MUTZ-3, a human cell line model for the cytokine-induced differentiation of dendritic cells from CD34+ precursors. Blood 100:701–703. https://doi.org/10.1182/blood.v100.2.701
22. Pigni M, Ashok D, Acha-Orbea H (2016) Derivation and utilization of functional CD8(+) dendritic cell lines. Methods Mol Biol 1423:39–49. https://doi.org/10.1007/978-1-4939-3606-9_3
23. Steiner Q-G, Otten LA, Hicks MJ et al (2008) In vivo transformation of mouse conventional CD8alpha+ dendritic cells leads to progressive multisystem histiocytosis. Blood 111:2073–2082. https://doi.org/10.1182/blood-2007-06-097576
24. Smita S, Ghosh A, Biswas VK et al (2021) Zbtb10 transcription factor is crucial for murine cDC1 activation and cytokine secretion. Eur J Immunol 51:1126–1142. https://doi.org/10.1002/eji.202048933
25. Smita S, Ahad A, Ghosh A et al (2018) Importance of EMT factor ZEB1 in cDC1 "MutuDC line" mediated induction of Th1 immune response. Front Immunol 9:2604. https://doi.org/10.3389/fimmu.2018.02604
26. Koga MM, Engel A, Pigni M et al (2021) IL10- and IL35-secreting MutuDC lines act in cooperation to inhibit memory T cell activation through LAG-3 expression. Front Immunol 12:607315. https://doi.org/10.3389/fimmu.2021.607315
27. Mpakali A, Georgaki G, Buson A et al (2024) Stabilization of the open conformation of insulin-regulated aminopeptidase by a novel substrate-selective small-molecule inhibitor. Protein Sci 33:e5151. https://doi.org/10.1002/pro.5151
28. Pigni M, Ashok D, Stevanin M, Acha-Orbea H (2018) Establishment and characterization of a functionally competent type 2 conventional dendritic cell line. Front Immunol 9:1912. https://doi.org/10.3389/fimmu.2018.01912
29. Saveanu L, van Endert P (2013) Preparing antigens suitable for cross-presentation assays in vitro and in vivo. Methods Mol Biol 960:389–400. https://doi.org/10.1007/978-1-62703-218-6_29
30. Giodini A, Rahner C, Cresswell P (2009) Receptor-mediated phagocytosis elicits cross-presentation in nonprofessional antigen-presenting cells. Proc Natl Acad Sci USA 106:

3324–3329. https://doi.org/10.1073/pnas.0813305106

31. Sengupta D, Graham M, Liu X, Cresswell P (2019) Proteasomal degradation within endocytic organelles mediates antigen cross-presentation. EMBO J 38:e99266. https://doi.org/10.15252/embj.201899266
32. Hubbard JJ, Pyzik M, Rath T et al (2020) FcRn is a CD32a coreceptor that determines susceptibility to IgG immune complex-driven autoimmunity. J Exp Med 217. https://doi.org/10.1084/jem.20200359
33. Weimershaus M, Mauvais F-X, Saveanu L et al (2018) Innate immune signals induce anterograde endosome transport promoting MHC class I cross-presentation. Cell Rep 24:3568–3581. https://doi.org/10.1016/j.celrep.2018.08.041
34. Leib J, Waeckel-Énée E, Fabrega S et al (2025) ERAP1-dependent extreme antigen processing efficacy can govern MHC class I expression hierarchy. J Immunol. https://doi.org/10.1093/jimmun/vkaf013
35. Kratzer R, Mauvais F-X, Burgevin A et al (2010) Fusion proteins for versatile antigen targeting to cell surface receptors reveal differential capacity to prime immune responses. J Immunol 184:6855–6864. https://doi.org/10.4049/jimmunol.0902555
36. Fuertes Marraco SA, Grosjean F, Duval A et al (2012) Novel murine dendritic cell lines: a powerful auxiliary tool for dendritic cell research. Front Immunol 3:331. https://doi.org/10.3389/fimmu.2012.00331
37. Karttunen J, Sanderson S, Shastri N (1992) Detection of rare antigen-presenting cells by the lacZ T-cell activation assay suggests an expression cloning strategy for T-cell antigens. Proc Natl Acad Sci USA 89:6020–6024. https://doi.org/10.1073/pnas.89.13.6020

Chapter 17

T Cell Assay to Track Presentation of an Influenza Hemagglutinin Epitope Presented by the MHC-II Protein HLA-DR1

Karthik M. Ramesh, Liying Lu, and Lawrence J. Stern

Abstract

Antigen presentation is a specialized function of the immune system where peptides generated from cellular or internalized proteins are presented on Major Histocompatibility Complex molecules on the surface of antigen-presenting cells for recognition by T cells. The binding of these MHC-peptide complexes to hypervariable receptors on the surface of a T cell leads to its activation. T cell activation induces increases in cell-surface activation markers such as CD69, which can be tracked by flow cytometry. Here we describe a method that uses flow cytometry and a recombinant T cell to track the presentation of an MHC-II HLA-DR1 restricted epitope generated from the influenza hemagglutinin protein.

Key words Antigen presentation, Major histocompatibility complex, Antigen-presenting cell, Peptide, Influenza, Hemagglutinin, HLA-DR, T cell, B cell, Flow cytometry

1 Introduction

Antigen presentation is a specialized function of the immune system performed by all nucleated cells where peptides are derived from self and foreign antigens and presented via major histocompatibility (MHC) molecules to T cells [1, 2]. The peptides derive from proteins expressed in the presenting cell or from extracellular proteins taken up by fluid-phase or receptor-mediated endocytosis (Fig. 1). MHC class I molecules (MHC-I), which are expressed by all nucleated cells, generally are loaded in the endoplasmic reticulum with peptides generated by cytosolic proteasomal proteolysis. After transport to the surface, the resultant MHC-I/peptide complexes are recognized by CD8+ T cells. MHC class II molecules (MHC-II) are expressed constitutively by professional antigen-presenting cells (APC) such as dendritic cells, macrophages, B cells, and thymic epithelia, which present antigens to CD4+ T cells. However, MHC-II expression can be induced on most cell

Efstratios Stratikos (ed.), *Antigen Processing and Presentation: Methods and Protocols*, Methods in Molecular Biology, vol. 3007, https://doi.org/10.1007/978-1-0716-5092-9_17,

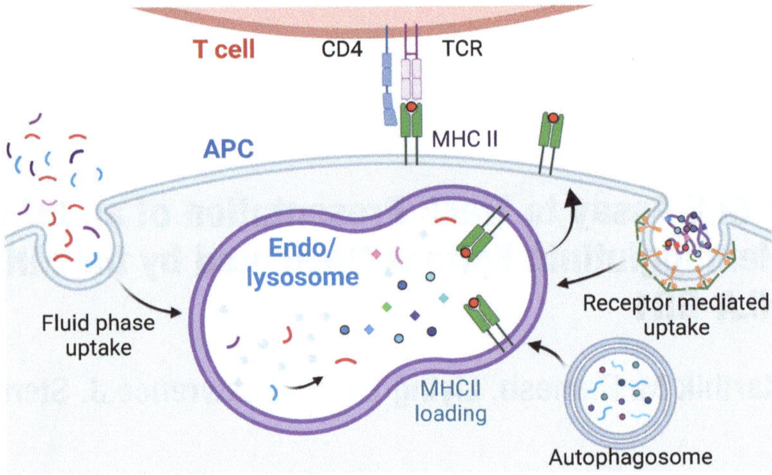

Fig. 1 Molecular basis of MHC-II antigen acquisition and presentation. Antigens are acquired by APCs by fluid phase or receptor-mediated endocytosis and trafficked to endosomal compartments for processing and loading onto MHC-II. Additionally, antigens can be brought in through recycling and autophagy pathways. Following this, peptide-loaded MHC-II are then presented on the surface of the APC to CD4 T cells

types by inflammatory cytokines such as interferon-g. MHC-II generally are loaded in endosomal/lysosomal compartments with peptides generated by proteolysis in those compartments, although loading with preprocessed peptides at the cell surface or in shallow recycling compartments also has been observed. After transport to the surface the resultant MHC-II/peptide complexes are recognized by CD4+ T cells. During their development in the thymus, cells assemble clonotypic receptors from variable and constant genes with hypervariable regions generated by random nucleotide addition or trimming. T cells with T cell receptors (TCRs) that recognize self-MHC proteins are positively selected, and those with TCRs that recognize self-peptides bound to these proteins are negatively selected, to generate a largely self-tolerant T cell repertoire capable of recognizing a wide variety of foreign peptides bound to self-MHC proteins, which is released from the thymus. Once T cells are activated by binding MHC-peptide complexes to their TCR, they express various effector functions such as producing cytokines, performing cytotoxic functions, or upregulating surface antigens for interactions with B-cells, macrophages, or many other immune and non-immune cell types [3]. T cell activation can be followed by tracking the expression of these upregulated cell-surface markers. The degree of T cell activation is correlated with the number of antigens detected on cell-surface or plate-bound peptide-MHC [4, 5].

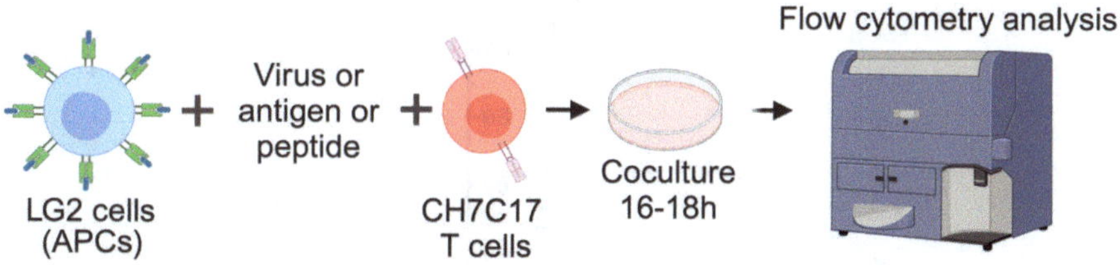

Fig. 2 Overview of the DR1/HA1.7 antigen presentation assay. APCs (LG2 cells) are treated with protein, peptide, or virus and co-cultured with T cells (CH7C17) for 16–18 h, following which the T cells are analyzed by flow cytometry for CD69 expression

Here we demonstrate an in vitro assay that uses a recombinant T cell to measure the degree of presentation of an epitope from influenza A virus displayed by an MHC-II protein on the surface of an antigen-presenting B cell (Fig. 2). We tracked the presentation of the influenza hemagglutinin epitope $HA_{306-318}$ (PKYVKQNTLKLAT, IEDB reference 48237) [6] characteristic of H3N2 strains such as A/Aichi/2/68 (NCBI Taxonomy ID: 387139) presented by HLA-DR1 (DRA1*01:01, DRB1*01:01) (IMGT reference HLA00662 HLA00664) on the B-lymphoblastoid cell line LG2 (Cellosaurus RRID: CVCL_C5UW) [7] using CH7C17 T cells (Cellosaurus RRID: CVCL_GX97). CH7C17 is a derivative of the TCRβ negative Jurkat derivative J.RT3-T3.5 (Cellosaurus RRID: CVCL_1316) that has been transfected with TCRα and TCRβ genes that code for the HA1.7 TCR [8] specific to the $HA_{306-318}$ presented on HLA-DR1 or HLA-DR4 [9]. Crystal structures are available for HLA-DR1 bound to $HA_{306-318}$ (PDB:1DLH) [10], HA1.7 TCR (PDB:4GKZ) [11] and the HLA-DR1/$HA_{306-318}$/HA1.7 complex (PDB:1FYT) [12]. The $HA_{306-318}$ epitope can be presented on other DR alleles [13] and has been used to track antigen presentation in many studies [14–34].

The assay can be used to track presentation of the $HA_{306-318}$ epitope derived from purified peptide across a concentration range of 4 nM to 2μM, recombinant hemagglutinin protein across a similar concentration range corresponding to approximately 250 ng/mL to 125μg/mL of monomer, respectively, or after virus infection (Figs. 3 and 4). Using a single 96-well plate format, the assay can be used to track up to 5 different antigens in duplicates across a range of eight concentrations for each antigen (Fig. 5). CH7C17 can be sorted into positive and negative populations based on the levels of CD69 expression using flow cytometry [35, 36], resulting in a robust measurement of cellular phenotypes indicating the levels of antigen presented.

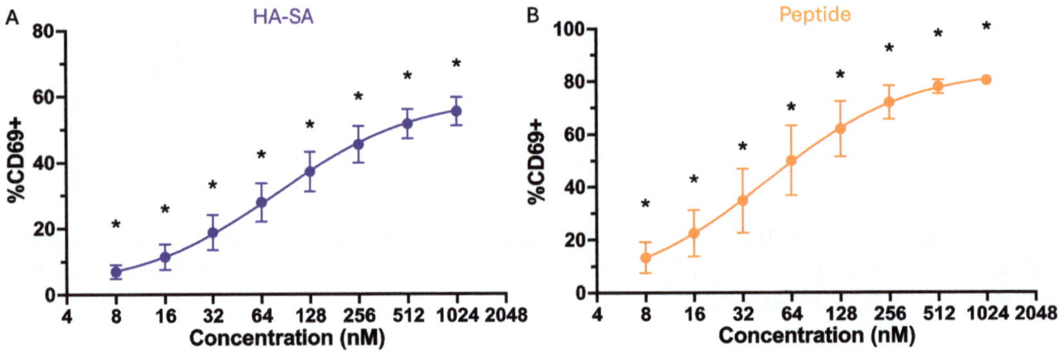

Fig. 3 CD69 levels correspond to antigen dose. (**a**) Dose response for HA-SA (protein). (**b**) Dose response for $HA_{306-318}$ (peptide). Data points in duplicates as shown in Fig. 5 (3 sets of duplicates across 3 experiments conducted on different days). Statistical significance assessed at each concentration against unstimulated using a two-tailed Student's *t*-test

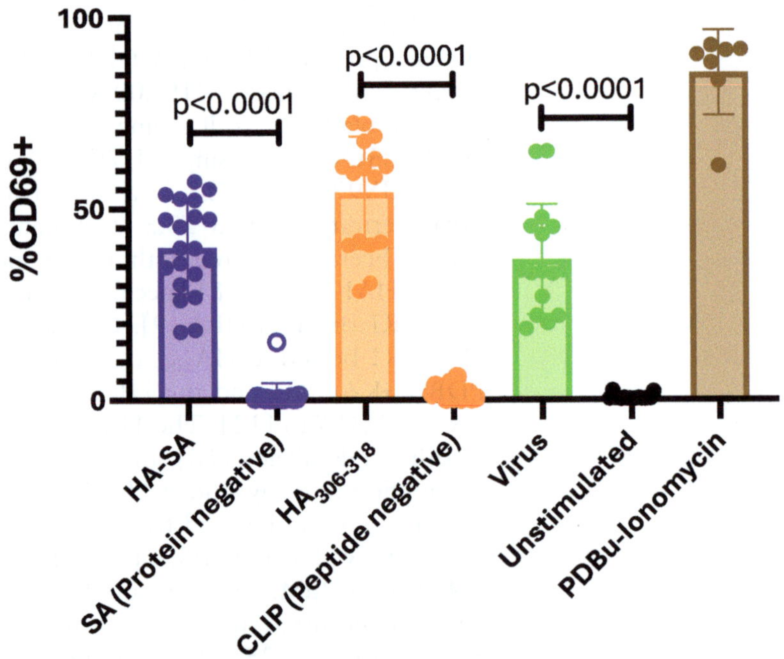

Fig. 4 Comparisons of T cell readouts of antigen-induced activation relative to assay controls. HA-SA antigen activates CH7C17 cells while SA alone does not (20 replicates across 10 experiments). Similarly, $HA_{306-318}$ peptide activates CH7C17 cells while the non-specific peptide CLIP does not (16 replicates across 8 experiments). H3N2 virus also activates CH7C17 cells when compared to unstimulated cells (16 replicates across 8 experiments). Positive control PDBu-Ionomycin treatment demonstrates T cell activation that bypasses antigen presentation (6 replicates across 3 experiments). Experiments were conducted on different days, with statistical significance assessed by a two-tailed Student's *t*-test

A	1	2	3	4	5	6	7	8
A	CH7C17 Unstained	CH7C17 Unstained	Antigen 1 1024	Antigen 1 1024	Peptide 1 1024	Peptide 1 1024	Virus 1:5	Virus 1:5
B	CH7C17 Live/Dead	CH7C17 Live/Dead	Antigen 1 512	Antigen 1 512	Peptide 1 512	Peptide 1 512	Virus 1:10	Virus 1:10
C	CH7C17 TCR	CH7C17 TCR	Antigen 1 256	Antigen 1 256	Peptide 1 256	Peptide 1 256	Virus 1:20	Virus 1:20
D	CH7C17 Tetramer	CH7C17 Tetramer	Antigen 1 128	Antigen 1 128	Peptide 1 128	Peptide 1 128	Virus 1:40	Virus 1:40
E	CH7C17 PDBu-Iono	CH7C17 PDBu-Iono	Antigen 1 64	Antigen 1 64	Peptide 1 64	Peptide 1 64	Virus 1:80	Virus 1:80
F	Coculture Streptavidin	Coculture Streptavidin	Antigen 1 32	Antigen 1 32	Peptide 1 32	Peptide 1 32	No virus	No virus
G	Coculture CLIP	Coculture CLIP	Antigen 1 16	Antigen 1 16	Peptide 1 16	Peptide 1 16		
H	Coculture Unstimulated	Coculture Unstimulated	Antigen 1 8	Antigen 1 8	Peptide 1 8	Peptide 1 8		

B	1	2	3	4	5	6	7	8
A	0	0	54.9	54.1	58.1	56.9	43.2	43.6
B	0	0	43.1	42.1	57.9	56.9	38	35.7
C	0	0	26.7	26.6	53.2	55	29.6	28.6
D	0	0	16.9	16.8	50.3	51	17.8	16.4
E	83.4	88	9.98	9.25	41.4	42.7	9.04	9.5
F	0.77	0.33	5.3	4.97	33.3	31.7	1.08	1.03
G	1.94	0.56	2.55	2.65	20.1	19.2		
H	0.56	0.63	1.45	1.55	11.5	11.4		

Fig. 5 Layout of an assay example. (**a**) Example 96-well plate layout for an experiment analyzing as antigens recombinant hemagglutinin-streptavidin oligomer (HA-SA) (antigen 1), purified synthetic $HA_{306-318}$ peptide (peptide 1), and an H3N2 influenza virus (virus) at various concentrations and dilutions. Shown here are: controls in columns 1 and 2; HA-SA protein loaded in columns 3 and 4, $HA_{306-318}$ peptide loaded in columns 5 and 6; H3N2 virus loaded in columns 7 and 8. Columns 9 through 12 are empty and not shown here. (**b**), Fractions of live CH7C17 cells in the CD69 positive gate, corresponding to the example assay plate showing increasing levels of CD69 expression with increasing antigen concentration

2 Materials

Prepare all cell culture media in sterile conditions and filter through 0.2μm tissue culture filter. Store reagents as indicated in each section.

2.1 Cell Culture

1. Antigen-presenting cells (LG2 cells) (*see* **Notes 1–4**).
2. T cells (CH7C17 cells, which are HA1.7 expressing Jurkat transfectants) (*see* **Notes 5–8**).
3. 25 cm^2 and 75 cm^2 tissue culture-treated flask (*see* **Notes 9** and **10**).
4. Heat Inactivation of FBS: Thaw a frozen FBS bottle 4 °C until thawed. Bring the thawed FBS to room temperature (RT) for 2 h. Heat the FBS in a water bath at 56 °C for 30 min. Swirl the bottle occasionally to ensure uniform heating. Once done,

wipe outer surfaces with 70% ethanol to prevent any contamination. Aliquot the FBS into 50 mL volumes and store at −20 °C. (*see* **Notes 9** and **10**).

5. Complete RPMI (cRPMI): Prepare by mixing 440 mL of RPMI 1640 containing L-Glutamine & 25 mM HEPES with 50 mL of heat-inactivated FBS, 5 mL of 100× Glutamax and 5 mL of 100× Pen-Strep. Store cRPMI at 4 °C (*see* **Notes 9** and **10**).

2.2 Antigen Presentation Co-culture

1. 96-well clear round bottom TC-treated cell culture microplate.
2. Dimethyl sulfoxide (DMSO).
3. 1× PBS (sterile).
4. Bovine serum albumin fraction V (7.5% solution): Prepare by dissolving 7.5 g of BSA in 100 mL of sterile 1× PBS. Filter through a sterile 0.2μm filter. Store at 4 °C. Alternatively, purchase a premade solution from Gibco.
5. PBS for viral infection (vPBS): Prepare 0.2% v/v Bovine albumin fraction V in 1× sterile PBS and store at 4 °C.
6. Influenza A virus A/Aichi/2/1968 (H3N2) (BEI resources NR3177). Store aliquoted at −80 °C (*see* **Notes 11** and **12**).
7. Influenza A virus Kilbourne F108: A/Aichi/2/1968 (HA, NA) x A/Puerto Rico/8/1934 (H3N2), Reassortant X-31b (BEI resources NR3483). Store aliquoted at −80 °C (*see* **Notes 11** and **12**).
8. Influenza hemagglutinin (HA): We purified HA in-house from transient transfection of HEK293 cells, following which the protein was biotinylated and oligomerized with streptavidin (HA-SA) (*see* **Note 12**).
9. $HA_{306-318}$ peptide (PKYVKQNTLKLAT, peptide positive control). Prepare 2 mg/mL stock in DMSO. Store aliquoted at −20 °C. (*see* **Notes 13** and **14**).
10. CLIP (VSKMRMATPLLMQ, peptide negative control). Prepare 2 mg/mL stock in DMSO. Store aliquoted at −20 °C (*see* **Note 15**).
11. Streptavidin (SA) (protein negative control). Prepare 2 mg/mL stock in water or 1× PBS and store aliquoted at −20 °C (*see* **Note 16**).
12. CellTrace™ Proliferation Kit (various fluorescent tags). Prepare by adding 50μL of DMSO to one vial and store aliquoted at −20 °C (*see* **Notes 17** and **18**).
13. Phorbol 12,13-dibutyrate (PDBu). Prepare 2 mM stock in DMSO and store aliquoted at −20 °C.
14. Ionomycin calcium salt. Prepare 3 mM stock in DMSO and store aliquoted at −20 °C.

2.3 Cell Staining

1. 96-well clear round bottom TC-treated cell culture microplate.
2. 1× PBS (sterile).
3. Fc receptor blocking agent (*see* **Note 19**).
4. Primary antibodies and secondary antibodies for surface staining and intracellular staining (as needed) (*see* **Notes 20–23**).
5. Isotype controls (as needed) (*see* **Notes 20–23**).
6. Live/dead viability kit. Prepare by adding 100μL of DMSO to one vial and store aliquoted at −20 °C (*see* **Notes 24–26**).
7. FACS buffer: Prepare by mixing sterile 1× PBS with 5% v/v FBS and filter through a sterile 0.2μm filter. Store at 4 °C. Optional: Add sodium azide to a final concentration of 0.02% for long-term storage (*see* **Note 26**).
8. Fixation buffer: Prepare 1–4% paraformaldehyde in 1× PBS solution and store aliquots at 4 °C. Alternatively purchase from Biolegend or BD Biosciences.
9. Intracellular staining (ICS) Fixation/Permeabilization Kit (available from Biolegend or BD Biosciences), which contains Fix/Perm solution and 10× Perm-Wash solution. Prepare 1× Perm-Wash solution by diluting 10× stock in water (*see* **Note 27**).

3 Methods

3.1 Cell Culture

1. LG2 and CH7C17 cells are cultured under the same conditions. Thaw a vial of frozen LG2 or CH7C17 cells in a 37 °C water bath by manually swirling the tube until a speck of ice remains (typically each vial contains five to ten million cells).
2. Immediately, transfer the contents of the vial into a sterile 15 mL tube containing 10 mL of cRPMI. (*see* **Notes 9** and **10**).
3. Centrifuge at 300×*g* for 5 min at 4 °C and discard the supernatant (*see* **Note 28**).
4. Resuspend the cells in 10 mL of fresh cRPMI and transfer to a T-25 culture flask.
5. Incubate at 37 °C incubator with 5% CO_2 for 2 days.
6. Centrifuge the cells at 300×*g* for 5 min at 4 °C and discard the supernatant.
7. Resuspend the cells in 5–10 mL of fresh cRPMI or 1× PBS and count using a hemocytometer.
8. Re-seed the cells at 10^5 cells/mL in 10–20 mL of cRPMI in a T-75 flask.
9. Maintain cells at numbers up to a maximum confluency of 10^6 cells/mL to retain maximum functionality (*see* **Notes 29–38**).

3.2 Labeling of Cells with CellTrace™ Stains

Here we are using the cell proliferation marker CellTrace™ Violet to label the CH7C17 cells and CellTrace™ Oregon Green to label the LG2 cells (*see* **Notes 17** and **18**) to enable identification of CH7C17 cells specifically to measure activation (Fig. 6).

1. Prepare stain by adding 50μL DMSO to one tube of CellTrace. Keep in dark on ice (*see* **Notes 17** and **18**).
2. Centrifuge the total required number of cells (1×10^5 cells per well is typical for this assay) in a 15 mL tube at 300×*g*, 4 °C and discard supernatant. Resuspend in PBS such that the final concentration is 1×10^6 cells/mL.
3. Add 1μL per million cells (which translate to 1μL/mL) and mix gently. Incubate at 37 °C for 20 min in the dark.
4. Quench the stain by adding 5× volume of pre-warmed cRPMI. Let it sit for 5 min. Centrifuge at 300×*g*, 4 °C and discard supernatant (*see* **Note 28**).
5. Resuspend cells in fresh cRPMI such that cell concentration is 1×10^6 cells/mL and let it sit for 10 min before using in the assay.

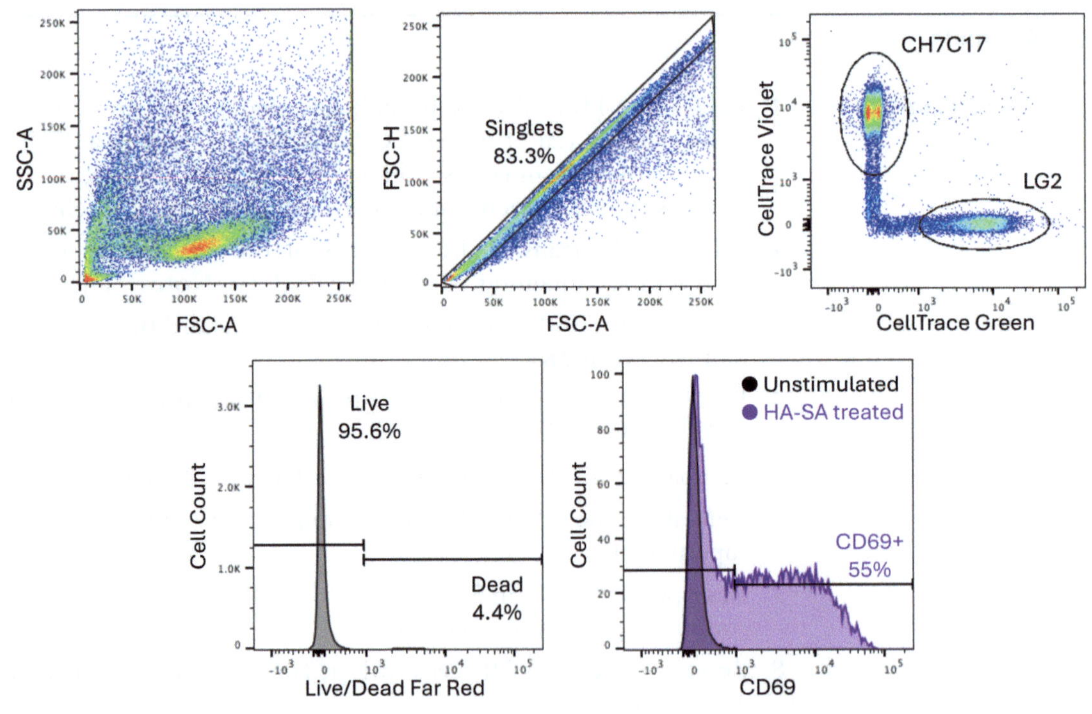

Fig. 6 Gating strategy. Cells are gated first on singlets following which CH7C17 cells and LG2 cells were separated based on their CellTrace™ stains. Live CH7C17 cells do not express the live/dead marker and these cells are gated and used to measure CD69 expression levels

3.3 Viral Infection of APCs

The virus in this example (Fig. 5) is loaded in columns 7 and 8, serially diluted in a v/v ratio of 1:5 down to 1:80.

1. Dispense 1 × 10^5 LG2 cells in 100μL cRPMI per well in columns 7 and 8 in a 96-well plate.
2. Centrifuge at 300×*g*, 4 °C for 3 min and discard supernatant (*see* **Note 28**).
3. Add 1× PBS and centrifuge at 300×*g*, 4 °C for 3 min. Repeat **step 1** more time. (*see* **Note 39**).
4. Resuspend cells in 100μL of virus working stock at indicated dilutions and incubate at 37 °C for 1 h (*see* **Note 40**).
5. Centrifuge at 300×*g*, 4 °C for 5 min and discard supernatant.
6. Resuspend in 100μL cRPMI and incubate at 37 °C for 24 h (*see* **Note 41**).
7. Following this, proceed to the addition of proteins, peptides, and controls to the APCs before the addition of T cells.

3.4 Protein and Peptide Treatment of APCs

In the example shown below (Fig. 5), the test protein, HA-SA, is loaded in columns 3–4 and the test peptide $HA_{306–318}$ is loaded in columns 5–6. Streptavidin protein loaded in wells F1 and F2 serves as a negative control for the test protein while CLIP loaded in wells G1 and G2 serves as a peptide negative control. All proteins and peptides were serially diluted from 1μM down to 8 nM.

1. Dispense 150μL of media to A1, A2 to D1, D2 of a 96-well tissue culture plate. These wells serve as the CH7C17 control wells for flow cytometry and will not receive any peptide or protein.
2. Dispense 200μL to E1, F1, G1 and H1.
3. Dispense 200μL in A3 to A6.
4. Dispense 100μL to all remaining wells except E2, F2, G2 and H2.
5. Prepare 2× the highest concentration of test protein, in this case 2μM of HA-SA, in A3 and A4 in 200μL (*see* **Notes 42** and **43**).
6. Serially dilute the test prep by taking 100μL from A3 and mixing it into B3. Repeat this process down the plate for both columns 3 and 4.
7. Discard the extra 100μL from H3 and H4.
8. Replicate this process for any additional antigens in other wells. In the example, 2μM of $HA_{306–318}$ was prepared in A5 and A6 in 200μL and serially diluted two-fold down the columns 5 and 6 in a similar manner as the HA-SA (*see* **Notes 43** and **44**).

9. Add Streptavidin to F1 at a final concentration of 2μM in the 200μL. Mix well and split 100μL into F2 (*see* **Notes 42** and **43**).
10. Add CLIP to G1 at a final concentration of 2μM in the 200μL. Mix well and split 100μL into G2 (*see* **Notes 43** and **44**).
11. Do not add any protein/peptide to E1 or E2 since these serve as the T cell activation control 2 h prior to the end of the incubation (*see* **Note 45**).
12. Do not add any protein/peptide to H1 or H2. These wells are the unstimulated control wells.

3.5 Setting up the Co-culture

1. Add 50μL of stained APCs (at 1×10^6 cells/mL) to all wells except the virally infected wells and the CH7C17 control wells.
2. Add 50μL of CH7C17 cells (at 1×10^6 cells/mL) to all the designated wells, including the CH7C17 control wells.
3. Gently shake the plate on a flat surface.
4. Incubate at 37 °C for 16–18 h and proceed for flow-cytometry staining (*see* **Notes 46** and **47**).
5. After 14 h, add PDBu at a final concentration of 20 nM and Ionomycin at a final concentration of 1μM to the designated wells and incubate at 37 °C for the remainder of the duration (*see* **Note 45**).

3.6 Flow Cytometry Staining

Antibodies and isotype controls are as needed. *See* **Notes 19** and **20** for the antibodies we used in the example assay.

1. If the existing total volume of liquid in wells is >200μL, proceed to **step 2** as is. If total volume is <200μL, make up volume to 200μL total per well with FACS buffer and proceed to **step 2** (*see* **Note 26**).
2. Centrifuge the assay plate at 300×*g*, 4 °C for 3 min and discard supernatant. Add 200μL FACS buffer to each well. Centrifuge at 300×*g*, 4 °C for 3 min and discard supernatant (*see* **Note 26**).
3. Add 50μL diluted Human Trustain FcX to all samples. Incubate 10 min at RT (*see* **Note 19**).
4. Wash with 200μL PBS. Centrifuge at 300×*g*, 4 °C for 3 min. Discard supernatant.
5. Add 50μL single stains (and if applicable, isotype control antibodies) to the designated wells and PBS to the unstained controls (*see* **Notes 20** and **21**).
6. Add 50μL surface primary antibodies + Live/Dead mix to target wells. Incubate for 25 min at 4 °C in the dark. (*see* **Notes 20–26**).

7. Wash with 200μL PBS. Centrifuge at 300×*g*, 4 °C for 3 min. Discard supernatant.
8. If applicable, add secondary antibody mix to the designated wells. Incubate for 25 min at 4 °C in the dark.
9. Wash with 200μL PBS. Centrifuge at 300×*g*, 4 °C for 3 min. Discard supernatant.
10. *If intracellular stains (ICS) are required, skip to **step 12**.* Resuspend in 100μL fixation buffer per well. Incubate for 15 min at 4 °C (*see* **Note 27**).
11. Add 100μL PBS, centrifuge at 300×*g*, 4 °C for 3 min and discard supernatant. Resuspend in 100–200μL PBS. Transfer samples to FACS tubes (if needed) and proceed to analysis. Alternatively, samples can be stored at 4 °C in the dark for up to 2 days (*see* **Notes 48** and **49**).
12. ***Steps 12–15** are applicable only if intracellular staining is required.* Resuspend in 100μL Fix/Perm buffer. Incubate for 15 min at 4 °C in the dark (*see* **Notes 27** and **48**).
13. Wash with 200μL 1× Perm-Wash buffer. Centrifuge at 300×*g*, 4 °C for 3 min. Discard supernatant. Repeat wash one more time.
14. Add 50μL ICS antibody mix (with Live/Dead if applicable) to target wells. Incubate for 25 min at 4 °C in the dark (*see* **Notes 27** and **48**).
15. Wash with 200μL 1x Perm-Wash buffer. Centrifuge at 300×*g*, 4 °C for 3 min. Discard supernatant and resuspend in 200μL PBS. Transfer samples to FACS tubes (if needed). Samples can be stored for 2 days at 4 °C in the dark or analyzed immediately for T cell activation levels as measured by CD69 expression (*see* **Notes 46–49**).

4 Notes

1. We used LG2 cells here as APCs but this protocol could be applied to other professional APCs such as B cells, dendritic cells, and macrophages.
2. LG2 cells are homozygous for HLA-A*02:01, B*27:05, C*07:01, DPA*01:01, DPB*02:01, DQA1*01:01, DQB*05:01, DRB1*01:01 and have been used as a source of HLA-DR1, DQ5, and DP2 [7]. We focused on a DR1-specific epitope. Other MHC-II alleles could also be tested for presentation similarly to DR1 using other B-lymphoblastoid or other antigen-presenting cells.

3. An alternative to providing the antigen through the APC would be to provide T cells with a purified MHC-II-peptide complex in the form of tetramers, wherein biotinylated MHC-II can be loaded with the peptide of interest and tetramerized using streptavidin [22, 37, 38].
4. MHC-II-peptide complexes bound to tissue culture wells or immobilized in the form of a micro- or nano-array can also be utilized to present antigen to T cells although this can depend on the spatial arrangement of the pMHC molecules [4, 5].
5. The HA1.7 TCR can recognize the epitope presented on either DR1 or DR4 (DRB1*04:01) [39].
6. CH7C17 cells retain the TCRβ from the TCRβ-negative mutant Jurkat-derived T cell line JRT [40] and this can pair with the HA1.7 TCRβ chain but this chimeric receptor does not recognize $HA_{306–318}$ [9] (also *see* **Note 37** and **38**).
7. Similarly, other T cell lines or clones as well as primary cells from patients could also be used for tracking antigen presentation similar to the CH7C17 cells that are described here.
8. T cell transfectants that express other TCRs recognizing the same or different epitopes can be generated to target other epitopes [41].
9. All tissue culture procedures must be performed aseptically.
10. All media components must be warmed up to 37 °C before use unless specified otherwise.
11. Handling of influenza viruses, transformed human cell lines (LG2, CH7C17) should be done in accordance with biosafety level 2 (BSL2) procedures. Follow waste disposal regulations when discarding materials contaminated with the virus. Before and after all procedures, all surfaces must be wiped down with bleach to prevent contamination and infection of the analyst.
12. We used influenza virus and influenza hemagglutinin as the antigens and the DR1-positive lymphoblastoid cell line LG2 because the $HA_{306–318}$ epitope is DR1 specific [10] and the CH7C17 cells express HA1.7 TCR that is specific to the $HA_{306–318}$ epitope presented on DR1 [6, 12].
13. The $HA_{306–318}$ peptide is a 13-mer and has been extensively studied. However, since MHC-II presents peptides varying from 9-mers to 20-mers other length variants could also be used. Variations of the $HA_{306–318}$ have been used in assays previously [42–45]. We find that the 13-mer sequence described here (synthesized by twenty-first Century Biochemicals) provides the best results in terms of binding to HLA-DR1.

14. The $HA_{306-318}$ epitope can also be presented by many other DR alleles [13] such as HLA-DRB1*07 [46], HLA-DRB1*04:03, HLA-DRB1*04:04, HLA-DRB1*04:05 [47], HLA-DRB5*01:02 [48, 49], HLA-DRB1*11:01 [50], HLA-DRB4*53 [51], HLA-DRB3*01:01 and HLA-DRB1*16:01 [42].
15. We selected CLIP (synthesized by twenty-first Century Biochemicals) as our negative control peptide. This is derived from the invariant chain chaperone and loaded in newly synthesized MHC-II molecules [1, 52]. Alternatively any peptide 9-mer or above restricted to the specific MHC-II allele of interest can be used as a negative control. We recommend CLIP since CLIP binds to most MHC-II proteins and can be used as a negative control regardless of the HLA allele.
16. We also selected streptavidin (SA) as our negative control for the protein since we oligomerized the HA using SA. However, any non-HA protein can serve as a negative control.
17. Since the assay involves the co-culturing of the cells and flow cytometry of the mixture, the analysis is facilitated by labeling one or both of the cell types with a marker to be able to differentiate them during flow cytometry. We labeled CH7C17 cells with CellTrace™ Violet and LG2 cells with CellTrace™ Oregon Green to enable identification of CH7C17 cells specifically to measure activation (Fig. 6). Alternatively, another label such as FITC could be used instead of the CellTrace™ labels or these labels could be omitted entirely in which case the %CD69 would reflect the entire population and not just the T cells.
18. Manufacturer recommendation for the CellTrace™ labels suggests resuspending each vial in 20μL of DMSO. We have observed no loss in labeling efficiency or fluorescence when resuspending each vial in 50μL of DMSO, reducing wastage.
19. In the example described here, we used Human Trustain FcX from Biolegend, which blocks Fc receptors on the cells and is diluted in FACS buffer. We recommend a 1:200 dilution from stock solution when staining up to 10^5 cells per well. However higher concentrations may be needed for higher cell counts. Alternate Fc block reagents may be used but would need to be tested for required working concentrations.
20. In the example described here, we used the following primary antibodies from Biolegend for surface staining: Anti-human CD69 PE (Clone FN50), Anti-human TCRαβ-APC (Clone IP26), Anti-human HLA-DR-FITC (Clone L243). We also used the following respective isotype controls: PE Rat IgG2a, κ Isotype Ctrl (Clone RTK2758), APC Mouse IgG1, κ Isotype Ctrl (Clone MOPC-21), FITC Mouse IgG2a, κ Isotype Ctrl (Clone MOPC-173).

21. Primary (and secondary antibodies if applicable) should be prepared in 1× PBS or FACS buffer. For single stain controls and isotype controls, we recommend a 1:400 dilution from the stock solution. All surface stain antibodies can be combined into a single mix. We recommend a 1:200 dilution from the stock solution when staining up to 10^5 cells per well. Alternate antibodies and fluorophores may be used, but would need testing to determine required working concentrations.
22. Fluorophore selection for the antibodies can be limited by the cytometer configuration available. In the example provided we selected fluorophores based on the cytometer configuration available to us at the time of designing this assay.
23. In the example shown here, we designed our panel such that tags on the surface markers for LG2 cells and CH7C17 cells did not have overlap in their excitation/emission spectra, simplifying the readout while retaining accuracy. The assay can be designed with alternate fluorophore tags as needed as long as they either do not overlap in their excitation/emission spectra or sufficient compensation is available.
24. The manufacturer recommends preparing Live/Dead stock by resuspending a single vial in 50μL of DMSO. However we observed no noticeable loss in labeling efficiency or fluorescence when using up to 100μL of DMSO. We recommend aliquoting and storing at −20 °C. Working solutions are prepared by diluting the stock in 1× PBS.
25. For Live/Dead staining, we recommend a 1:1000 dilution from the stock solution to stain up to 10^5 cells. Cells can also be stained for Live/Dead markers separately prior to surface staining. However we observed no issues when adding Live/Dead stain to the surface staining solutions so as to reduce a centrifugation and resuspension step.
26. We observed non-specific binding of the Live/Dead stain when using FACS buffers containing sodium azide. In such cases, it is recommended to stain cells with Live/Dead stain diluted in 1× PBS prior to staining for surface markers or intracellular markers.
27. Fluorescent-tagged antibodies for intracellular staining (ICS) are prepared in 1× Perm-wash buffer. All ICS antibodies can be combined into a single mix. We recommend a 1:200 dilution from the stock solution similar to surface marker antibodies.
28. We recommend performing all centrifugation steps involving live cells at 300×*g*, 4 °C. However if excessive cell death is observed, the centrifugation speed and/or time can be modified as needed.

29. We find that both CH7C17 and LG2 cells must be passaged at least twice after revival to get the best results in the assay.
30. For the best performance of the CH7C17 cells, split cells a day before they are needed.
31. Both CH7C17 and LG2 cells can be maintained at numbers up to a maximum confluency of 10^6 cells/mL to retain maximum functionality. This translates to splitting cells every 2–3 days depending on their count.
32. Cells tend to clump together when growing and this is normal. Pipetting will help disperse the cells.
33. Centrifuging and resuspending the culture before splitting is not necessary if the cells have high viability and do not show excessive debris. Healthy cultures can be split by diluting directly into fresh cRPMI.
34. LG2 cells can be maintained up to 8–10 passages while CH7C17 cells can be maintained up to 12–15 passages without significant loss in functionality. However, this number is only estimated and needs to be tested for specific applications.
35. To test LG2 cells for functionality, measure the expression of HLA-DR1 levels on the surface. Alternatively, HLA-DR1 can be purified from LG2 cells and measured using an ELISA. We have typically observed, on average, 1μg of HLA-DR1 per million healthy cells.
36. To test CH7C17 cells for functionality, measure the levels of HA1.7 TCR on the surface using HLA-DR1/$HA_{306–318}$ tetramers as well as total TCR levels using a pan-TCR$\alpha\beta$ antibody. We find >80% of the cells expressing a TCR retain high expression of HA1.7 in healthy cells. Experimental testing has shown 60% of the cells retain high expression of the TCR at up to 15 passages (also *see* **Notes 37** and **38**).
37. Due to the transfected nature of the TCR, CH7C17s cannot be maintained in culture indefinitely without losing the HA1.7 TCR. Experimental testing has shown 60% of the cells retain high expression of the TCR at up to 15 passages. However, this number may vary depending on growth conditions.
38. If loss of functionality is observed, the recommendation is to sort for high-HA1.7 expressing cells by labeling cells with the DR1/$HA_{306–318}$ tetramers. Tetramers are used instead of an anti-TCR or anti-CD3 antibody because of the native Jurkat TCRα chain, which can hybridize with the HA1.7 TCRβ chain to express a TCR, which is not specific to the HA epitope.
39. It is important to remove all traces of FBS from the cells since FBS has been known to inhibit virus growth [53, 54].

40. Working stocks of viruses must be made fresh before use by diluting viral stocks to the desired dilution in vPBS. We find it easier to prepare dilutions of the virus in tubes and dispense them in the plate as needed.
41. Viral infection requires an additional 24 h preceding protein and peptide treatment in the assay since the APCs will need time to replicate the virus and present antigens.
42. The plate is prepared with 100μL of 2× concentration in each well so that it is halved when the cells are added for a final volume of 200μL. Adjust this plate layout and concentrations as necessary but remember to prepare 2× concentration to accommodate for the cells (Fig. 5a).
43. The assay here is described for a final volume of 200μL per well. The concentrations shown here are the final concentration for the assay after addition of all the cells (Fig. 5a).
44. For working concentrations of peptides, dilute 1:10 in warm cRPMI prior to starting the co-culture instead of diluting smaller volumes in the plate.
45. PDBu and Ionomycin can activate T cells in as little as 30 min of incubation. We have observed robust responses as little as 2 h and up to 6 h prior to staining.
46. We selected CD69 as the readout for this assay, which is an early activation marker with upregulation as early as 4 h [55, 56]. However, we have observed peak expression at 16–18 h for CH7C17 cells with no significant improvement up to 24 h of incubation.
47. While we selected CD69 as a readout due to its sensitivity in the low-concentration ranges, CD25 could be used as an alternate readout specifically for the CH7C17 cells [57]. CD3 downregulation, which reflects internalization of shedding of engaged receptors, also could be used [34, 58].
48. Fixation also neutralizes any infectious agents and is recommended for all assays. When performing ICS, Fix/Perm buffer also fixes the cells and the surface-stained antibodies, so additional fixation is not required at the end of the assay.
49. We recommend analyzing the cells immediately. However, if fixation alters readout, it could be skipped. In cases where cells are not fixed, they should not be stored for analysis on a later date.

Conflict of Interest

The authors have no conflicts of interest to declare that are relevant to the content of this chapter.

References

1. Roche PA, Furuta K (2015) The ins and outs of MHC class II-mediated antigen processing and presentation. Nat Rev Immunol 15:203–216. https://doi.org/10.1038/nri3818
2. Pishesha N, Harmand TJ, Ploegh HL (2022) A guide to antigen processing and presentation. Nat Rev Immunol 22:751–764. https://doi.org/10.1038/s41577-022-00707-2
3. Jensen PE (2007) Recent advances in antigen processing and presentation. Nat Immunol 8: 1041–1048. https://doi.org/10.1038/ni1516
4. Stone JD, Demkowicz WE, Stern LJ (2005) HLA-restricted epitope identification and detection of functional T cell responses by using MHC-peptide and costimulatory microarrays. Proc Natl Acad Sci USA 102:3744–3749. https://doi.org/10.1073/pnas.0407019102
5. Deeg J, Axmann M, Matic J et al (2013) T cell activation is determined by the number of presented antigens. Nano Lett 13:5619–5626. https://doi.org/10.1021/nl403266t
6. Lamb JR, Eckels DD, Lake P et al (1982) Human T-cell clones recognize chemically synthesized peptides of influenza haemagglutinin. Nature 300:66–69. https://doi.org/10.1038/300066a0
7. Gorga JC, Horejsi V, Johnson DR et al (1987) Purification and characterization of class II histocompatibility antigens from a homozygous human B cell line. J Biol Chem 262: 16087–16094
8. Hewitt CR, Lamb JR, Hayball J et al (1992) Major histocompatibility complex independent clonal T cell anergy by direct interaction of Staphylococcus aureus enterotoxin B with the T cell antigen receptor. J Exp Med 175:1493–1499. https://doi.org/10.1084/jem.175.6.1493
9. Wedderburn LR, Searle SJ, Rees AR et al (1995) Mapping T cell recognition: the identification of a T cell receptor residue critical to the specific interaction with an influenza hemagglutinin peptide. Eur J Immunol 25:1654–1662. https://doi.org/10.1002/eji.1830250627
10. Stern LJ, Brown JH, Jardetzky TS et al (1994) Crystal structure of the human class II MHC protein HLA-DR1 complexed with an influenza virus peptide. Nature 368:215–221. https://doi.org/10.1038/368215a0
11. Holland CJ, Rizkallah PJ, Vollers S et al (2012) Minimal conformational plasticity enables TCR cross-reactivity to different MHC class II heterodimers. Sci Rep 2:629. https://doi.org/10.1038/srep00629
12. Hennecke J, Carfi A, Wiley DC (2000) Structure of a covalently stabilized complex of a human alphabeta T-cell receptor, influenza HA peptide and MHC class II molecule, HLA-DR1. EMBO J 19:5611–5624. https://doi.org/10.1093/emboj/19.21.5611
13. Roche PA, Cresswell P (1990) High-affinity binding of an influenza hemagglutinin-derived peptide to purified HLA-DR. J Immunol 144: 1849–1856. https://doi.org/10.4049/jimmunol.144.5.1849
14. Nanaware PP, Jurewicz MM, Leszyk JD et al (2019) HLA-DO modulates the diversity of the MHC-II self-peptidome. Mol Cell Proteomics 18:490–503. https://doi.org/10.1074/mcp.RA118.000956
15. Gómez-Morón A, Requena S, Roda-Navarro P, Martín-Cófreces NB (2023) Activation kinetics of regulatory molecules during immunological synapse in T cells. Methods Cell Biol 178: 149–171. https://doi.org/10.1016/bs.mcb.2022.10.014
16. Zarutskie JA, Busch R, Zavala-Ruiz Z et al (2001) The kinetic basis of peptide exchange catalysis by HLA-DM. Proc Natl Acad Sci USA 98:12450–12455. https://doi.org/10.1073/pnas.211439398
17. Mamedov A, Vorobyeva N, Filimonova I et al (2019) Protective allele for multiple sclerosis HLA-DRB1*01:01 provides kinetic discrimination of myelin and exogenous antigenic peptides. Front Immunol 10:3088. https://doi.org/10.3389/fimmu.2019.03088
18. Aldinucci A, Rizzetto L, Pieri L et al (2010) Inhibition of immune synapse by altered dendritic cell actin distribution: a new pathway of mesenchymal stem cell immune regulation. J Immunol 185:5102–5110. https://doi.org/10.4049/jimmunol.1001332
19. Painter CA, Negroni MP, Kellersberger KA et al (2011) Conformational lability in the class II MHC 310 helix and adjacent extended strand dictate HLA-DM susceptibility and peptide exchange. Proc Natl Acad Sci USA 108: 19329–19334. https://doi.org/10.1073/pnas.1108074108
20. Stratikos E, Wiley DC, Stern LJ (2004) Enhanced catalytic action of HLA-DM on the exchange of peptides lacking backbone hydrogen bonds between their N-terminal region and the MHC class II alpha-chain. J Immunol 172:1109–1117. https://doi.org/10.4049/jimmunol.172.2.1109

21. Anderson MW, Gorski J (2005) Cooperativity during the formation of peptide/MHC class II complexes. Biochemistry 44:5617–5624. https://doi.org/10.1021/bi048675s

22. Ye M, Kasey S, Khurana S et al (2004) MHC class II tetramers containing influenza hemagglutinin and EBV EBNA1 epitopes detect reliably specific CD4(+) T cells in healthy volunteers. Hum Immunol 65:507–513. https://doi.org/10.1016/j.humimm.2004.02.019

23. Danke NA, Kwok WW (2003) HLA class II-restricted CD4+ T cell responses directed against influenza viral antigens postinfluenza vaccination. J Immunol 171:3163–3169. https://doi.org/10.4049/jimmunol.171.6.3163

24. Gebe JA, Falk BA, Rock KA et al (2003) Low-avidity recognition by CD4+ T cells directed to self-antigens. Eur J Immunol 33:1409–1417. https://doi.org/10.1002/eji.200323871

25. Figueredo Dos Santos C, Tilkin-Mariame AF, De Préval C, Lakhdar-Ghazal F (2002) Influence of histidine beta81 of HLA-DR101 on peptide binding and presentation to T-cell receptor. Hum Immunol 63:459–466. https://doi.org/10.1016/s0198-8859(02)00394-4

26. Mack J, Falk K, Rötzschke O et al (2001) Synthesis of linear and comb-like peptide constructs containing up to four copies of a T cell epitope and their capacity to stimulate T cells. J Pept Sci 7:338–345. https://doi.org/10.1002/psc.327

27. Korb LC, Mirshahidi S, Ramyar K et al (1999) Induction of T cell anergy by low numbers of agonist ligands. J Immunol 162:6401–6409

28. Natarajan SK, Stern LJ, Sadegh-Nasseri S (1999) Sodium dodecyl sulfate stability of HLA-DR1 complexes correlates with burial of hydrophobic residues in pocket 1. J Immunol 162:3463–3470

29. Rajnavölgyi E, Horváth A, Gogolák P et al (1997) Characterizing immunodominant and protective influenza hemagglutinin epitopes by functional activity and relative binding to major histocompatibility complex class II sites. Eur J Immunol 27:3105–3114. https://doi.org/10.1002/eji.1830271205

30. Tsitoura DC, Holter W, Cerwenka A et al (1996) Induction of anergy in human T helper 0 cells by stimulation with altered T cell antigen receptor ligands. J Immunol 156:2801–2808

31. Sadegh-Nasseri S, Stern LJ, Wiley DC, Germain RN (1994) MHC class II function preserved by low-affinity peptide interactions preceding stable binding. Nature 370:647–650. https://doi.org/10.1038/370647a0

32. Cameron TO, Cohen GB, Islam SA, Stern LJ (2002) Examination of the highly diverse CD4 (+) T-cell repertoire directed against an influenza peptide: a step towards TCR proteomics. Immunogenetics 54:611–620. https://doi.org/10.1007/s00251-002-0508-y

33. Cochran JR, Cameron TO, Stone JD et al (2001) Receptor proximity, not intermolecular orientation, is critical for triggering T-cell activation. J Biol Chem 276:28068–28074. https://doi.org/10.1074/jbc.M103280200

34. Cochran JR, Cameron TO, Stern LJ (2000) The relationship of MHC-peptide binding and T cell activation probed using chemically defined MHC class II oligomers. Immunity 12:241–250. https://doi.org/10.1016/s1074-7613(00)80177-6

35. Givan AL (2011) Flow cytometry: an introduction. Methods Mol Biol 699:1–29. https://doi.org/10.1007/978-1-61737-950-5_1

36. McKinnon KM (2018) Flow cytometry: an overview. Curr Protoc Immunol 120:5.1.1–5.1.11. https://doi.org/10.1002/cpim.40

37. Dileepan T, Malhotra D, Kotov DI et al (2021) MHC class II tetramers engineered for enhanced binding to CD4 improve detection of antigen-specific T cells. Nat Biotechnol 39:943–948. https://doi.org/10.1038/s41587-021-00893-9

38. James EA, LaFond R, Durinovic-Bello I, Kwok W (2009) Visualizing antigen specific CD4+ T cells using MHC class II tetramers. J Vis Exp. https://doi.org/10.3791/1167

39. Hennecke J, Wiley DC (2002) Structure of a complex of the human α/β T cell receptor (TCR) HA1.7, influenza hemagglutinin peptide, and major histocompatibility complex class II molecule, HLA-DR4 (DRA*0101 and DRB1*0401) : insight into TCR cross-restriction and alloreactivity. J Exp Med 195:571–581. https://doi.org/10.1084/jem.20011194

40. Weiss A, Stobo JD (1984) Requirement for the coexpression of T3 and the T cell antigen receptor on a malignant human T cell line. J Exp Med 160:1284–1299. https://doi.org/10.1084/jem.160.5.1284

41. Song I, Gil A, Mishra R et al (2017) Broad TCR repertoire and diverse structural solutions for recognition of an immunodominant CD8 (+) T cell epitope. Nat Struct Mol Biol 24:395–406. https://doi.org/10.1038/nsmb.3383

42. Zeliszewski D, Gaudebout P, Golvano JJ et al (1994) Molecular basis for degenerate T-cell

recognition of one peptide in the context of several DR molecules. Hum Immunol 41:28–33. https://doi.org/10.1016/0198-8859(94)90080-9

43. Tuosto L, Karr RW, Fu XT et al (1994) Different regions of the N-terminal domains of HLA-DR1 influence recognition of individual peptide-DR1 complexes. Hum Immunol 40: 312–322. https://doi.org/10.1016/0198-8859(94)90031-0
44. Newton-Nash DK, Eckels DD (1993) Differential effect of polymorphism at HLA-DR1 beta-chain positions 85 and 86 on binding and recognition of DR1-restricted antigenic peptides. J Immunol 150:1813–1821
45. Ffrench RA, Tang XL, Anders EM et al (1989) Class II-restricted T-cell clones to a synthetic peptide of influenza virus hemagglutinin differ in their fine specificities and in the ability to respond to virus. J Virol 63:3087–3094. https://doi.org/10.1128/JVI.63.7.3087-3094.1989
46. Krieger JI, Karr RW, Grey HM et al (1991) Single amino acid changes in DR and antigen define residues critical for peptide-MHC binding and T cell recognition. J Immunol 146: 2331–2340
47. Rothbard JB, Busch R, Bal V et al (1989) Reversal of HLA restriction by a point mutation in an antigenic peptide. Int Immunol 1: 487–495. https://doi.org/10.1093/intimm/1.5.487
48. Yamamoto K, Fukui Y, Esaki Y et al (1994) Functional interaction between human histocompatibility leukocyte antigen (HLA) class II and mouse CD4 molecule in antigen recognition by T cells in HLA-DR and DQ transgenic mice. J Exp Med 180:165–171. https://doi.org/10.1084/jem.180.1.165
49. Brawley JV, Concannon P (1996) Modulation of promiscuous T cell receptor recognition by mutagenesis of CDR2 residues. J Exp Med 183:2043–2051. https://doi.org/10.1084/jem.183.5.2043
50. McKinney JS, Fu XT, Swearingen C et al (1994) Individual effects of the DR11-variable beta-chain residues 67, 71, and 86 upon DR (alpha,beta 1*1101)-restricted, peptide-specific T cell proliferation. J Immunol 153: 5564–5571
51. Hewitt CR, Feldmann M (1989) Human T cell clones present antigen. J Immunol 143:762–769
52. Castellino F, Zhong G, Germain RN (1997) Antigen presentation by MHC class II molecules: invariant chain function, protein trafficking, and the molecular basis of diverse determinant capture. Hum Immunol 54:159–169. https://doi.org/10.1016/s0198-8859(97)00078-5
53. Iki S, Yokota S, Okabayashi T et al (2005) Serum-dependent expression of promyelocytic leukemia protein suppresses propagation of influenza virus. Virology 343:106–115. https://doi.org/10.1016/j.virol.2005.08.010
54. Qin Z, Ju H, Liu Y et al (2013) Fetal bovine serum inhibits hepatitis C virus attachment to host cells. J Virol Methods 193:261–269. https://doi.org/10.1016/j.jviromet.2013.06.024
55. Rea IM, McNerlan SE, Alexander HD (1999) CD69, CD25, and HLA-DR activation antigen expression on CD3+ lymphocytes and relationship to serum TNF-α, IFN-γ, and sIL-2R levels in aging. Exp Gerontol 34:79–93. https://doi.org/10.1016/S0531-5565(98)00058-8
56. Cibrián D, Sánchez-Madrid F (2017) CD69: from activation marker to metabolic gatekeeper. Eur J Immunol 47:946–953. https://doi.org/10.1002/eji.201646837
57. Stone JD, Cochran JR, Stern LJ (2001) T-cell activation by soluble MHC oligomers can be described by a two-parameter binding model. Biophys J 81:2547–2557. https://doi.org/10.1016/S0006-3495(01)75899-7
58. José ES, Borroto A, Niedergang F et al (2000) Triggering the TCR complex causes the downregulation of nonengaged receptors by a signal transduction-dependent mechanism. Immunity 12:161–170. https://doi.org/10.1016/S1074-7613(00)80169-7

Chapter 18

Transient Transfection and T Cell Activation in the Assessment of Endoplasmic Reticulum Aminopeptidase 1 and 2 Peptide Trimming Function

Emma Reeves and Edward James

Abstract

Endoplasmic reticulum aminopeptidases 1 and 2 (ERAP1 and ERAP2) play a critical role in antigen processing by trimming N-terminally extended peptides within the ER, thereby shaping the repertoire of peptides presented by Major Histocompatibility Complex Class I (MHC I) molecules. Polymorphic variants in these enzymes give rise to functionally distinct allotypes, influencing peptide trimming efficiency and specificity, modulating CD8+ T cell responses in both health and disease. This chapter outlines a cellular model system for assessing ERAP1 and ERAP2 peptide trimming activity, using peptide-specific T cell activation as a surrogate readout. By employing transient transfection and co-culture with either T cell hybridoma (B3Z) or cytotoxic T lymphocytes (CTL), this approach enables the evaluation of peptide processing efficiency of ERAP1/2 based on the presentation and recognition of optimally generated MHC I-restricted epitopes.

Key words ERAP1, ERAP2, T cell activation, T cell hybridoma, Transfection, Antigen processing, Antigen presentation

1 Introduction

The generation of antigenic peptides for presentation by Major Histocompatibility Complex class I (MHC I) molecules is a fundamental process in the activation of cytotoxic CD8+ T cells and a key element of the adaptive immune response. Among the key players in this pathway are the endoplasmic reticulum aminopeptidases 1 and 2 (ERAP1 and ERAP2), which are responsible for trimming N-terminally extended precursor peptides within the ER to generate optimal length epitopes for MHC I binding [1]. These aminopeptidases are important not only for generating stable and high-affinity peptide-MHC I complexes [2] but also for shaping the peptide repertoire, immunodominance hierarchy and specificity of CD8+ T cell responses [3–5].

Efstratios Stratikos (ed.), *Antigen Processing and Presentation: Methods and Protocols*, Methods in Molecular Biology, vol. 3007, https://doi.org/10.1007/978-1-0716-5092-9_18,

ERAP1, and to a lesser extent ERAP2, exhibit polymorphic variation that results in distinct allotypes, conferring functional differences in enzymatic activity and substrate specificity [6–8]. Genetic variation in ERAP1/2 has been implicated in a range of immune-mediated conditions, including autoimmune and autoinflammatory disease, viral infections and cancer, underscoring their functional relevance in shaping the immunopeptidome and influencing immune surveillance and tolerance mechanisms. Understanding how ERAP1 and ERAP2 allotypes influence antigen presentation is therefore critical in elucidating their role in immune regulation and disease pathogenesis.

Direct measurement of ERAP1/ERAP2 activity in cells can be technically challenging due to the complexity of intracellular peptide processing, with in vitro assays often used to ascertain enzymatic activity and kinetics towards specific substrates [6]. However, these assays do not account for the interplay of antigen processing and presentation (APP) components in a physiological system. To overcome these limitations, surrogate assays that link peptide processing by ERAP1/ERAP2 to functional T cell readouts have proven valuable in ascertaining the functional activity of these aminopeptidases [8, 9]. T cell activation assays, described in this chapter, rely on the successful generation and presentation of specific peptide-MHC I complexes, which are detected through the activation of peptide-specific T cells.

A particularly useful tool in T cell activation assays is the use of LacZ-inducible T cell hybridomas, which provide a sensitive and quantifiable readout of T cell receptor (TCR) engagement and subsequent activation. These hybridomas have been engineered to express β-galactosidase under the control of the nuclear factor of activated T cells (NFAT)-responsive promoter of the IL-2 gene [10]. The production of β-galactosidase upon recognition of peptide-MHC I complexes can be easily measured using colorimetric or fluorogenic substrates, allowing for precise monitoring of specific antigen processing and presentation [11].

In this chapter, we describe a robust and adaptable model system that utilizes transient transfection of ERAP1/ERAP2, MHC I and ER-targeted antigen precursors into an ERAP1/2 mammalian knock-out cell system. Co-culture with either LacZ inducible T cell hybridoma or cytotoxic T lymphocytes enables the indirect assessment of ERAP1 and ERAP2 activity by measurement of T cell activation in response to the final, optimally trimmed peptide presented by the relevant MHC I. While the model described here primarily utilizes the murine H2-K^b restricted B3Z T cell hybridoma, which recognizes the ovalbumin-derived epitope, $OVA_{257\text{-}264}$ SIINFEKL (SL8), it is easily adaptable to other peptide MHC I systems. Use of the BE7A2Z T cell hybridoma, which recognizes the human papillomavirus (HPV) 16 E7-derived epitope, HPV-16 $E7_{82\text{-}90}$ LLMGTLGIV (LV9) restricted to

HLA-A*02:01, extends this model into the human HLA allele context and has been used to determine the trimming activity of ERAP1 allotype combinations identified in HPV+ head and neck cancers [9].

This methodology offers several advantages: it preserves the physiological context of peptide trimming within the ER, allows for the assessment of ER trimming by use of knock-out cell line models, and provides a functional immune readout reflecting the biological significance and consequence of ERAP1 and ERAP2 peptide processing. Furthermore, the methodology can be adapted for the investigation of other components of APP and for the screening of pharmacological modulators of ERAP1 and ERAP2.

Overall, this approach bridges the gap between biochemical characterization of ERAP1 activity and the immunological impact of diverse trimming functions due to polymorphism on T cell responses.

2 Materials

2.1 General Reagents and Consumables

1. 1.5 mL centrifuge tubes, sterile.
2. 15 mL centrifuge tubes, sterile.
3. 10 cm tissue culture plates.
4. 6-well tissue culture plates.
5. 96-well flat-bottom tissue culture plates.
6. 96-well round-bottom tissue culture plates.
7. 3 mL pastettes, sterile.
8. Phosphate-buffered saline tablets (PBS, Sigma), 1 tablet per 200 mL dH_2O.
9. Fetal bovine serum (FBS), heat-inactivated.
10. Complete+ RPMI media: RPMI supplemented with 10% FBS, 2 mM L-glutamine, 2 mM penicillin/streptomycin, 1 mM Hepes, 1 mM sodium pyruvate, with the addition of 50 μM 2-mercaptoethanol. *See* **Note 1**.
11. Complete DMEM media: DMEM supplemented with 10% FBS, 2 mM L-glutamine, 2 mM penicillin/streptomycin, 1 mM Hepes and 1 mM sodium pyruvate.
12. Cell culture incubator, 37 °C/5% CO_2.
13. Water bath set to 37 °C.
14. Microscope.
15. Hemocytometer and trypan blue solution, 0.4% (Thermo-Fisher Scientific).

Table 1
Components of transient transfection

DNA component	Plasmid concentration (μg/μL)	Amount to add into transfection (μL)	Final amount (μg) added to transfection
ERAP1 or ERAP2-pcDNA3 or empty-pcDNA3 (control)	0.5	1	0.5
Relevant MHC (e.g. H2-K^b)—pcDNA3	0.5	0.5	0.25
Peptide minigene (e.g. AIVMK-SIINFEKL and SIINFEKL)—pcDNA3	0.5	0.5	0.25

2.2 Cell Transfection and Harvesting

1. HEK293T ERAP1 and ERAP2 knock-out cell line (293T-E1E2KO) or other cell line of choice. 293T-E1E2KO are cultured in DMEM media in 10 cm culture dishes.
2. Serum-free DMEM media, 97 μL per transfection condition, warmed at 37 °C.
3. FuGENE®6 transfection reagent (Promega), stored at 4 °C. *See* **Note 2**.
4. Plasmid DNA (*see* Table 1) cloned into a relevant mammalian expression vector, e.g. pcDNA3, at 1 μg/μL stock in TE buffer and stored at 4 °C. It is also best practice to store an aliquot at −20 °C long term.
5. 1× Tris/EDTA (TE) buffer: 10 mM Tris-HCl (pH 8.0) + 0.1 mM EDTA.

2.3 T Cell Activation: T Cell Hybridoma

1. LacZ-inducible T cell hybridoma. This protocol uses B3Z (H2-K^b/SIINFEKL specific) cultured in complete+ RPMI; however, others can be used such as BE7A2Z (HLA-A*02:01-LLMGTLGIV specific), depending on availability and requirement.
2. Antigen-presenting cells expressing the relevant MHC I allele. This protocol uses K89 L-cells stably expressing H2-K^b, cultured in complete+ RPMI media in 10 cm tissue culture dish.
3. T75 tissue culture flask, culturing B3Z T cell hybridoma in 30 mL complete+ RPMI.
4. 1 mM EDTA in 1× PBS, sterile filtered. Use 4 μL per 10 cm dish of K89 cells to detach cells.
5. Multichannel pipette with 200 μL capacity.
6. Reagent reservoir for multichannel pipette.

7. 10 mg synthetic SL8 peptide, lyophilized >95% HPLC grade resuspended at 10 mM in DMSO, aliquoted and stored at − 20 °C.
8. Chlorophenol red-β-D-galactopyranoside (CPRG) buffer: 91 mg CPRG powder (Roche), 1.25 mL NP40, 9 mL 1 M $MgCl_2$ made to 1 L with 1× PBS. Store at 4 °C protected from light. *See* **Note 3**.
9. Microplate absorbance plate reader, such as BioRad iMark, with coverage of 595–655 nm wavelength.

2.4 T Cell Activation: CTL Co-culture

1. CTL, PBMC or other source of T cell pool.
2. Brefeldin A solution (1000×). Use at 1 μL per 1 mL of cells, store undiluted at 4 °C.
3. Phorbol 12-myristate-13-acetate (PMA).
4. Ionomycin.
5. FACS wash buffer: PBS supplemented with 5% FCS, sterile filtered and stored at 4 °C.
6. Fluorochrome-conjugated antibodies specific for cell surface markers; CD3 and CD8 as well as IFNγ.
7. BD Cytofix/Cytoperm™ Fixation/Permeabilization kit (BD Biosciences) consisting of Fixation/Permeabilization buffer and BD Perm/Wash (10× concentrate containing FBS and saponin) to be diluted 1:10 in dH_2O prior to use.
8. 5 mL round-bottom test tubes suitable for flow cytometry.
9. BD FACS Canto flow cytometer or equivalent machine.
10. Flow cytometry analysis software of choice (e.g. FlowJo™).

3 Methods

3.1 Transfection of Cells

3.1.1 Cell Harvesting and Plating for Transfection (Day 1)

1. Harvest, count and plate 293T-E1E2KO cells, or an alternative cell line model, in 2 mL complete DMEM culture media (Subheading 2.1) in a 6-well tissue culture plate at 1×10^5 cells/mL. *See* **Note 4**.
2. Incubate the cells overnight (37 °C/5%CO_2) to ensure adherence to the tissue culture plate and to achieve 50–80% confluency prior to plasmid transfection.

3.1.2 Transfection of Cells (Day 2)

FuGENE®6 transfection reagent is used at 3:1 FuGENE®6:DNA ratio, e.g. 3 μL FuGENE®6: 1 μg total DNA. This amount can be scaled up or down if necessary (e.g. 2 μg DNA), ensuring the same ratio. *See* **Note 5**.

1. Check cells to ensure 50–80% confluence in 6-well plate (*see* **Note 6**). This is essential for optimal transfection efficiency.

2. Allow FuGENE®6 transfection reagent and serum-free DMEM to warm to room temperature.
3. Dilute relevant stock plasmids (1 μg/μL) to a working concentration of 0.5 μg/μL in TE buffer. *See* **Note 7**.
4. Calculate the relative amount of DNA plasmid required for each transfection based on the ratio in Table 1 and a total of 1 μg DNA for 6-well plate transfection.
5. Add 97 μL serum-free DMEM to a sterile 1.5 mL Eppendorf tube in the tissue culture hood.
6. Carefully add 3 μL FuGENE®6 transfection reagent to the serum-free media, close the lid and flick the tube gently 4–5 times (*see* **Note 8**).
7. Incubate the FuGENE®6 and serum-free media mix for 5 min at room temperature in the tissue culture hood: *see* **Notes 9** and **10**.
8. Add the relevant amount of DNA (as calculated according to Table 1) to the FuGENE®6/serum free media in the Eppendorf. Depending on the number of transfections required, the DNA can be added in any order to the FuGENE®6/serum free media.
9. Flick the tube gently to mix the contents of the tube to promote FuGENE®6-DNA complex formation.
10. Incubate the transfection mix for a minimum of 15 min, ensuring no longer than 45 mins as this might affect transfection efficiency, at room temperature in the tissue culture hood.
11. Remove cells from the incubator 2 min prior to adding the transfection mix and label accordingly. It is not necessary to remove the media from the cells prior to adding the transfection mix. *See* **Note 11**.
12. Slowly add the transfection mix in a dropwise manner to each well of the 6-well plate, moving the plate left to right/up and down whilst adding the mix. Ensure all transfection mix is added to the cells (>100 μL). *See* **Note 12**.
13. Incubate the transfected cells for 24 h at 37 °C/5%CO_2.

3.1.3 Harvesting Transfected Cells (Day 3)

1. Add 4 mL fresh complete DMEM to a 15 mL falcon tube for each transfection condition.
2. Using a 3 mL pastette, harvest all 293T-E1E2KO cells from the surface of each well of the 6-well culture plate, adding them into the corresponding 15 mL falcon tube with DMEM.
3. Mix the transfected cells with the media and take 30 μL for counting.
4. Centrifuge the transfected cells (6 mL total) at 400*g*× 5 min at room temperature to pellet the transfected cells.

5. Using the 30 μL aliquot of cells, count cells using hemocytometer to determine the total number of cells.
6. Resuspend cells at the desired concentration for downstream assays.

Transfected cells can be used for downstream assays to determine peptide-specific T cell activation as a surrogate of ERAP1/2 peptide trimming activity and generation of final peptides from N-terminally extended precursors.

3.2 Determining T Cell Activation Using Peptide-HLA-Specific, LacZ-Inducible T Cell Hybridomas

3.2.1 Harvesting T Cell Hybridoma

1. T cell hybridoma (e.g. B3Z) is a suspension cell line and should be cultured in T75 flasks in 30 mL complete+ RPMI media. Using a 10 mL pipette, mix the cells into a single cell suspension. *See* **Note 13**.
2. Remove 30 μL T cell hybridoma from the flask for counting.
3. Count cells using hemocytometer to determine the number of cells per mL.
4. Calculate the number of wells that require co-culture with T cell hybridoma and how many total cells are required, with the addition of 5 wells to ensure enough cell suspension. This will usually be 8 wells per transfection condition, using a concentration of 1×10^5 cells in 100 μL per well (concentration of 1×10^6 cells/mL). *See* **Note 14**.
5. Remove the required cell suspension to achieve the desired total cell number and centrifuge the cells at 400*g*× 5 min to pellet the cells.
6. Resuspend the pellet of cells at 1×10^6 cells/mL in complete + RPMI.
7. Proceed to setting up the co-culture.

3.2.2 Co-culture of Transfected Cells with T Cell Hybridoma

Generation of overnight co-culture of transfected antigen-presenting cells and the corresponding peptide-/HLA-specific T cell hybridoma starting at an effector:target ratio of 1:1.

1. Resuspend transfected 293T-E1E2KO cells at a final concentration of 1×10^6 cells/mL in fresh DMEM media (following on from Subheading 3.1.3).
2. Add 200 μL transfected 293T-E1E2KO cells to the second well of a flat-bottom 96-well tissue culture plate (e.g. B2, C2, D2) for each transfection condition. This should give a concentration of 2×10^5 cells in each well.
3. Add 100 μL fresh DMEM media to the subsequent 7 wells of each row per transfection condition.
4. Using a multichannel pipette set to 100 μL, mix the cells in the first well and remove 100 μL with the multichannel pipette, to leave 100 μL in the first well.

5. Add the remaining 100 μL into the second well (media only) and mix.
6. Continue to remove 100 μL from each well and add it into the subsequent well, creating a 1:2 dilution across the row per transfection condition. The starting concentration of cells in well 1 will be 1×10^5 cells total.
7. Discard the media from well 7, leaving well 8 as a media-only control. *See* **Note 15**.
8. Add 100 μL T cell hybridoma (1×10^5 cells total) from Subheading 3.2.1 to all transfected cells and the media-only control. There is no need to mix the cells together within the well.
9. Incubate the co-culture overnight at 37 °C/5%CO_2.

3.2.3 Detecting Activation of T Cell Hybridoma Through LacZ Transcription

1. Remove the 96-well plate containing the transfected cells and T cell hybridoma co-culture from the incubator and centrifuge at 500*g*× 2 min to pellet the cells.
2. Flick off all the supernatant (note: cells will stick to the base of the well), ensuring minimal media is left in the well.
3. Add 100 μL CPRG reagent (*see* Subheading 2.3) to each well but DO NOT MIX. Try to minimize the number of air bubbles in the CPRG wells as this will affect the plate readings (*see* **Note 16**). This step is not required to be sterile and can be done on the general lab bench.
4. Incubate the plate at room temperature. Color change of CPRG should be visible in the positive controls (e.g. ERAP1-WT) shortly after the addition of CPRG, depending on B3Z sensitivity (*see* Subheading 3.3).
5. Using an appropriate microplate absorbance plate reader (e.g. BioRad iMark), measure the CPRG color change at wavelengths 595 nm with subtraction of 655 nm, ensuring plate mixing prior to reading.

   Details of plate reader protocol:

   Reading mode: Endpoint

   Read Speed: Step

   Wavelength: Dual readings at 595 nm (measurement) and 655 nm (reference)

   Mix time: 3 s

   Mix speed: Medium
6. Continue to acquire readings over 8 h depending on the rate of color change. *See* **Note 17**.

3.3 Testing T Cell Hybridoma Sensitivity

This protocol uses antigen-presenting cells with the relevant HLA-allele and synthetic final peptide sequence to determine the level of sensitivity of the T cell hybridoma to a specific concentration of peptide.

1. Harvest antigen-presenting cells (APCs), e.g. K89 L-cells stably expressing H2-K^b for presentation of SIINFEKL peptide. For K89 cells:
   (a) Remove and discard all media from 10 cm tissue culture dish.
   (b) Add 4 mL 1 mM EDTA in PBS (room temperature) to the cells and return to 37 °C/5%CO_2 for 3–5 min.
   (c) During the incubation, add 6 mL complete+ RPMI to a 15 mL falcon tube.
   (d) Using a pastette, remove all cells from the dish and add them into the 6 mL RPMI media.
   (e) Mix the cells with the pastette and remove a small aliquot (30–50 μL) for counting.
   (f) Centrifuge cells at 400*g*× 5 min.
2. Count the APCs (K89 cells).
3. Harvest and count T cell hybridoma as in Subheading 3.2.1.
4. Resuspend both K89 cells and T cell hybridoma at 1×10^6 cells/mL in RPMI.
5. Add 200 μL (2×10^5 cells) K89 cells (or equivalent APCs) to the second well of a flat-bottom 96-well culture plate.
6. Add 100 μL K89 cells to the subsequent 7 wells to achieve 8 wells across one row of the plate. Repeat if different starting concentrations of peptide are required.
7. Dilute the 10 μM stock of peptide (e.g. SL8) 1:100 in PBS or culture media to achieve 100 nM stock.
8. Add 4 μL 100 nM stock peptide to the first well of K89 cells (200 μL).
9. Using 200 μL pipette, set the pipette to 100 μL and mix the K89 cells and peptide in the first well.
10. Remove 100 μL from the first well and add it into the second well containing 100 μL K89 cells.
11. Mix the well and remove 100 μL. Continue this across the next 5 wells.
    (a) Leave the last well blank as a K89-only control.
    (b) Remove the final 100 μL from well 7 to waste.
    (c) The final starting concentration is 1 nM and titrated 1:2 across the row (*see* **Note 18**).

12. Add 100 μL (1×10^5) T cell hybridoma (e.g. B3Z) to each well. There is no need to mix this.
13. Incubate at 37 °C/5%CO_2 overnight.
14. Add CPRG, incubate on the bench and take intermittent plate readings as detailed in Subheading 3.2.3. *See* **Note 17 and 19**.

3.4 Determining T Cell Activation Through IFNγ Secretion of CTL Co-cultured with APC

3.4.1 Co-culture of Transfected APCs with Specific CTL

1. Resuspend harvested transfected cells from Subheading 3.1 at 1×10^6 cells/mL complete DMEM.
2. Obtain relevant CTL (methods not detailed) for determining T cell activation. Resuspend CTL at 1×10^7 cells/mL in relevant media for CTL culture.
3. Set up the co-culture of samples in a final volume of 200 μL/well in a 96-well round-bottom tissue culture plate according to Table 2. The addition of BFA at 1:1000 × stock (0.2 μL) enhances IFNγ detection. *See* **Note 20**.
4. Set up the following:
   (a) One well per transfected APC condition as in Table 2 (no PMA/Ionomycin in these wells).
   (b) Control wells formed of the following:
      (i) No APCs (100 μL media only in place of APCs).
      (ii) No CTL (100 μL media only in place of CTL).
      (iii) No APCs (100 μL media only in place of APCs) + 1×10^6 CTL cells in 100 μL with the addition of PMA/Ionomycin (Table 2) as a positive control for T cell activation.
5. Incubate the co-culture at 37 °C/5% CO_2 for 4–6 h (*see* **Note 21**).
6. Proceed to the flow cytometry staining protocol.

3.4.2 Flow Cytometry Staining for IFNγ Detection

1. Remove the co-culture 96-well plate from the incubator and centrifuge plate at 500*g*× 2 min.

Table 2
APC and CTL co-culture components

Component	Volume/concentration	Stock concentration
Transfected antigen-presenting cells	1×10^5 cells in 100 μL	1×10^6 cells/mL
Cytotoxic T lymphocytes	1×10^6 cells in 100 μL	1×10^7 cells/mL
Brefeldin A (BFA)	0.2 μL	1000×
PMA/Ionomycin (+ve control only)	50 ng/mL PMA 1 μM Ionomycin	
Total	200.2 μL	

2. Carefully flick off all supernatant from the cells.
3. Add fluorochrome-conjugated cell surface marker antibodies anti-CD3 and anti-CD8 to the cells, diluted in 50 μL FACS buffer at the relevant concentrations specified for the antibody. *See* **Note 22**.
4. Incubate surface antibodies on ice/4 °C and in the dark for 30 min. *See* **Note 23**.
5. Add 100 μL/well FACS wash buffer and centrifuge the plate at 500*g* ×2 min.
6. Wash the cells twice using 100 μL/well FACS wash buffer and centrifugation at 500*g* ×2 mins to pellet the cells.
7. Gently flick off all supernatant.
8. Resuspend the cell pellets in 100 μL/well of fixation/permeabilization buffer and incubate on ice/4 °C in the dark for 20 min.
9. Add 100 μL 1× BD Perm/Wash buffer to the cells and centrifuge at 500*g* ×2 min.
10. Gently flick off all supernatant and resuspend cells in another 100 μL 1× BD Perm/Wash and centrifuge at 500*g*× 2 min.
11. Gently flick off all supernatant.
12. Resuspend the fixed/permeabilized cells in 50 μL 1× BD Perm/Wash buffer containing the optimal concentration of fluorophore-conjugated IFNγ antibody. *See* **Note 24**.
13. Incubated on ice in the dark for 30 min.
14. Wash the cells twice using 100 μL/well 1× BD Perm/Wash buffer and centrifugation at 500*g*× 2 min.
15. Resuspend cells in 200 μL FACS wash buffer and add them into FACS tubes.
16. Acquire data using BD FACS Canto II or a similar machine.
    (i) Set voltage and FSC/SSC using the unstained control sample.
    (ii) Use cell surface single stain controls to determine compensation parameters.
    (iii) Use unstained/negative controls to set the gating strategy.
    (iv) Use positive control (PMA/Ionomycin) to confirm positive IFNγ gate.

4 Notes

1. 2-mercaptoethanol (2-ME: also known as β-mercaptoethanol) is an important addition in the culture media for T cell hybridoma and murine cells at 50 μM. 2-ME reduces oxidative stress of the local environment through the removal of free radicals that can be toxic to the cell in culture, ensuring optimal cell survival and growth.
2. FuGENE®6 is a non-liposomal transfection reagent mix in ethanol that should be stored at 4 °C with the lid closed tightly. Do not freeze or store below 4 °C.
3. CPRG is a substrate for β-galactosidase and is a colorimetric measure of β-galactosidase produced in response to NFAT elements from the IL-2 promoter in response to T cell activation and will change color from yellow to red/purple upon release of chlorophenol red. CPRG powder should be stored at −20 °C. CPRG solution is light sensitive and should be placed immediately into a container protected from light and stored at 4 °C.
4. It is important to determine the optimal plating density of the user prior to transfection and T cell activation assay. Briefly, plate cells at different densities ranging from 0.5×10^5 to 5×10^5 cells/mL on day 1, transfect with 1 μg pcDNA3 construct containing GFP on day 2 and determine the percentage GFP-positive cells by flow cytometry to identify the optimal transfection efficiency plating density.
5. FuGENE®6 transfection has been optimized at 3:1 ratio of FuGENE®6 to DNA. For 6-well plates 1 μg (3 μL FuGENE®6) or 2 μg (6 μL FuGENE®6) in 97 μL serum-free media is optimal. Transfection can be scaled up to 10 cm culture dishes using 500 μL serum-free media, 6 μg total DNA and 18 μL FuGENE®6 reagent. Alternatively, transfection can be scaled down to 96-well transfections using 5 μL serum-free media, 0.1 μg DNA and 0.3 μL FuGENE®6 reagent.
6. Transfection efficiency relies on optimal cell confluency of 50–80% after overnight incubation due to the length of transfection incubation (24–48 h). Transfection efficiency significantly reduces when cells are not confluent enough, or alternatively when cells are too confluent, leading to stress.
7. Plasmid DNA should be free of protein and RNA contaminants, with a A260/A280 ratio of 1.7–2.0. DNA of sufficient quality can be obtained by plasmid purification kits such as Qiagen maxiprep spin kit (Qiagen) or PureYield Plasmid Purification (Promega). Ensure extracted DNA is air dried for no

more than 10 min and resuspended in an appropriate volume of TE buffer before determining concentration.

8. It is crucial to add FuGENE®6 directly into the serum-free media, ensuring it does not come into contact with the plastic surfaces of the 1.5 mL tube. Contact of undiluted FuGENE®6 reagent with plastic can inactivate the reagent and reduce transfection efficiency.
9. Ensure FuGENE®6 reagent is returned to 4 °C immediately after use to maximize longevity of reagent and optimal transfection efficiency.
10. This 5-min incubation is time sensitive; avoid adding plasmid DNA earlier or later than this time point.
11. The final volume of mixture can vary depending on plasmid DNA added.
12. It is not necessary to remove/replace culture media or serum and/or wash cells prior to the addition of transfection: DNA mix.
13. B3Z T cells should be kept at a concentration of $0.5–1.2 \times 10^6$ cells/mL to maintain optimal sensitivity to SL8 peptide. B3Z cells can become semi-adherent due to culturing adaptations.
14. Obtaining the correct number of B3Z T cell hybridoma for co-culture can also be achieved by removing a set volume of cells (e.g. 25 mL) from culture flask, counting cells and resuspending the cell pellet at 1×10^6 cells/mL.
15. Other controls can be added in subsequent wells if needed.
16. Air bubbles in the CRPG wells can affect the output readings. If air bubbles form, use a p200 pipette tip to remove.
17. Measure CPRG color change at time frames dependent on the rate of color change. Depending on B3Z T cell hybridoma sensitivity, the rate of color change may vary between experiments.
18. Use 1 nM as a starting concentration to determine T cell hybridoma sensitivity; however, this can be increased (up to 10 μM) and/or decreased (down to 10pM) depending on the results of the initial sensitivity test.
19. Depending on T cell hybridoma sensitivity, it may be necessary to subclone the cell line to obtain a more sensitive clone. Briefly, count cells and plate into four 96-well flat-bottom culture plates at a density of 1 cell/well in 200 μL. Incubate at 37 °C/5%CO_2 until single-cell-derived colonies are visible under the microscope (7–10 days); approximately 1/3 of the plate should contain cell colonies. Test the sensitivity of these clones by co-culturing 100 μL cells clones with 100 μL K89 APCs +/− 10 μM SL8 peptide and proceed with CPRG. Select

clones where CPRG changes color indicative of T cell activation and the peptide control CPRG remains unchanged. Grow these into 12-well and subsequently 6-well culture plates before re-testing sensitivity. Choose 1–2 clones to grow into T25 flasks.

20. BFA is an inhibitor of protein transport from ER to the Golgi, leading to accumulation of proteins in the ER. The use of BFA during IFNγ stimulation enhances the detection of intracellular IFNγ by FACS staining.
21. Optimization: if no response is observed after 6 h of incubation, extend the incubation time to 16–24 h.
22. It is advised to titrate these antibodies prior to use, starting with the manufacturer's recommendations, to determine the optimal antibody concentration for use in these experiments.
23. It is recommended that staining of surface antibodies be carried out on live, unfixed cells prior to fixation/permeabilization and staining for IFNγ to achieve the best fluorescent signals.
24. Optimization: fluorophore-conjugated IFNγ antibody incubation can be extended to 45–60 min to improve the fluorescent signal.

Conflict of Interest

The authors have no conflicts of interest to declare that are relevant to the content of this chapter.

References

1. Serwold T, Gonzalez F, Kim J, Jacob R, Shastri N (2002) ERAAP customizes peptides for MHC class I molecules in the endoplasmic reticulum. Nature 419:480–483
2. Hammer GE, Gonzalez F, James E, Nolla H, Shastri N (2007) In the absence of aminopeptidase ERAAP, MHC class I molecules present many unstable and highly immunogenic peptides. Nat Immunol 8:101–108
3. Hammer GE, Gonzalez F, Champsaur M, Cado D, Shastri N (2006) The aminopeptidase ERAAP shapes the peptide repertoire displayed by major histocompatibility complex class I molecules. Nat Immunol 7:103–112
4. Kemming J et al (2019) ERAP1 allotypes shape the epitope repertoire of virus-specific CD8(+) T cell responses in acute hepatitis C virus infection. J Hepatol 70:1072–1081
5. York IA, Brehm MA, Zendzian S, Towne CF, Rock KL (2006) Endoplasmic reticulum aminopeptidase 1 (ERAP1) trims MHC class I-presented peptides in vivo and plays an important role in immunodominance. Proc Natl Acad Sci USA 103:9202–9207
6. Hutchinson JP et al (2021) Common allotypes of ER aminopeptidase 1 have substrate-dependent and highly variable enzymatic properties. J Biol Chem 296:100443
7. Raja A, Kuiper JJW (2023) Evolutionary immuno-genetics of endoplasmic reticulum aminopeptidase II (ERAP2). Genes Immun 24:295–302
8. Reeves E, Edwards CJ, Elliott T, James E (2013) Naturally occurring ERAP1 haplotypes encode functionally distinct alleles with fine substrate specificity. J Immunol 191:35–43

9. Reeves E et al (2019) HPV epitope processing differences correlate with ERAP1 Allotype and extent of CD8(+) T-cell tumor infiltration in OPSCC. Cancer Immunol Res 7:1202–1213
10. Karttunen J, Sanderson S, Shastri N (1992) Detection of rare antigen-presenting cells by the lacZ T-cell activation assay suggests an expression cloning strategy for T-cell antigens. Proc Natl Acad Sci USA 89:6020–6024
11. Sanderson S, Shastri N (1994) LacZ inducible, antigen/MHC-specific T cell hybrids. Int Immunol 6:369–376

Chapter 19

Methods for Analyzing Human T Cell Activation Through TCR or CAR Engagement

Loredana Saveanu and Irini Evnouchidou

Abstract

There are various ways to activate human T cells through their T Cell Receptor (TCR) or through overexpressed Chimeric Antigen Receptors (CARs), which are one of the most promising treatments in cancer immunotherapy. Here, we describe some basic methods to activate the human TCR or CAR using Antigen Presenting Cells presenting their natural antigen or superantigen, or stimulation with soluble antibodies or antibody-covered beads. There are several methods to analyze this activation and we describe here cytokine detection by ELISA, phosphoprotein detection by Western blot and expression of activation molecules at the cell surface by flow cytometry.

Key words TCR, CAR, T cell activation, ELISA, Western blot, flow cytometry

1 Introduction

TCR engagement is the first step that leads to activation of downstream pathways in a T cell. This activation will ultimately lead to expression and secretion of various cytokines by the T cell, with IL-2 being one of the most used for analysis since it is a highly secreted cytokine by both human T cell lines such as the Jurkat T cell line and by human primary T cells [1]. Critical steps of the TCR activation pathways include phosphorylation of key proteins such as the TCR CD3ζ chain, Lck, ZAP-70, LAT and PLCγ [2]. Interestingly, the same pathways are activated by the CARs that are emerging as a promising T cell immunotherapy tool in cancer treatment and more recently even in other pathologies such as heart disease, autoimmunity and senescence [3]. It seems that the main difference in downstream signaling between TCR and CAR activation is the intensity of the signal [4, 5]. Signaling phosphorylation kinetics can be followed and quantified by western blot and can mirror differences in the intensity of the activation signal as well as the speed at which this signal is propagated. Finally, T cell

Efstratios Stratikos (ed.), *Antigen Processing and Presentation: Methods and Protocols*, Methods in Molecular Biology, vol. 3007,
https://doi.org/10.1007/978-1-0716-5092-9_19,

activation leads to upregulation of cell surface expression of various activation molecules including CD69, CD25 (IL-2R) and CD137 (4-1BB) [6]. This provides another way to detect T cell activation through flow cytometry. It is preferable to combine several methods to follow TCR or CAR activation using diverse stimuli and various amounts of target antigen or kinetics in order to have a clearer picture of potential differences attributed to a specific protein knock-out or overexpression in T cell studies.

2 Materials

2.1 IL-2 Secretion Measurement by ELISA

2.1.1 T Cell Activation by SEE-loaded Raji B Cells

1. RPMI complete medium: RPMI 1640 supplemented with GlutaMAX, 10% FCS (fetal calf serum), 100 U/mL penicillin and 100 μg/mL streptomycin.
2. 96-well flat-bottom plates.
3. 96-well round-bottom plates.
4. SEE (Staphylococcal Enterotoxin E) (Toxin Technology).
5. 15 mL falcon tubes.

2.1.2 IL-2 ELISA

1. Human IL-2 DuoSet ELISA (R&D Systems).
2. ELISA Plates: e.g., Nunc MaxiSorp.
3. Plate sealers: e.g., Microseal B PCR plate seal (Bio-rad).
4. 25 ml reagent reservoir.
5. Multichannel pipette.
6. Squirt bottle.
7. PBS pH 7.4.
8. Wash buffer: 0.05% Tween-20 in PBS pH 7.4.
9. Blocking buffer: 1% BSA in PBS pH 7.4.
10. Reagent diluent: 0.1% BSA, 0.05% Tween-20 in PBS pH 7.4.
11. Substrate solution: TMB substrate reagent set (BD Biosciences).
12. Stop solution: 2 N H_2SO_4.
13. Infinite 200 microplate reader (Tecan).

2.2 TCR Signaling After Activation with Antibodies

2.2.1 TCR Activation with Antibodies

1. RPMI complete medium: RPMI 1640 supplemented with GlutaMAX, 10% FCS (fetal calf serum), 100 U/mL penicillin and 100 μg/mL streptomycin.
2. PBS pH 7.4.
3. Anti-CD3 Ab (OKT3).
4. Anti-CD28 Ab (CD28.2).

2.2.2 Western Blot for TCR or CAR Signaling Proteins

1. Lysis buffer: 50 mM Tris, 150 mM NaCl, pH 7.4, 1% CHAPS, Complete protease inhibitor (Roche), phosphatase inhibitor cocktails 2 and 3 (Sigma-Aldrich).
2. Laemmli buffer: DTT 300 mM, Bromophenol Blue 0.3%, Tris pH 6.8 150 mM, SDS 6.8%, glycerol 30%.
3. Protein ladder, e.g., Precision Plus Protein Dual Color Standards (Bio-rad).
4. TGX Stain-free Criterion 4–15% acrylamide gels (Bio-Rad).
5. Nitrocellulose membranes (Bio-rad).
6. Tris/Glycine/SDS buffer (Bio-rad).
7. Tris-buffered saline (TBS; 10×): 1.5 M NaCl, 0.2 M Tris-HCl, pH 7.4.
8. TBS-Tween-20 0.1% (TBST).
9. Blocking buffer: 5% non-fat milk in TBST.
10. Anti-pLAT rabbit (Millipore).
11. Anti-pSrc (D49G4).
12. Anti-pZAP70 (65E4).
13. Anti-CD247 (pY142) (K25-407.69).
14. Anti-pPLCγ1 rabbit (Cell Signaling).
15. Loading control: Anti-vinculin (hVIN-1).
16. Trans-Blot Turbo Transfer System (Bio-rad).
17. Clarity Western ECL Substrate (Bio-rad).
18. ChemiDoc Imaging system (Bio-rad).
19. ImageLab software (Bio-rad).

2.3 CAR T Cell Activation by Flow Cytometry

2.3.1 CAR T Cell Activation

1. RPMI complete medium: RPMI 1640 supplemented with GlutaMAX, 10% FCS (fetal calf serum), 100 U/mL penicillin and 100 μg/mL streptomycin.
2. 96-well round-bottom plates.

2.3.2 Flow Cytometry for T Cell Activation Molecules

1. 96-well V-bottom plates.
2. 25 mL reagent reservoir.
3. Multichannel pipette.
4. FACS buffer: 0.5% BSA, 0.05% NaN_3 in PBS (store at 4 °C up to 6 months).
5. Small FACS tubes 1.2 mL.
6. Viability dye, e.g., Ghost Dye Violet 510 (Tonbo Biosciences).
7. Anti-CD69-APC (clone FN50).
8. Anti-CD25-PerCP-Cy5.5 (clone BC96).

9. Anti-CD137-PE-Cy7 (clone 4B4-1).
10. Anti-G4S CAR linker biotinylated (Cell Signaling) (*see* **Notes 1** and **2**).
11. Streptavidin-PE.

3 Methods

3.1 IL-2 Secretion Measurement by ELISA

3.1.1 T Cell Activation by SEE-Loaded Raji B Cells

1. Resuspend Raji B cells at 10^6 cells/ml in RPMI complete medium and seed 50 μL in 96-well flat-bottom plates in triplicates per condition.
2. Resuspend Jurkat T cells at 10^6 cells/mL in RPMI complete medium and add 100 μL to the Raji B cells.
3. Prepare 5 tenfold serial SEE dilutions from 1 mg/mL stock starting at 1:10,000 (100 ng/mL). Add 50 μL SEE per dilution and incubate for 6 h at 37 °C, 5% CO_2. (*see* **Notes 3** and **4** for alternative activation).
4. Centrifuge the plate at 4 °C, 5 min at 500×*g* and collect the supernatants in 96-well round-bottom plates.
5. Store at—20 ° C, till analysis by ELISA.

3.1.2 IL-2 ELISA

1. Dilute the capture antibody in PBS at 4 μg/mL final concentration in PBS. Coat a 96-well ELISA plate with 100 μL per well of the capture antibody. Seal the plate and incubate overnight at room temperature.
2. Wash the plate three times with Wash Buffer using a squirt bottle. After the last wash, invert the plate and blot against clean absorbent paper to remove any remaining buffer.
3. Add 300 μL blocking buffer to each well and incubate at room temperature for at least 1 h. At this point, prepare standards.
4. Prepare seven IL-2 standards using twofold serial dilutions in Reagent Diluent. Prepare 500 μL of high standard at 1000 pg/mL and 6 tubes with 250 μL Reagent Diluent. Transfer serially 250 μL of the high standard to the second tube, then 250 μL of the second tube to the third tube and so on, till you obtain all dilutions.
5. Wash plate three times as in **step 2**.
6. Add to each well 100 μL sample diluted in Reagent Diluent (*see* **Note 5**) or standards. Cover with an adhesive strip and incubate 2 h at room temperature or overnight at 4 °C.
7. Wash the plate three times as in **step 2**.
8. Add to each well 100 μL of the detection antibody diluted to 100 ng/mL in Reagent Diluent. Cover with an adhesive strip and incubate 2 h at room temperature.

9. Wash the plate three times as in **step 2**.
10. Add to each well 100 μL streptavidin-HRP diluted 40-fold in Reagent Diluent. Incubate for 20 min at room temperature in the dark. Place TMB substrate solution at room temperature (*see* **Note 6**).
11. Wash the plate three times as in **step 2**.
12. Add to each well 100 μL of freshly prepared TMB substrate solution. Incubate for 15–20 min at room temperature in the dark (*see* **Note 7**).
13. Add to each well 50 μL stop solution (*see* **Note 8**).
14. Measure absorbance at 450 nm on an Infinite 200 microplate reader (*see* **Note 9**).
15. Draw a standard curve using the values obtained from the standards and fit curve with the one-site binding model on Prism including calculation of unknowns from the standard curve.

3.2 TCR or CAR Signaling After Activation with Antibodies

3.2.1 TCR Activation with Antibodies

1. Resuspend Jurkat T cells at 10^6 cells/ml in complete RPMI medium on ice. Keep 1 mL of the cell suspension in an eppendorf tube on ice.
2. Add to the rest of the cells 125 ng/mL anti-CD3ε and 250 ng/mL anti-CD28 and split the cell suspension in 4 eppendorf tubes, adding 1 mL/tube on ice (*see* **Notes 10** and **11**).
3. Directly transfer the cells to a water bath at 37 °C for 1.5, 5, 15 and 30 min.
4. At the end of each incubation, transfer the tube immediately to ice and add 500 μL cold PBS to stop the activation.
5. Centrifuge all samples at 500×*g*, at 4 °C for 5 min.
6. Aspirate supernatant, add 1 mL PBS and repeat centrifugation.
7. Aspirate supernatant and store pellets at −80 °C.

3.2.2 Western Blot for TCR or CAR Signaling Proteins

1. Prepare cell lysates from pellets: Resuspend each pellet in 20 μL cold lysis buffer. Incubate 20 min on ice and centrifuge for 10 min at 4 °C at 10,000×*g*. Transfer supernatant to a new tube.
2. Add 10 μL sample loading buffer to the supernatant and boil for 10 min at 95 °C.
3. Load 10 μL per sample on the gel.
4. Run the gel till the gel front has reached the lower part of the gel.
5. Transfer on a nitrocellulose membrane for 10 min at 20 V.
6. Block the membrane with 5% non-fat milk for 1 h at room temperature or overnight at 4 °C.

7. Prepare primary antibody dilutions in 5% BSA in TBS-Tween 0.1%.
8. Wash once the membrane in TBS-Tween 0.1% and cut the membrane at the corresponding molecular weights.
9. Incubate with primary antibodies overnight at 4 °C (*see* **Note 12**).
10. Wash the membrane 4 times in TBS-Tween 0.1% for 5 min per wash (*see* **Note 13**).
11. Prepare secondary antibody dilutions in TBS-Tween 0.1%.
12. Wash the membrane 4 times in TBS-Tween 0.1% for 5 min per wash.
13. Add HRP substrate and incubate the membrane for 5 min in the dark.
14. Acquire chemiluminescent signal on a Chemi-doc Imaging System (*see* **Note 14**).
15. Analyze and quantify bands of interest using the Fiji software.

3.3 CAR-T Cell Activation by Flow Cytometry

3.3.1 CAR-T Cell Activation

1. Plate Raji B cells so as to be able to activate Jurkat CAR-T cells with Raji B cells at different E:T ratios (4:1, 2:1, 1:1). Add 25,000, 50,000 and 100,000 cells/well in round-bottom plates. Add medium so that the total volume reaches 100 μL.
2. Add Jurkat CAR-T cells at 100,000 cells/well and incubate overnight at 37 °C, 5% CO_2.
3. The next day, centrifuge the plate and collect supernatant for ELISA. Store in round-bottom plates at −20 °C. Perform ELISA as in Subheading 3.1.2.

3.3.2 Flow Cytometry for T Cell Activation Molecules

1. Transfer cells in a V-bottom plate and centrifuge 2 min at 700×*g* at 4 °C.
2. Aspirate supernatant and resuspend in PBS. Repeat centrifugation.
3. Stain with 100 μL viability Ghostdye Violet 510 (1:1000 dilution in PBS) for 20 min at 4 °C.
4. Prepare Ab mix: CD69-APC, CD25-PerCP-Cy5.5, CD137-PE-Cy7 (all diluted at 1:100 in FACS buffer) and biotin-anti-G4S CAR linker (diluted at 1:50 in FACS buffer).
5. Wash cells once in 150 μL FACS buffer and stain with 50 μL of Ab mix at 4 °C for 30 min in the dark (*see* **Note 15**).
6. Prepare streptavidin-PE (diluted at 1:200 in FACS buffer).
7. Wash cells twice with FACS buffer and stain with streptavidin-PE at 4 °C for 30 min in the dark.
8. Wash once and analyze by flow cytometry (Fig. 1).

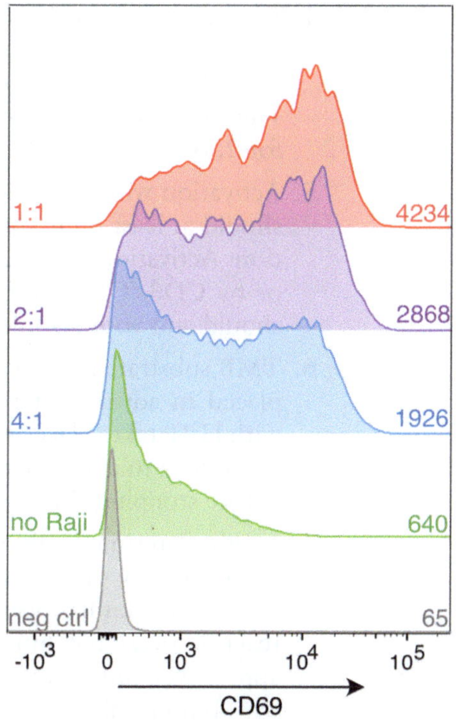

Fig. 1 Jurkat CAR-T cells were activated with Raji B cells at different effector: target (E:T) ratios for 24 h and the CD69 activation marker was measured by flow cytometry. Lower E:T ratios lead to higher CD69 activation since more targets are available per effector T cell

9. For compensation you need unstained Jurkat T cells, dead Jurkat T cells (killed at 95 °C for 10 min) stained with viability dye and compensation beads stained with the Abs.

4 Notes

1. Depending on the CAR sequence different Abs could be used: Abs against a linker in the CAR, e.g., G4S, Whitlow/218, Abs against the specific scFv or antigen recognized by the CAR scFv, e.g., b
iotinylated CD19.
2. For normal T cell activation, the anti-CAR Ab can be replaced with anti-CD3 (clone OKT3).
3. For activation by Daju-A2 cells loaded with MART1 peptide: Seed 15 × 10^3 Daju-A2 cells per well in a 96-well flat-bottom plate overnight. The day after, add MART1 peptide (ProImmune) at a final concentration of 10 μM as well as 200 × 10^3 Jurkat cells expressing TCR-MelanA2. Harvest supernatants after overnight incubation at 37 °C, 5% CO_2.

4. For activation by CD3/CD28 antibodies: Plate 150 × 10^3 Jurkat T cells per well in a 96-well flat-bottom plate in 100 μL. Add 100 μL of anti-CD3/anti-CD28 antibodies at 1 μg/mL final concentration and incubate at 37 °C, 5% CO_2 for 2, 6 or 24 h.
5. Activation with SEE-loaded Raji B cells is very strong, so you should at least dilute to half your supernatant to avoid saturation. Activation by Daju-A2 cells loaded with MART1 peptide or by CD3/CD28 antibodies is weaker and in that case you should add your sample undiluted for the ELISA.
6. TMB substrate solution is kept at 4 °C; therefore, it should be placed in advance at room temperature so that the reaction with HRP takes place at room temperature. Lower temperature could lead to a longer time needed to obtain the desired color before stopping the reaction.
7. Incubation time could vary depending on various factors such as the efficacy of the washes, the precise room temperature or the time the kit has been in use. It is better to check visually the reaction every few minutes and stop it when all the standards appear well but there is still no visible background for the well with buffer only.
8. Sometimes there can be bubbles in the wells after the addition of the stop solution and this could be problematic for absorbance detection. To avoid this, you can either press your multichannel pipette till the first level or you can briefly centrifuge the plate after addition of the stop solution.
9. For more precision, you should use a wavelength correction set to 540 or 570 nm.
10. It is very important for point 0 not to add any Ab in that tube because even when keeping the tube on ice, we have observed sometimes a slight activation compared to not adding the Abs at all.
11. CAR-T cells can be activated with CD19 microbeads (Miltenyi).
12. Phospho-antibodies often give weak signals, so it is preferable to always incubate them with the membrane overnight. For the same reason, it is not advised to freeze and reuse phospho-antibody dilutions; always use fresh antibody dilutions.
13. Longer incubation time during washes could help to reduce background.
14. In case there is no signal for some antibody, reacquire that membrane separately and use signal accumulation mode if needed.
15. Resuspend cells well with a multichannel pipette to ensure homogeneous staining.

Conflict of Interest

The authors have no conflicts of interest to declare that are relevant to the content of this chapter.

References

1. Hughes CC, Pober JS (1996) Transcriptional regulation of the interleukin-2 gene in normal human peripheral blood T cells. Convergence of costimulatory signals and differences from transformed T cells. J Biol Chem 271:5369–5377. https://doi.org/10.1074/jbc.271.10.5369
2. Alcover A, Alarcón B, Di Bartolo V (2018) Cell biology of T cell receptor expression and regulation. Annu Rev Immunol 36:103–125. https://doi.org/10.1146/annurev-immunol-042617-053429
3. Larson RC, Maus MV (2021) Recent advances and discoveries in the mechanisms and functions of CAR T cells. Nat Rev Cancer 21:145–161. https://doi.org/10.1038/s41568-020-00323-z
4. Salter AI, Ivey RG, Kennedy JJ et al (2018) Phosphoproteomic analysis of chimeric antigen receptor signaling reveals kinetic and quantitative differences that affect cell function. Sci Signal 11. https://doi.org/10.1126/scisignal.aat6753
5. Salter AI, Rajan A, Kennedy JJ et al (2021) Comparative analysis of TCR and CAR signaling informs CAR designs with superior antigen sensitivity and in vivo function. Sci Signal 14: eabe2606. https://doi.org/10.1126/scisignal.abe2606
6. Wolfl M, Kuball J, Ho WY et al (2007) Activation-induced expression of CD137 permits detection, isolation, and expansion of the full repertoire of CD8+ T cells responding to antigen without requiring knowledge of epitope specificities. Blood 110:201–210. https://doi.org/10.1182/blood-2006-11-056168

Correction to: Expression, Purification, and Functional Analysis Methods of Antigen-Processing Aminopeptidases ERAP1, ERAP2, and IRAP

Anastasia Mpakali

Correction to:
Chapter 9 in: Efstratios Stratikos (ed.), *Antigen Processing and Presentation: Methods and Protocols*, Methods in Molecular Biology, vol. 3007,
https://doi.org/10.1007/978-1-0716-5092-9_9

The original version of Chapter 9 – "**Expression, Purification, and Functional Analysis Methods of Antigen-Processing Aminopeptidases ERAP1, ERAP2, and IRAP**" was inadvertently published with Figures 1 and 3 swapped in the previously published version.

The book has been updated with these changes.

The updated version of this chapter can be found at
https://doi.org/10.1007/978-1-0716-5092-9_9

Efstratios Stratikos (ed.), *Antigen Processing and Presentation: Methods and Protocols*, Methods in Molecular Biology, vol. 3007,
https://doi.org/10.1007/978-1-0716-5092-9_20,

Index

Efstratios Stratikos (ed.), *Antigen Processing and Presentation: Methods and Protocols*, Methods in Molecular Biology, vol. 3007, https://doi.org/10.1007/978-1-0716-5092-9,

MIX

FSC C105338

If you have any concerns about our products,
you can contact us on
ProductSafety@springernature.com

In case Publisher is established outside the EU,
the EU authorized representative is:
Springer Nature Customer Service Center GmbH
Europaplatz 3, 69115 Heidelberg, Germany

Printed by Libri Plureos GmbH
in Hamburg, Germany